COMPUTER-AIDED PROBLEM SOLVING FOR SCIENTISTS AND ENGINEERS

COMPUTER-AIDED PROBLEM SOLVING FOR SCIENTISTS AND ENGINEERS

Sundaresan Jayaraman
Georgia Institute of Technology

McGRAW-HILL, INC.
New York St. Louis San Francisco Auckland Bogotá
Caracas Hamburg Lisbon London Madrid Mexico Milan Montreal
New Delhi Paris San Juan São Paulo Singapore Sydney Tokyo Toronto

**COMPUTER-AIDED PROBLEM SOLVING FOR
SCIENTISTS AND ENGINEERS**

Copyright © 1991 by McGraw-Hill, Inc. All rights reserved. Printed in the United States of America. Except as permitted under the United States Copyright Act of 1976, no part of this publication may be reproduced or distributed in any form or by any means, or stored in a data base or retrieval system, without the prior written permission of the publisher.

1 2 3 4 5 6 7 8 9 0 DOC DOC 9 0 9 8 7 6 5 4 3 2 1

ISBN 0-07-032353-4

This book was set in Times Roman by Beacon Graphics Corporation.
The editors were B. J. Clark and David A. Damstra;
the production supervisor was Louise Karam.
The cover was designed by David Romanoff.
The photo researcher was Rita Geffert.
R. R. Donnelley & Sons Company was printer and binder.

Library of Congress Cataloging-in-Publication Data

Jayaraman, Sundaresan.
    Computer-aided problem solving for scientists and engineers /
Sundaresan Jayaraman.
      p.   cm.
    Includes index.
    ISBN 0-07-032353-4
    1. Science—Methodology.   2. Engineering—Methodology.   3. Problem
solving.   I. Title.
Q175.J349   1991
507.2—dc20   90-27252

Acknowledgments

dBASE IV is a registered trademark of Ashton-Tate Corporation; Excel is a registered trademark of Microsoft Corp.; IBM is a registered trademark of International Business Machine Corporation; Lotus 1-2-3 is a registered trademark of Lotus Development Corp.; Macintosh is a registered trademark of Apple Computer, Inc.; MathCAD is a registered trademark of Mathsoft Inc.; Mathematica is a registered trademark of Wolfram Research; QUATTRO, QUATTRO PRO, and Reflex are registered trademarks of Borland International, Inc.; TK SOLVER PLUS is a registered trademark of Universal Technical Systems, Inc.; Unix is a registered trademark of AT&T Technologies; Windows is a registered trademark of Microsoft, Corp.; WordPerfect is a registered trademark of WordPerfect Corporation.

ABOUT THE AUTHOR

SUNDARESAN JAYARAMAN is an associate professor at the Georgia Institute of Technology in Atlanta, Georgia, where he teaches undergraduate and graduate courses on computer applications in engineering, process control, and mechanics of textile structures. His current research interests are in the areas of computer-integrated manufacturing, knowledge-based systems for engineering, and engineering design of textile structures and processes.

Dr. Jayaraman received his Ph.D. degree from North Carolina State University, Raleigh, and the M.Tech and B.Tech degrees from the University of Madras, India. He was involved in the design and development of TK!Solver, the first equation-solving program from Software Arts, Inc., and the predecessor of TK Solver Plus. Prior to joining Georgia Tech, Dr. Jayaraman worked as a product manager at Lotus Development Corp., Cambridge, Massachusetts, and Software Arts, Inc., Wellesley, Massachusetts.

Dr. Jayaraman is a recipient of the 1989 Presidential Young Investigator Award from the National Science Foundation for his research in the area of computer-aided manufacturing. In addition to other publications and presentations, Dr. Jayaraman has coauthored *The TK!Solver Book: A Guide to Problem Solving in Science, Engineering, Business and Education,* published by Osborne/McGraw-Hill, Berkeley, California, in 1984. He is a member of several professional societies including the Fiber Society, AAAI, ASEE, IEEE Computer Society, and ACM.

To my wife, Sumathi

CONTENTS IN BRIEF

PREFACE		xix
APPLICATION EXAMPLES AND CASE STUDIES		xxiii
LIST OF QUICK REFERENCE COMMAND TABLES		xxiv

PART 1   COMPUTING AND PROBLEM SOLVING

Chapter 1	Introduction to Computer Hardware and Software	1
Chapter 2	Problem Solving in Science and Engineering	27

PART 2   PROGRAMMING

Chapter 3	Fundamentals of Computer Programming	41

PART 3   PROBLEM-SOLVING TOOLS: SPREADSHEETS

Chapter 4	Spreadsheets for Problem Solving	71
Chapter 5	Spreadsheet Commands and Features	111
Chapter 6	Spreadsheet Application Examples (Case Studies)	161

PART 4   PROBLEM-SOLVING TOOLS: EQUATION SOLVERS

Chapter 7	Equation Solving	207
Chapter 8	Equation Solving and Engineering Design and Analysis of Beams	261
Chapter 9	Equation-Solving Application Examples	313

PART 5   PRODUCTIVITY TOOLS

Chapter 10	Database Management	363
Chapter 11	Word Processing	395

INDEX	425

CONTENTS

PREFACE — xix
APPLICATION EXAMPLES AND CASE STUDIES — xxiii
LIST OF QUICK REFERENCE COMMAND TABLES — xxiv

PART 1 COMPUTING AND PROBLEM SOLVING

1 Introduction to Computer Hardware and Software — 1

1.1 ROLE OF COMPUTERS — 1
1.2 ANATOMY OF A COMPUTER — 3
    1.2.1 Computer and the Human Mind: Information Processing — 4
1.3 COMPUTER HARDWARE — 4
    1.3.1 Configuration of a Personal Computer — 5
1.4 INFORMATION PROCESSING — 13
    1.4.1 Number Representation — 13
    1.4.2 Conversions between Number Systems — 14
    1.4.3 Binary Arithmetic — 16
    1.4.4 Boolean Algebra and Logic Operations — 19
1.5 COMPUTER SOFTWARE — 19
    1.5.1 Operating System — 20
    1.5.2 Major Operating Systems — 21
    1.5.3 Application Software — 22
1.6 THE EVOLUTION OF COMPUTERS:
    A HISTORICAL PERSPECTIVE — 23
    1.6.1 The Age of Mechanical Calculators — 23
    1.6.2 The Electronic Calculator — 24
    1.6.3 The Minicomputer — 24
    1.6.4 The Personal Computer — 25
SUMMARY — 25

2	**Problem Solving in Science and Engineering**	**27**
2.1	WHAT IS PROBLEM SOLVING?	27
	2.1.1 Solving the Car Problem	27
	2.1.2 Prime Number Algorithm	28
2.2	A FRAMEWORK FOR QUANTITATIVE PROBLEM SOLVING	30
	2.2.1 Analysis of the Problem	30
	2.2.2 Devise a Solution Strategy	31
	2.2.3 Test the Algorithm	31
	2.2.4 Interpretation of Results	31
	2.2.5 Refine Solution Strategy	31
	2.2.6 Solve Problems	31
2.3	ILLUSTRATION OF PROBLEM-SOLVING METHODOLOGY	32
	2.3.1 Analysis of the Problem	32
	2.3.2 Devise a Solution Strategy	33
	2.3.3 Test the Algorithm	33
	2.3.4 Interpretation of Results	33
	2.3.5 Refine Solution Strategy	33
	2.3.6 Solve Problems	34
	2.3.7 Role of Computers in Problem-Solving	35
2.4	COMMUNICATING WITH COMPUTERS: A HIERARCHY OF LANGUAGES	36
	2.4.1 Search for a Common Language	36
	2.4.2 Problem Solving, Programming, and Application Software	39
	SUMMARY	40

PART 2 PROGRAMMING

3	**Fundamentals of Computer Programming**	**41**
3.1	STEPS IN PROGRAM DEVELOPMENT	41
3.2	ROOTS OF A QUADRATIC EQUATION	42
	3.2.1 Developing the Algorithm	42
	3.2.2 Translating the Algorithm into Program Code	43
	3.2.3 Executing the Program	48
	3.2.4 Improving and Generalizing the Program	50
3.3	REPETITIVE COMPUTATIONS AND THE LOOP CONSTRUCT	54
	3.3.1 Ohm's Law	54
3.4	ARRAY DATA STRUCTURE AND APPLICATION OF LOOPS	56
	3.4.1 A Sorting Algorithm	56
	3.4.2 Implementing the Sorting Algorithm	58

	3.5 STATISTICAL COMPUTATIONS AND SUBROUTINES	62
	3.5.1 A Program for Simple Statistics	62
	3.5.2 Developing the Algorithm	63
	3.5.3 Implementing the Algorithm	65
	3.6 GOOD PROGRAMMING PRACTICES	67
	3.6.1 A Modular Approach to Programming	68
	SUMMARY	69

PART 3   PROBLEM-SOLVING TOOLS: SPREADSHEETS

4 Spreadsheets for Problem Solving   71

4.1 TRAJECTORY OF A PROJECTILE	71
4.2 WHAT IS A SPREADSHEET?	72
4.2.1 Spreadsheet Basics	74
4.2.2 The Spreadsheet Window	75
4.2.3 Moving around the Spreadsheet	77
4.3 A STRUCTURED APPROACH TO PROBLEM SOLVING WITH SPREADSHEETS	78
4.3.1 Developing the Spreadsheet Template	80
4.3.2 Using the Template: What-If Analysis	82
4.4 THE PROJECTILE EXAMPLE	83
4.4.1 Creating a Spreadsheet Template	83
4.4.2 Using the Template	90
4.5 THE COMMAND STRUCTURE	91
4.5.1 Invoking a Command	92
4.5.2 Increasing the Cell Width	93
4.5.3 Inserting Rows and Columns	94
4.5.4 Escaping from Trouble	102
4.5.5 Printing the Template	103
4.5.6 Saving the Template	106
4.5.7 Exiting the Spreadsheet	107
4.6 A TOUCH OF HISTORY	109
SUMMARY	110

5 Spreadsheet Commands and Features   111

5.1 AN EXAMPLE FROM THERMODYNAMICS	111
5.1.1 Problem Analysis and Template Layout	112
5.1.2 Creating the Template	113
5.2 REPETITIVE COMPUTATIONS AND THE COPY COMMAND	119
5.2.1 Generalizing the Template: Looking for Relationships	119
5.2.2 The Copy Command	119

5.3 IMPROVING THE TEMPLATE APPEARANCE: THE RANGE
    FORMAT COMMAND 127
    5.3.1 The Range or Block Command 128
5.4 GENERATING TABLES AND GRAPHS 132
    5.4.1 Table of Results 133
    5.4.2 Creating and Displaying Graphs 136
5.5 OTHER SPREADSHEET COMMANDS 148
    5.5.1 The Erase and Delete Commands 149
    5.5.2 The Window or Layout Command 152
    5.5.3 The File Command 156
SUMMARY 160

6 Spreadsheet Application Examples (Case Studies) 161

6.1 RELIABILITY ANALYSIS OF ELECTRONIC COMPONENTS 161
    6.1.1 Problem Analysis and Template Layout 162
    6.1.2 Creating the Template 168
    6.1.3 Solving the Problem 179
6.2 THE ITERATION FEATURE IN MASS
    BALANCE CALCULATIONS 180
    6.2.1 Problem Analysis and Template Layout 182
    6.2.2 Circular References and Iteration 183
    6.2.3 Creating the Template 185
    6.2.4 Avoiding Circular References and Iterative Solutions 188
    6.2.5 Generalizing the Template 190
6.3 BUILT-IN FUNCTIONS AND ENGINEERING ECONOMICS 191
    6.3.1 Time Value of Money and Evaluation Criteria 192
    6.3.2 An Example: Acquisition of Engineering Workstations 196
    6.3.3 Creating and Using the Template 199
6.4 LIMITATIONS OF SPREADSHEETS 202
SUMMARY 204

PART 4 PROBLEM-SOLVING TOOLS: EQUATION SOLVERS

7 Equation Solving 207

7.1 HELICAL SPRING DESIGN 207
7.2 WHAT IS AN EQUATION SOLVER? 211
    7.2.1 Basics of an Equation Solver 211
    7.2.2 The TK Sheets 212
7.3 A STRUCTURED APPROACH TO PROBLEM SOLVING
    WITH EQUATION SOLVERS 216
    7.3.1 Developing the TK Model 217
    7.3.2 Using the TK Model: What-If and How-Can Analyses 221

7.4	THE HELICAL SPRING DESIGN EXAMPLE	223
	7.4.1 Creating a TK Model	223
	7.4.2 Using the Model: Perform What-If and/or How-Can Analysis	238
7.5	ITERATIVE SOLVING	241
	7.5.1 Need for Iterative Solving	242
	7.5.2 Using the Iterative Solver	243
	7.5.3 Other Conditions Requiring Iterative Solution	248
7.6	TK'S COMMAND STRUCTURE	248
	7.6.1 The Slash Command Menu	248
	7.6.2 Printing the Sheets	251
	7.6.3 Saving the Model	253
	7.6.4 TK Sheets and the Select Command	258
	7.6.5 Resetting and Exiting TK	259
7.7	A TOUCH OF HISTORY	259
	SUMMARY	260

8 Equation Solving and Engineering Design and Analysis of Beams   261

8.1	ENGINEERING DESIGN	261
	8.1.1 Characteristics of the Design Process	261
	8.1.2 Engineering Design of Beams	262
	8.1.3 Problem Analysis and Model Design	263
8.2	CREATING THE MODEL	265
	8.2.1 Defining Unit Conversions	268
	8.2.2 Assigning Values and Solving the Model	271
8.3	SHEARING FORCE AND BENDING MOMENT DIAGRAMS	275
	8.3.1 Repetitive Computations and List Solving	276
	8.3.2 Tables and Graphs	281
8.4	CONTROLLING THE DISPLAY: NUMERIC FORMATTING	288
8.5	REPRESENTING DESIGN DATA: THE USER-DEFINED FUNCTION FEATURE	293
	8.5.1 Functions in TK	295
	8.5.2 Using Wood Properties Database in the Model	300
	8.5.3 Incorporating User-Defined List Functions in Rules	303
8.6	CONTROLLING THE FLOW OF COMPUTATION: CONDITIONAL RULES	305
	8.6.1 The Given Function	305
	8.6.2 Selecting Sound Designs: Decision-Making Statements	307
	SUMMARY	312

9 Equation-Solving Application Examples   313

9.1	ELECTRIC CIRCUIT ANALYSIS AND RULE FUNCTIONS	313
	9.1.1 Problem Analysis and Model Design	314

9.1.2 Creating the Model: The Rule Function Feature	315
9.1.3 The Wheatstone Bridge	321
9.2 ELEMENTS OF A TRIANGLE AND RULE FUNCTIONS	326
9.2.1 Mathematical Relationships	326
9.2.2 Creating the Model	326
9.2.3 Utilizing the Model	329
9.3 FACTORIALS AND PROCEDURE FUNCTIONS	335
9.3.1 Mathematical Relationships and Problem Analysis	335
9.3.2 Creating the Model: The Procedure Function Feature	335
9.3.3 Computing Factorials: A Recursive Procedure Function	339
9.4 GREATEST COMMON DIVISOR AND PROCEDURE FUNCTION	345
9.4.1 Mathematical Relationships	346
9.4.2 Tracing the Procedure: The Debug Function	348
9.5 STOICHIOMETRY: INTEGRATING DECLARATIVE AND PROCEDURAL COMPUTATIONS	348
9.5.1 Protecting the Environment: Processing Industrial Waste	348
9.5.2 Creating the Model	349
9.5.3 Another Chemical Reaction	356
9.6 TK'S LIBRARY OF FUNCTIONS	360
SUMMARY	361

PART 5   PRODUCTIVITY TOOLS

10   Database Management   363

10.1 DATABASE CONCEPTS	363
10.1.1 What Is a Database?	363
10.1.2 Database Terminology	364
10.2 DESIGNING AND CREATING A DATABASE	366
10.2.1 Analyzing Data and User Needs	366
10.2.2 Implementing the Database Design	367
10.2.3 Entering Data	367
10.2.4 Saving the Database	368
10.2.5 Editing the Database	368
10.2.6 Utilizing the Database	368
10.2.7 Data Security and Integrity	369
10.2.8 Characteristics of a Database Management Program	369
10.3 A DATABASE FOR THE PERIODIC TABLE OF ELEMENTS	371
10.3.1 Data and Needs Analysis	371
10.3.2 Implementing the Database Design	372

	10.3.3 Entering Data	376
	10.3.4 Editing the Database	376
	10.3.5 Utilizing the Database	380
	10.4 RELATIONAL DATABASE CONCEPTS	392
	SUMMARY	394
11	**Word Processing**	**395**
	11.1 WHAT IS WORD PROCESSING?	395
	11.2 EVOLUTION OF A DOCUMENT: THE FIVE-STEP PROCESS	396
	11.2.1 Creating Text	396
	11.2.2 Editing a Document	403
	11.2.3 Improving the Document Appearance: Formatting a Document	413
	11.2.4 Saving the Document	418
	11.2.5 Printing the Document	418
	11.3 ADVANCED CONCEPTS	419
	SUMMARY	423
	INDEX	425

PREFACE

Problem solving in science and engineering has been dramatically influenced by the extensive proliferation of personal computers equipped with innovative software packages. Today's professional no longer relies solely on programming in a conventional language (e.g., FORTRAN, Pascal, C, BASIC) for problem solving. Instead, personal computers and powerful software packages are gradually replacing mainframe computers as problem-solving tools.

These advancements in personal computer hardware and software have influenced the contents of the introductory computing course at universities in science and engineering programs. Traditionally, the first course has been a programming course typically taught using a mainframe computer. Currently, students are introduced to problem-solving tools (e.g., spreadsheets and equation solvers) and productivity tools (e.g., word processors, graphics, and database management programs) in the first course. Some universities are expanding the scope of the programming course to cover these tools, while others are instituting an independent course for this purpose. Either way, the ultimate goal is that students be able to use these tools in both their advanced courses and careers.

Despite the increasing use of software tools, there is no sole source of information that addresses quantitative problem solving in science and engineering in the light of the software available for personal computers. *Computer-Aided Problem Solving for Scientists and Engineers* is a first attempt to fill this gap and has been written with the following objectives:

1 To introduce the basics of computing hardware and software

2 To develop a structured methodology for problem solving that can be adopted with or without software tools

3 To illustrate the scope and power of computer-aided problem solving by covering software tools typically used in science and engineering: spreadsheets, equation solvers, database management programs, and word processors

4 To introduce programming principles so that the computer can be used effectively to solve problems even in the absence of specific software tools

These objectives and the final selection of book topics have evolved over time and are based on both classroom experience in an introductory computing course and real-world applications of personal computing.

STYLE AND ORGANIZATION

As an introductory text aimed at students and professionals with varying backgrounds, *Computer-Aided Problem Solving for Scientists and Engineers* elucidates the fundamental concepts behind problem solving using software tools with the help of broad-based science and engineering examples. Each example is comprehensive and includes an explanation of the basic concepts along with the necessary equations, formulas, and assumptions.

The underlying concepts of each category of software tools (e.g., spreadsheets) are introduced through a specific application example using a hands-on approach and by developing a problem-solving methodology. Although the specific sequence of commands may vary from program to program (e.g., in spreadsheets from Lotus 1-2-3 to Excel), the introduced concepts and problem-solving methodology will be equally applicable for any program in that category. To enhance the usefulness of the book, especially in a laboratory setting, the commonly used commands for the following popular software packages are provided in quick reference tables:

- **Spreadsheets:** Lotus 1-2-3, Quattro, and Excel
- **Equation solving:** TK Solver Plus
- **Word processing:** WordPerfect and Microsoft Word
- **Database management:** dBASE IV and Reflex

Likewise, in the coverage of programming principles, equivalent BASIC and FORTRAN statements for the various constructs are provided.

The presentation style is pragmatic and applications-oriented. Each chapter begins with an overview of its contents and concludes with a summary of the material covered. The 11 chapters in the text have been grouped into 5 parts:

Part 1	Computing and Problem Solving
Part 2	Programming
Part 3	Problem-Solving Tools: Spreadsheets
Part 4	Problem-Solving Tools: Equation Solvers
Part 5	Productivity Tools

An advantage of this grouping is that each part is self-contained, and depending on the specific course objectives, certain parts may be skipped without loss of continuity. For example, if programming is not part of the course, Part 2 may be skipped. However, Part 1 introduces the basics of computing and problem solving, and may be regarded as the foundation for the remainder of the book.

The modular nature of topics enables the book to be used by practicing professionals, especially those wanting to harness the power of personal computer hardware and software in their workplace.

Software Coverage

Generally (or 80 percent of the time), we use only about 40 percent of a software's features and/or commands. This "80/40 rule" is not different from the everyday example of a videocassette recorder on which only a fraction of its capabilities is typically used. Likewise, the objective in the text is not to cover *every* feature of each program, which would require over 400 pages per program, but to cover major features used regularly by the typical scientist and/or engineer. The selection of topics under each software category is based on both experiences in the real world and with students at universities. Though the book is not intended to replace the manuals accompanying the programs, the information presented can be gainfully used to acquire proficiency in the key features of the various programs.

USING THE TEXT: FOR THE INSTRUCTOR

This book is intended primarily for use in the introductory computing course at universities. Consequently, no experience with computing is assumed. The distribution of class hours for the various topics will depend on the individual instructor and course objectives. However, the following table presents some possible combinations for experimentation:

Part of book	With programming		Without programming	
	30-h class	45-h class	30-h class	45-h class
1	6	6	6	8
2	8	11	0	0
3	5	9	8	14
4	6	12	11	16
5	5	7	5	7
Total hours:	30	45	30	45

In the table, the schedule listed under "With programming" refers to a course that covers software tools in addition to a programming language, while the "Without programming" schedule refers to a stand-alone software tools course.

The class hours should be complemented by laboratory sessions. The example in the introductory chapter of the software category (e.g., projectile trajectory in spreadsheets) can be used for the essential hands-on initiation to a specific program. The instructor can modify and enhance the wide range of science and engineering examples used throughout the text and also create additional discipline-specific examples to suit the course needs.

The book can also be used in any course that utilizes one or more of the software tools covered in the text. For example, a junior-level fluid mechanics course may require the use of equation-solving tools covered in Part 4.

ACKNOWLEDGMENTS

I thank the hundreds of students who have taken my introductory computing course at the Georgia Institute of Technology for providing the necessary motivation to undertake the writing of this book. The material in this text has been tested in various forms in the course. Feedback from the students has helped in defining and refining the contents of both the course and the text. Thanks are also due Georgia Tech for providing the environment for the development of the course and the book.

The reviewers' input during the development of the text has been valuable. For this, I thank the following individuals: Steve Bayerlein, University of Idaho; John Biddle, Cal State Polytechnic University; James McDonough, University of Cincinnati; and Donald Woolston, University of Wisconsin—Madison.

I thank my graduate students who provided continuous help and support in one form or another during the preparation of the manuscript. In particular, Rajeev Malhotra, K. Srinivasan, and Jill Davis deserve special thanks for their comments on early versions of the chapters. My colleague, Dr. Howard Olson, who read the first few chapters, deserves thanks for his input. Thanks are also due Dr. Phiroze Dastoor, my research colleague, who read the completed manuscript and provided a number of valuable suggestions, and Dawn Heilman, my former student, for her help with the figures and command structure of the latest version of dBase IV. Dr. Milos Konopasek deserves thanks for his comments on the TK chapters.

Thanks go to B. J. Clark at McGraw-Hill for his enthusiastic sponsorship of the project. In addition, I thank the following individuals at McGraw-Hill for their efforts in making the book a reality: David Damstra, senior editing supervisor; Judy Pietrobono, reviews coordinator; Marci Nugent, copy editor; Louise Karam, production supervisor; David Romanoff, cover designer; Rita Geffert, photo researcher; and Karen Jackson and her marketing team.

Lastly, I thank my family: Sumathi for enduring the seemingly never-ending book "excuse" and lost times; my daughter Tharuni Aparna for putting up with my absence during the first eight months of her coming into this wonderful world; and my parents, brother, and sister for their constant encouragement and enthusiastic support for all my activities. A very special thanks to my parents and my wife, Sumathi, for the countless hours spent in reading and correcting the proofs. But for their efforts, the tight production schedule could not have been realized.

Sundaresan Jayaraman

APPLICATION EXAMPLES AND CASE STUDIES

Example	Section	Page Number
Chemical Engineering		
Mass Balance in a Chemical Reactor	6.2	180
Chemistry		
Periodic Table of Elements	10.3	371
Stoichiometry	9.5	348
Civil Engineering		
Beam Design and Analysis	8.1	261
Electrical Engineering		
Circuit Analysis	9.1	313
Ohm's Law	3.3	54
Engineering Economics		
Time Value of Money and Investment Options	6.3	192
Mathematics		
Elements of a Triangle	9.2	326
Factorials and Number of Combinations	9.3	335
Greatest Common Divisor	9.4	345
Prime Numbers	2.1	28
Roots of a Quadratic Equation	3.2	42
Sorting a Set of Numbers	3.4	56
Mechanical Engineering		
Spring Design and Analysis	7.1	207
Physics		
Laws of Motion	2.3	32
Path of a Projectile	4.4	83
Statistics		
Mean, Standard Deviation, and Variance	3.5	62
Reliability Analysis	6.1	161
Thermodynamics		
Behavior of an Ideal Gas	5.1	111

LIST OF QUICK REFERENCE COMMAND TABLES

Table		Page Number
Spreadsheets (Lotus 1-2-3, Quattro, Excel)		
4.1	Cursor movement commands	79
4.3	Changing cell width commands	95
4.4	Insert row and column commands	97
4.5	Print commands	107
4.6	File save and retrieve commands	107
4.7	Quit or exit commands	109
5.1	Formula display commands	116
5.2	Recalculation commands	118
5.3	Copy commands	123
5.5	Numeric format commands	133
5.6	Cell name commands	137
5.7	Move commands	139
5.10	Graph commands	147
5.11	Graph name commands	147
5.12	Erase commands	150
5.13	Delete commands	153
5.14	Window commands	155
5.15	Titles commands	157
5.16	File combine commands	159
6.6	Label name commands	185
6.7	Iterations commands	188
Equation Solver (TK Solver Plus)		
7.1	Slash commands	250
7.2	Storage or file commands	258
Database Programs (dBase IV, Reflex)		
10.2	Create and use commands	376
10.3	Records commands	376
10.4	File commands	379
10.5	Edit commands	380
10.6	Sort commands	383
10.7	Search commands	386
10.9	Print commands	390
Word Processors (WordPerfect, Microsoft Word)		
11.1	Cursor movement commands	399
11.2	Text delete commands	403
11.3	File commands	406
11.4	Search string commands	407
11.5	Text copy and move commands	409
11.6	Page layout or format commands	414
11.7	Display control commands	418
11.8	Print commands	420
11.9	Spell checker and thesaurus commands	422

CHAPTER 1

INTRODUCTION TO COMPUTER HARDWARE AND SOFTWARE

In this chapter, we discuss the anatomy of a computer in detail and look at how information is represented in the computer. We cover number representation schemes, binary arithmetic, and the principles of boolean algebra. The concepts of system and application software are introduced. We also examine the wide range of applications in which the computer is being used — from synthesizing music to designing a suspension bridge. Finally, we take a brief look at the evolution of computers.

1.1 ROLE OF COMPUTERS

Computers have become an integral part of everyday life. Whether it is withdrawing money from an automatic teller machine, making airline reservations, designing a suspension bridge, estimating the stresses in an elastic beam, solving a set of nonlinear simultaneous equations, computing the orbital path of the space station, controlling machines in manufacturing, synthesizing music, writing a project report, or preparing the company's profit and loss statement, we make use of computers. In fact, computers have become as ubiquitous as automobiles.

Computers in Product Development Figure 1.1 shows a typical product development cycle in the real world of manufacturing. Beginning with engineering design, the cycle proceeds to the analysis of the design, which is followed by the development of the prototype. The prototype is then tested under various operating conditions, e.g., stresses, strains, and environment, to evaluate its performance. The test data are analyzed and the cycle of design-analysis-prototyping-testing is repeated until performance criteria are met. Not only is this iterative process time-consuming, it also

2 INTRODUCTION TO COMPUTER HARDWARE AND SOFTWARE

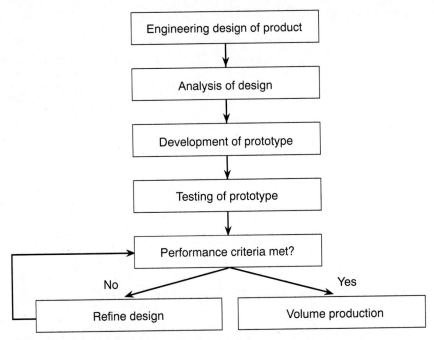

FIGURE 1.1 The product development cycle in manufacturing.

requires large amounts of human, fiscal, and other resources. Here is where the computer comes in.

Computers have had a significant influence on the workings of scientists and engineers involved in product development. It is no longer necessary to build and test hundreds of prototypes for each design of a product or system. Engineering designs can be created with the help of computer-aided design (CAD) software and analyzed using computer-aided engineering (CAE) tools. Systems from various design alternatives can be modeled and simulated on the computer for performance evaluation. The "best" design can be selected even before a *single* prototype is developed!

Other Applications Computing power has been harnessed for activities ranging from the design of the space shuttle to the design of nuclear power plants; from the design of very large scale integrated circuits to the design of a city's water and sewage system; from forecasting a hurricane to simulating the potential impact of a devastating earthquake; from disseminating information within a manufacturing facility to communicating between plants on different continents.

Computing power includes the physical hardware and the necessary sequence of instructions or software for carrying out the task. The success, or failure for that matter, of any computer application, be it in product development, manufacturing, or problem solving, depends largely on the synergistic relationship between hardware and software. Consequently, understanding the basics of computer hardware and software is the first step toward the successful use of computers. We will now begin our journey through the fascinating world of computers by taking a detailed look at the basic components of a computer.

1.2 ANATOMY OF A COMPUTER

A computer is essentially a device that accepts information, stores it, processes it, and communicates the results in a useful form. It accomplishes this by breaking a task or problem down into logical operations that can be performed on **binary** numbers, which are fixed-length strings of 0s and 1s. All information within the computer—whether text or numbers—is represented by binary numbers. Likewise, whatever the size, a computer contains devices for its five main functions: input, storage, arithmetic, control, and output.

Figure 1.2 shows the typical anatomy of a computer. The input device is used for entering information into the computer. Examples of an input device include the keyboard, mouse, light pen, punched card, and tape. The entered information is stored in the computer's memory unit for further processing. The central processing unit (CPU) is the brain of the computer and is responsible for processing the information. It is made up of the arithmetic-logic unit (ALU) and the control unit (CU). All the arithmetic and logic operations are performed in the ALU, while the CU regulates the various operations and flow of information in the computer. Information is also stored externally on floppy and hard disks and on magnetic tapes.

The output device is used to communicate the results of the computations. The video monitor and printer are examples of output devices. These are also known as *peripheral devices,* or simply *peripherals,* since they are not central to the computation function. Together, these components make up the computer *hardware.*

The hardware can do nothing by itself except "boost" the individual's ego on whose desk it sits. The hardware requires a set of instructions, commonly referred to as the *program* or *software,* that directs the computer to perform specific tasks. Just as a car without gas is of no real utility, hardware without software does not serve the intended purpose. There are two major classes of software: system and application. *System software* controls the operations of the computer and manages the flow of information

FIGURE 1.2 The anatomy of a computer.

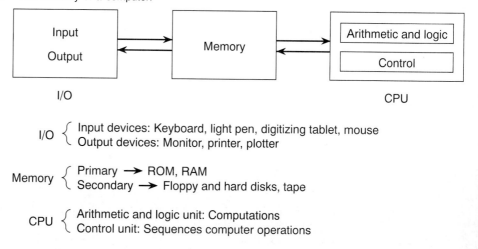

between the various components shown in Figure 1.2. *Application software* refers to programs designed for specific applications: calculating the torsion in a rod, displaying the graph of monthly product sales, retrieving information from an engineering database, or editing a term paper.

1.2.1 Computer and the Human Mind: Information Processing

The computer is analogous to the human mind. The eyes, ears, and other sensory organs are the human input devices, i.e., the means for gathering data for subsequent processing. The brain is the CPU that processes the sensory data with the help of programming in the form of knowledge and past experience stored in the human memory. The results of the processing can be conveyed orally, in writing, or through other means; thus, the mouth and hands are examples of human output devices.

For example (see Figure 1.3), when solving a crossword puzzle, the eyes scan the blanks in the puzzle and the clues. The blanks and the clues represent *data*—in the present form, they do not represent the solution to the puzzle. This data is conveyed to the brain, which attempts to interpret it and develop a solution. This is the computation or information processing phase and is carried out until either a solution is reached or it is determined that one cannot be found. The solution represents *information* because it is a meaningful and useful interpretation of data. The solution is entered in the blanks (in the puzzle) by hand, and the cycle of input, computation, and output continues with the next set of blanks and clue. Just as the task of solving a puzzle has been broken down into its elements by the human mind, the computer breaks down a task and arrives at a solution. Thus, in many ways, information processing by the computer is similar to human information processing.

1.3 COMPUTER HARDWARE

Traditionally, computers have been classified into four major categories on the basis of cost, size and computational power. These are: **supercomputers**, **mainframes**, **minicomputers**, and **microcomputers**. Supercomputers are the most powerful and expensive of computers; they are typically used in complex scientific and engineering applications—nuclear physics, computational fluid dynamics, weather forecasting,

FIGURE 1.3 Information processing in solving a crossword puzzle.

Organ	Action	Phase
Eyes	Scan the blanks and read the clue.	Input
Brain	Interpret clue.	Computation
Memory	Provide knowledge for interpretation.	Computation
Brain	Determine if solution is reached.	Computation
Brain	If necessary, decide on continuation.	Computation
Hand	Enter solution in blanks.	Output
Mouth	Express satisfaction or frustration.	Output

etc. Cray Research is a pioneer in this field and others include systems from Control Data Corporation and NEC of Japan. The new SX-X line of supercomputers from NEC is said to be eight times faster than the most powerful CRAY, performing about 20 billion operations a second!

Until the late seventies, computers were synonymous with mainframe computers. These are general-purpose machines used for a broad range of information processing chores by multiple users. Examples include the IBM 4341, IBM 360/370 series, CDC Cyber series, and UNISYS 2200.

Minicomputers, or minis, are smaller versions of mainframes both in memory capacity and in the number of users that can be handled simultaneously. Examples include the IBM AS/400, DEC VAX and Data General Eclipse.

Microcomputers constitute the most recently developed class of computers. The microcomputer, or micro, derives its name from the fact that the CPU is a single integrated circuit on a silicon chip — the microprocessor. Though micros are much smaller in size than mainframes, they are more powerful than some of the systems of the past. The wide range of application software readily available for use on these systems has contributed to their extensive proliferation in the eighties. The boundary between minis and micros is rapidly disappearing. The so-called supermicros and workstations, such as those from Sun, Apollo, and even DEC, are transcending the boundary lines.

1.3.1 Configuration of a Personal Computer

A personal computer is a stand-alone microcomputer that is affordable for the individual user (see Figure 1.4). As shown in Figure 1.2, the personal computer hardware can be broadly grouped into three categories: CPU, memory, and input-output (I/O). We will take a closer look at the characteristics of each of these subsystems.

Central Processing Unit (CPU) The CPU of a personal computer is a microprocessor. The computing power of the microprocessor is dictated by the number of *bits* that can be processed in parallel and the clock speed. The term *bit* stands for *binary* dig*it* (0 or 1). It represents the basic unit of information that can be accepted, stored, processed, and output by the computer. Eight bits make a *byte* (see Figure 1.5). This relationship is analogous to the penny and dollar. Though the penny is the basic unit of currency, the dollar is a more convenient measure. Typically, a collection of 4 bytes is referred to as a *word*. In older systems, 2-byte words were common. The greater the number of bits a microprocessor can handle in parallel, the faster the information can be processed. The first generation of microprocessors could handle 4 bits in parallel and were known as 4-bit microprocessors. This has evolved through the 8- and 16-bit microprocessor to the present-day 32-bit microprocessor.

The system or CPU clock is responsible for synchronizing the internal operations of the CPU. *Clock speed* refers to the frequency of electronic pulses generated by the system clock. Clock speeds have increased from about 0.4 MHz in the 4-bit microprocessors to about 33 MHz in the newer microprocessors.

Motorola and Intel are the two main U.S. manufacturers of microprocessors. Motorola's family of processors is used in the personal computers produced by Apple

(a)

(b)

FIGURE 1.4 (a) IBM PS/2 computer (courtesy of International Business Machines Corporation). (b) Apple Macintosh SE/30 (courtesy of Apple Computer, Inc.).

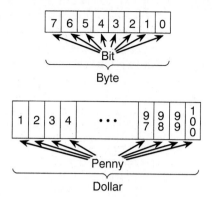

FIGURE 1.5 Bits and bytes and pennies and dollars.

and in workstations from vendors such as Sun Microsystems. The latest of these processors is the MC-68040, which incorporates 1.2 million transistors—the equivalent of five complex processor chips on a single sliver of silicon. Intel's microprocessors are used in IBM and IBM-compatible personal computers. The Intel 80486 is the most recent and powerful microprocessor from Intel, incorporating approximately 1.2 million transistors on a single chip.

An emerging trend in microprocessors is the RISC (reduced-instruction-set computer) chip in which many of the processor commands are incorporated in the chip itself, thus speeding up the computations. Intel's i860 is an example of a RISC chip.

Memory The personal computer has two kinds of memory: primary and secondary. Primary memory, or main memory, is located alongside the CPU on the motherboard inside the computer unit. For this reason, it is also referred to as *internal memory*. On the other hand, secondary memory is external to the computer motherboard.

Primary Memory There are two kinds of primary memory: read-only memory (ROM) and random-access memory (RAM). The former contains information entered by the manufacturer and can be "read" only by the CPU (or user). The information cannot be modified or erased, and it is not lost when the power is shut off. Typically the ROM contains system programs burned or programmed in by the chip manufacturer. ROMs containing programs for specific purposes are called PROMs (programmable read-only memory). Since the programs are stored in memory, they are executed much faster than if they were loaded into RAM from an external source. An EPROM (erasable PROM) is a type of PROM whose contents can be erased by nondestructive means, e.g., ultraviolet (uv) light. EPROMs are generally used for developing PROMs.

RAM is the working memory of the personal computer: new information can both be read from and written into RAM. For this reason it is also called *read-write memory*. RAM is expensive and it is volatile. The contents are lost when the power is turned off. A RAM chip holds the bits of information in "cells." However, information is transferred into and out of RAM in the form of bytes. In other words, RAM can be thought of as a matrix of bytes where each byte is identified by a unique address; this is similar to the mailboxes in a post office where each box has a unique number. RAM is said to be *byte-addressable*.

The capacity of RAM is denoted by the number of bits that can be stored. Memory size is measured in bytes and is expressed in kilobytes, or K for short. 1K of memory represents 2^{10}, or 1024, bytes. Note that K does not correspond to the everyday interpretation as exactly 1000. Typical RAM sizes are 256K, 512K, 640K, 1024K (same as 1 megabyte, or MB). Thus, a personal computer with 640K RAM has 640×1024, or 655,360 bytes, which is equal to $655,360 \times 8$, or 5,242,880 bits. The computer memory can be expanded by the addition of extra memory boards (see Figure 1.6). The density of the RAM chip, i.e., number of bits that can be packed, has increased considerably with the advent of large-scale and very large scale integration (LSI and VLSI) technologies. This has also contributed to the reduction in memory costs.

Secondary Memory Since information in RAM will be erased when the power is turned off, it is necessary to have a facility to store the information (e.g., a model created for circuit analysis) in the RAM for later use. The personal computer has a secondary storage device, usually a disk drive. Secondary storages devices, also referred to as *external devices,* are cheaper than RAM, and there are no limitations on the size of the storage. Floppy disks, hard disks, optical disks, and cartridge tapes are examples of secondary storage media.

Floppy disks come in several sizes: the 5.25- and 3.5-in disks are common. The latter is being used in the current generation of personal computers. The floppy disk consists of a mylar disk coated with magnetic material and placed in a protective jacket. Information is recorded on one or both sides of the diskette in circular rings called **tracks**. Each track is divided into sections called **sectors** as shown in Figure 1.7. Tracks and sectors are marked magnetically when the diskette is formatted or readied for use the first time.

FIGURE 1.6 A memory board (courtesy of International Business Machines Corporation).

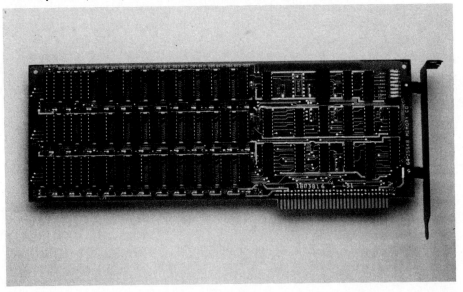

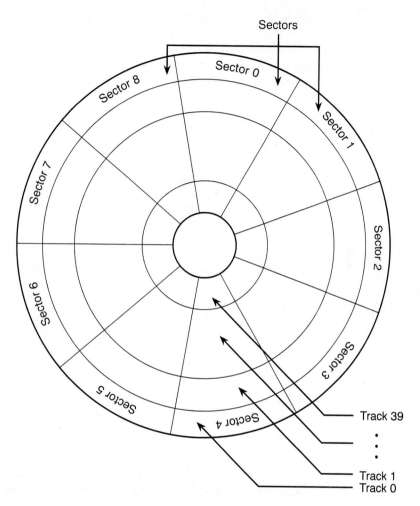

FIGURE 1.7   Information storage on a floppy diskette.

Information is transferred to and from the diskette a sector at a time by means of two read-write heads in the disk drive. The diskette is rotated at about 300 revolutions per minute (r/min), and the heads move in tandem radially across the disk to access a specific sector of the circular track. The storage capacity of the diskette depends on the quality of the media, the format used for recording information, and the design of the head. The capacity of the diskette is given by the following relationship:

$$\text{Capacity/disk} = \text{number of sides/disk} \times \text{number of tracks/side} \\ \times \text{number of sectors/track} \times \text{number of bytes/sector}$$

Thus, a double-sided, double-density disk with 40 tracks, 9 sectors, and 1024 bytes/track can store 720K of information (2 × 40 × 9 × 1024). Newer high-density disks can store up to 1.44 MB of information. The performance of the floppy disk drive is denoted by the response time to a read-write request, which can range from 200 to 500 milliseconds (ms). This performance is acceptable for small programs with a minimum of disk access but creates excessive delays for larger applications. For such applications, the hard disk drive offers about 10 to 20 times better performance.

The hard disk drive has four or six movable read-write heads and two or three disks in a sealed unit. The disk is not removable and is thus referred to as a *fixed disk*. Since the sealed unit offers better protection against dust, smoke, and other impurities, the read-write heads are designed as a precision unit to be positioned on an air cushion or to fly at heights that are very close to the disk surface but are not in contact as is the floppy read-write head. Moreover, the head reads up to 600 tracks/in. Therefore, higher density of recording and faster access speeds are possible in a hard disk. Typical hard disk drives used in personal computers can store 10, 20, 30, 40, 71, and 120 MB of information, and access times are from 28 down to 18 ms. Since the information stored on any segment of a floppy or hard disk can be accessed directly, they are called *direct-access* storage devices.

Since the hard disk can store large amounts of information, the cost of losing all that information is also very high. Information can be lost when the read-write head "crashes" (i.e., touches and damages the magnetic coating) due to dirt or smoke particles. It is essential to periodically make backup copies of the information on the hard disk. A streaming tape drive is an ideal backup device as the tape can store about 60 MB of information. The information stored on a tape cartridge can be accessed only sequentially, i.e., the read-write head must pass over the portions of the tape preceding the desired block of information. The streaming tape is known as a *sequential-access* storage device, and it is slower than a floppy drive in accessing information.

Another storage device with rising use and popularity is the optical disk drive. Its origins can be traced back to laser video disk technology. In the first-generation devices, also known as CD ROMs, information could be written once and read many times in a special disk drive attached to the personal computer. In the present generation, exemplified by the NeXT computer, optical disk storage technology is used: Information can be written to and read optically from special disks.

Input-Output Devices The third component of personal computer hardware comprises the input-output devices, i.e., the devices responsible for transferring information to and from the computer (see Figure 1.8).

Input Devices The **keyboard** is the standard input device. In addition to the keys for numbers and alphabets, there are special-purpose keys on the keyboard that include arrow keys and function keys. As the name implies, function keys are used in the software to carry out preassigned tasks.

The **mouse** is a hand-held mechanical or optical device whose movement on a horizontal plane is monitored and which is used to point to and select any point on the display screen. With the advent of graphical user interfaces (GUI), the use of the mouse as an input device is rapidly growing.

A **light pen** has a light-sensitive cell at its tip, and when it comes in contact with the display screen, it "picks" or identifies a point (e.g., a menu item) on the screen. It is commonly used in conjunction with computer-aided design and drafting programs.

FIGURE 1.8 (a) Keyboard and mouse; (b) optical scanner; (c) digitizing tablet; (d) various monitor terminals; (e) laser printer; (f) plotter. (Parts a, b, d, and e courtesy of Apple Computer, Inc.; parts c and f courtesy of Hewlett-Packard Company.)

A **digitizing tablet** is a drawing surface that is sensitive to the touch of a stylus. The movement of the stylus is translated to the movement of the cursor on the display screen. It is used for drawing figures and digitizing shapes and patterns.

Other input means include image-scanning and speech input devices, signals from instruments and machines, and information from other computers and networks.

Output Devices The **video display terminal** (VDT) or **monitor** is the most common output device used with personal computers. A portion of the computer's memory is directly mapped to the monitor so that the memory contents can be viewed on the screen. A display adapter card is used to manage the display of information on the screen. Images appearing on the monitor are formed by an array of dots or picture elements (*pixels*). The quality of resolution is directly proportional to the number of pixels that can be addressed by the computer. A low-resolution display can have about 128×48 pixels while a high-resolution monitor can have 1024×1024 pixels. The higher the resolution, the clearer and crisper the images. Information can be displayed in one of two modes: text or graphics. In the text or character mode, a group of pixels is addressed to form a character on the screen, whereas in the graphics mode the pixels are addressed individually. The particular display mode depends on the hardware-software configuration.

There are two types of monitors: monochrome and color. Both monitors require the use of display adapter cards for transferring information from the computer's memory to the screen. The monochrome monitor displays only one color, usually green or amber, while the color monitor can display a wide spectrum of colors.

A **flat panel display** is a thin flat screen monitor used in portable personal computers. The newest 14-in monitor is a 1.5-in-thick panel capable of displaying 16 colors. The display uses thin-film transistor technology whereby the screen is an array of more than 1.5 million transistors controlling the same number of individual color dots. As a result, it offers greater reliability and longer life than a conventional monitor, albeit being considerably more expensive.

The purpose of information processing is to store and exchange information. For this, a hard or printed copy of the information should be generated. The **printer** is a hard-copy output device commonly used with personal computers. Printers can be broadly classified as *impact* or *nonimpact*. Dot matrix and letter-quality printers fall into the first category while laser printers fall into the latter.

In the **dot matrix printer,** the image is formed by the impact of the printhead wires on the paper. The greater the number of wires on the head, the higher the density of dots formed and better the quality of the output. The dot matrix printer is relatively inexpensive, and the print output quality is adequate for drafts and routine applications. It is a bidirectional printer, i.e., it prints when the printhead moves from left to right and from right to left. Typical printing speeds are in the range of 90 to 220 characters per second (cps).

The **letter-quality printer,** or daisy wheel printer, is slower and more expensive than a dot matrix printer. Its output is similar to the one produced on a typewriter and of a higher quality than that of a dot matrix printer. The daisy wheel printer produces output at the rate of 50 to 70 cps.

A **laser printer** produces presentation-quality text and graphical output at speeds of about 400 to 500 cps. The process is similar to photocopying; a laser is used to position dots of electrostatically charged toner on a rotating drum. The toner is then transferred to the paper and is set by heat.

An **ink-jet printer** is another output device in which a jet of ink is sprayed on the paper to form the images. The output quality is better than that of dot matrix printers. It is increasingly used for multicolor printing.

For graphical output, the **pen plotter** is the commonly used output device. The pen is moved in response to a signal giving a set of coordinates; it produces dots very close to each other so that the resulting image appears solid, i.e., continuous, on the paper. Typically, there are several pens in different colors. The pen plotter is extremely slow and is not well suited for text output.

Other output forms include voice and direct digital signals to control machines and transfer information to networks and computers.

1.4 INFORMATION PROCESSING

Since the personal computer is a binary information processor, all information, whether numbers or text, must be represented inside the computer in binary form. The commonly used scheme for representing information in the personal computer is called the American Standard Code for Information Interchange (ASCII) system. In this scheme, each character is represented by an 8-bit ASCII code. A maximum of 2^8, or 256, individual characters including numbers, all uppercase and lowercase alphabets, punctuation marks, and special characters can be represented by this scheme. A portion of the ASCII code is shown in Figure 1.9.

1.4.1 Number Representation

A number system is a scheme for representing numeric values. In the **decimal system,** combinations of digits from 0 through 9 are used. The relative positions of the digits dictate the actual value of the combination. The *radix,* or *base,* of the decimal system is 10, and the positional values are all factors of the radix, viz., 1, 10, 100, 1000, 10000, etc. Thus, the digit combination $(009)_{10}$ has a value of 9, while $(900)_{10}$ has a value of 900. The subscript 10 denotes the number system, i.e., decimal. However, it is omitted in everyday practice. The number representation in the decimal system is shown in Figure 1.10a.

In the **binary system,** combinations of the two digits 0 and 1 are used to represent numbers; the radix is 2, and the positional values are all factors of 2, viz., 1, 2, 4, 8, 16, 32, 64, etc. The digit combination $(011)_2$ has a value of 3, while $(110)_2$ has a value of 6. As shown in Figure 1.10b, the value of a number represented in the binary system is determined using a scheme similar to that for the decimal system.

In general, if d_i, d_j, d_k, d_l, d_m, etc., symbolically represent the digits in the number system, and the radix or number base is R, then the value of representation can be determined as follows:

$$d_i d_j d_k d_l d_m = d_i \times R^4 + d_j \times R^3 + d_k \times R^2 + d_l \times R^1 + d_m \times R^0$$

Character	ASCII code
0	0011 0000
1	0011 0001
2	0011 0010
3	0011 0011
4	0011 0100
5	0011 0101
6	0011 0110
7	0011 0111
8	0011 1000
9	0011 1001
A	0100 0001
B	0100 0010
C	0100 0011
D	0100 0100
E	0100 0101
F	0100 0110
G	0100 0111
H	0100 1000
I	0100 1001
J	0100 1010
K	0100 1011
L	0100 1100
M	0100 1101
N	0100 1110
O	0100 1111
P	0101 0000
Q	0101 0001
R	0101 0010
S	0101 0011
T	0101 0100
U	0101 0101
V	0101 0110
W	0101 0111
X	0101 1000
Y	0101 1001
Z	0101 1010
Blank	0010 0000
;	0010 1100
:	0011 1010
?	0011 1111

FIGURE 1.9 An illustration of the ASCII code.

1.4.2 Conversions between Number Systems

It is easy to convert from one number system to the other. Converting from the binary to the decimal system is similar to determining the value of the binary representation (see Figure 1.10b). To convert a decimal number to its binary equivalent, divide the number repeatedly by the radix, i.e., 2, and track the remainders. The following example illustrates the conversion of $(19)_{10}$ to its binary equivalent.

Decimal system: Radix = 10
Representation: $(2459)_{10}$
Positional value: Radix$^{\text{digit position}}$
Value: Multiply each digit by its positional value and add the products.

Digits	2	4	5	9
Position	3	2	1	0
Positional value	10^3	10^2	10^1	10^0
Value	2×10^3 +	4×10^2 +	5×10^1 +	9×10^0
=	2000 +	400 +	50 +	9
=	2459			

(a)

Binary system: Radix = 2
Representation: $(1010)_2$
Positional value: Radix$^{\text{digit position}}$
Value: Multiply each digit by its positional value and add the products.

Digits	1	0	1	0
Position	3	2	1	0
Positional value	2^3	2^2	2^1	2^0
Value	1×2^3 +	0×2^2 +	1×2^1 +	0×2^0
=	8 +	0 +	2 +	0
=	10			

(b)

FIGURE 1.10 Number representation schemes. (a) Decimal system. (b) Binary system.

Thus $(19)_{10} = (10011)_2$. Note that the last remainder becomes the most significant (or left-most) digit.

Two other number representation systems used are the octal and hexadecimal systems. These two provide a convenient and easily convertible representation for binary numbers. In the octal system, the binary digits are grouped into sets of three (radix = 2^3 or 8), while in the hexadecimal system they are grouped into sets of four (radix = 2^4 or 16). In the octal system, digits used for representation are from 0 through 7, and in the hexadecimal system the digits include 0 through 9 and A through F.

1.4.3 Binary Arithmetic

Arithmetic operations (addition, subtraction, multiplication, and division) can be carried out on binary numbers in the same way as decimal numbers. Since there are only two digits, the operations are simpler.

Binary Addition When you add two binary digits, there are just the following four combinations:

$$
\begin{array}{cccc}
0 & 0 & 1 & 1 \\
+0 & +1 & +0 & +1 \\
\hline
0 & 1 & 1 & 10
\end{array}
$$

↑_____Carry out to next position

The result of adding two 1s is 2, which is 10 in binary notation. The 1 in $(10)_2$ represents the carry digit, and it is added to the next higher binary position. This is very similar to the decimal addition of 8 and 4:

$$
\begin{array}{r}
8 \\
+\ 4 \\
\hline
12
\end{array}
$$

↑_____Carry 1

The 1 represents the carry digit, while 2 is recorded in the units position. Since the 1 is in the tens position, its value is 10.

Another Example Let's consider the following binary addition:

$$
\begin{array}{rr}
\text{Carry:} & 1111101 \\
 & 010110101 \\
+ & 1011101 \\
\hline
 & 100010010
\end{array}
$$

The equivalent decimal example is:

$$
\begin{array}{rr}
\text{Carry:} & 1 \\
 & 181 \\
+ & 93 \\
\hline
 & 274
\end{array}
$$

Binary Subtraction In subtracting two binary digits, there are just the following four combinations:

$$
\begin{array}{cccc}
0 & 0 & 1 & 1 \\
\underline{-0} & \underline{-1} & \underline{-0} & \underline{-1} \\
0 & -1 & 1 & 0
\end{array}
$$

The result of subtracting 1 from 0 is −1. However, as in decimal subtraction, "borrowing" is permitted from a higher bit if one exists. For example, subtracting 3 from 12, in decimal notation, is carried out as follows:

$$
\begin{array}{rr}
\text{Borrow:} & 1 \\
& 12 \\
& \underline{-\ 3} \\
& 9
\end{array}
$$

Since we cannot subtract 3 from 2, we borrow a 1 from the tens position, which gives the units position a value of 12; the result of the subtraction is 9. The equivalent binary subtraction is as follows:

$$
\begin{array}{rr}
\text{Borrow:} & 11 \\
& 1100 \\
& \underline{-0011} \\
& 1001
\end{array}
$$

Since we cannot subtract 1 from 0 and there is a higher bit, we attempt to borrow from it. However, it happens to be 0; so, we try the next higher bit. It is a 1. This 1 becomes a $(10)_2$ in the first position, and the result of the subtraction is 1. When borrowing, a 1 is left in the second position, and the result of the second bit subtraction is 0. Thus the result is $(1001)_2$, which is the same as the decimal subtraction.

Another Example Let's consider the following binary subtraction:

$$
\begin{array}{rr}
\text{Borrow:} & 1\ 11 \\
& 010110101 \\
& \underline{-\ \ \ 1011101} \\
& 001011000
\end{array}
$$

The equivalent decimal example is:

$$
\begin{array}{rr}
\text{Borrow:} & 11 \\
& 181 \\
& \underline{+\ 93} \\
& 88
\end{array}
$$

Two's Complement Microcomputers can perform only binary additions and not subtractions. However, a subtraction can be converted into an addition using the concept of the complement of a number. The **one's complement** of a binary number is obtained by replacing the 0 digits with 1 digits and the 1 digits with 0 digits. The **two's complement** of a binary number is derived by adding 1 to its one's complement. In some computers in the fifties and sixties, one's-complement arithmetic was used. In present-day systems, however, two's-complement arithmetic is used. The following is an example:

$$
\begin{array}{ll}
\text{Binary number:} & 101001 \\
\text{One's complement:} & 010110 \\
& +1 \\
\text{Two's complement:} & 010111
\end{array}
$$

Subtraction Using the Two's Complement Binary subtraction can be performed by adding the two's complement of the subtrahend to the minuend. The final carry digit must be discarded. Recall that the subtrahend is the number that is subtracted from the minuend. Let's perform the earlier binary subtraction 1100 − 0011 using this method; the two's complement of the subtrahend, 0011, is 1101.

$$
\begin{array}{ll}
\text{Minuend:} & 1100 \\
\text{Two's complement of subtrahend:} & +\,1101 \\
& 11001 \\
& \uparrow \\
& \text{Discard}
\end{array}
$$

The result is $(1001)_2$, which is the same as before.

Let's consider another example of subtracting a larger number from a smaller one, i.e., $(00101)_2 - (10110)_2$:

$$
\begin{array}{ll}
\text{Minuend:} & 00101 \\
\text{Two's complement of subtrahend:} & +01010 \\
& 01111
\end{array}
$$

The final carry is 0. This means the result is a negative number and is in its two's-complement form. The final answer can be obtained by taking the two's complement of the result and adding a negative sign. Thus, the two's complement of 01111 is 10001, and the answer is $-(10001)_2$ or −17. Note that the equivalent decimal operation is 5 − 22, which is −17.

To sum up, if the final carry is 1, the answer is positive. If it is 0, the answer is a negative number and is in its two's-complement form.

Binary Multiplication Let's consider the product of two binary numbers, 0101 and 1010. The operation to be performed is similar to decimal multiplication, and it is in fact easier since there are only two digits, 0 or 1.

```
        0101
    ×   1010
        0000
        0101
        0000
        0101
      0110010    Answer
```

The equivalent decimal operation is 5 × 10, and the answer is 50.

Binary Division Let's consider the division of two binary numbers, viz., 11001 by 101:

```
                  101    Quotient
    Divisor 101)11001    Dividend
                  101
                00101
                  101
                00000    Remainder
```

The equivalent decimal operation is 25/5, and the answer is 5.

1.4.4 Boolean Algebra and Logic Operations

Boolean algebra provides the basis for the decision making and logic operations carried out in the personal computer. The binary digits 0 and 1 are used to define the logical decisions. The boolean or logical operators OR, AND, and XOR (exclusive OR) combine two binary digits and produce a single-digit result. The boolean operator NOT complements a binary digit. In boolean algebra, 1 stands for TRUE, and 0 stands for FALSE. The logic operations for the various combinations of input values are defined using a truth table. This is shown in Table 1.1 where A and B are logic or boolean variables with values of 0 or 1. The .OR. operator is denoted by + or ∨, the .AND. operator by . or ∧, the .XOR. operator by ⊕ or ⩗, and the .NOT. operator by −.

De Morgan's Theorem The boolean operations can be combined to produce any desired output from a set of given inputs using De Morgan's theorem. The theorem can be expressed in one of two ways:

$$\overline{A \wedge B} = \overline{A} \vee \overline{B}$$

$$\overline{A \vee B} = \overline{A} \wedge \overline{B}$$

1.5 COMPUTER SOFTWARE

Software consists of programs that enable the hardware to perform the desired tasks. As shown in Figure 1.11, there are several layers of software between the user and the hardware. Systems software consists of programs that serve as an interface

TABLE 1.1 TRUTH TABLE FOR BOOLEAN OPERATIONS

Logic variables		Boolean operation				
A	B	A.OR.B	A.AND.B	A.XOR.B	.NOT.A	.NOT.B
0	0	0	0	0	1	1
0	1	1	0	1	1	0
1	0	1	0	1	0	1
1	1	1	1	0	0	0

FIGURE 1.11 Layers of system software buffer the user from the personal computer hardware.

```
                User
    ┌──────────────────────┐
    │ Application program  │
    ├──────────────────────┤
    │  Operating system    │
    │  Command processor   │
    ├──────────────────────┤
    │  Diagnosis routines  │
    ├──────────────────────┤
    │   Bootstrap loader   │
    ├──────────────────────┤
    │      Hardware        │
    └──────────────────────┘
```

between the application program and the hardware. Application software consists of programs for carrying out tasks such as word processing, database management, and equation solving.

1.5.1 Operating System

The **operating system** is the most important component of system software, and it regulates the operations in the personal computer: input-output, memory access, and arithmetic and logic computations. The operating system is analogous to the *traffic cop* or *signal* at an intersection on the road. Just as the cop or signal ensures orderly flow of vehicles and pedestrians, the operating system regulates the flow of information and instructions between the CPU, memory, and peripheral devices. Otherwise, the action of the input device to store some information in memory may be rudely interrupted by the CPU's attempt to retrieve information from memory, resulting in a collision with disastrous consequences to the operation of the computer.

Booting the Personal Computer The personal computer's ROM contains two system software programs: **bootstrap loader** and **diagnosis routines.** The former starts up the computer when the power is turned on. The diagnostic routines test the major components of the system, and they display appropriate error messages if any. Then the operating system is read into RAM, and the computer is said to be booted when the operating system prompt is displayed on the monitor. At the same time, control of the system is turned over to the **command processor,** the program responsible for interpreting commands from input devices such as the keyboard. These commands could include loading and running programs, formatting diskettes, checking the contents of the disk, etc. The **input-output manager,** another part of the operating system, is responsible for regulating all input and output activities: accessing disk drives, displaying information on the monitor, printing, etc.

1.5.2 Major Operating Systems

Since the system software works closely with the computer hardware, it is tailored to that class of microprocessors. A single operating system that is independent of the microprocessor hardware has not emerged as a dominant system in the personal computer world. The UNIX operating system is the closest to such a *universal* system. However, it has not gained a strong foothold in the world of personal computers yet. There are several drawbacks associated with the absence of such a standard: increased cost of learning and training to users as they move from one hardware system to another, compatibility, i.e., certain application programs that run under one operating system may not run under another, and problems with moving information generated under one system to the other.

The operating system software for personal computers is typically supplied on disks; for this reason, it is often called the **disk operating system,** or **DOS** for short.

The **CP/M** (control program for microprocessors) operating system designed by Gary Kildall for Zilog's Z-80 and Intel's 8080 microprocessors was the popular operating system in the 1970s. However, its decline as a standard began with the advent of the IBM personal computer in 1981.

Microsoft's partnership with IBM has led to the emergence of **MS-DOS** (Microsoft Disk Operating System) as the de facto standard operating system for a large proportion of personal computers: about 20 million machines. It is designed to work with Intel's family of microprocessors: I-8088, 8086, 80286, and 80386. More recently, Microsoft has released Windows 3.0 for the 80286 and 80386 class of personal computers. This extension to DOS provides for a multitasking environment on the personal computer.

The Apple II family of personal computers uses **Apple DOS** and **ProDOS**, while the Apple Macintosh has an operating system of its own.

The **UNIX** operating system originally developed at AT&T's Bell Laboratories is available on several personal computers under different names, e.g., XENIX, AIX. This powerful operating system, originally available on minicomputers and workstations, offers several advantages in multiuser and multitasking environments, i.e., where several users are on a single system or several different tasks are being performed at any given time, respectively. Application programs developed for the MS-DOS are currently being ported to run under UNIX, and this will contribute to its acceptance in the personal computer world. UNIX is the de facto standard in the world of engineering and scientific workstations.

1.5.3 Application Software

The proliferation of personal computers witnessed in the decade of the eighties is a direct result of the ready availability of easy-to-use software for a wide range of applications, from preparing income tax returns to designing bridges. The typical personal computer user is not necessarily a programmer and desires flexibility, tolerance, and guidance during the interaction with the computer. The application programs fill this need to a large degree with their excellent human-computer interfaces. These programs enable the user to harness the power of the personal computer without direct and ongoing assistance from programmers. For this reason, this era is also known as the age of the microcomputer revolution.

Program Development Software As the name implies, program development software is used for developing programs for specific applications. These include assembler languages and higher level programming languages. Assembler languages are used to write code for enhancing processor performance in specific tasks. Each assembler language is unique and its complexity demands in-depth study. Programming languages, on the other hand, are easier to learn and are widely used for applications development.

BASIC (*B*eginner's *A*ll-purpose *S*ymbolic *I*nstruction *C*ode) is the most commonly used programming language on the personal computer. However, it is not suitable for large programming projects. The newer versions of BASIC incorporate concepts of structured programming (concepts associated with other advanced programming languages) which are essential for large-scale software development efforts.

FORTRAN (*FOR*mula *TRAN*slator), a very popular language on mainframes and minis, is used for scientific and engineering applications. Pascal provides a structured programming environment and other advanced concepts (such as strong data typing), which makes it a popular vehicle for application development. While the growth of Pascal is leveling off, it is used as the first language in many computer science curricula. The C programming language is rapidly gaining prominence as a tool for both applications and systems programming, though its strengths lie in the latter. For example, the UNIX operating system is implemented in C.

Application Programs Programs for creating documents, solving mathematical models, analyzing balance sheets, creating databases, generating presentation-quality graphs, and playing games are all referred to as application software. They are designed with a specific application in mind. For example, **word processors** are designed for processing words, i.e., creating reports and documents; **equation solvers** for numeric problem solving; **spreadsheets** for financial analysis and problem solving; **database management** tools for creating and using databases; **graphics** packages for visual display of information; **communications** programs for information transfer; and **utility** programs for disk and file maintenance.

The programs in each of these general categories can be used for a wide variety of applications. For example, a spreadsheet program can be used for preparing the company's profit and loss statement, estimating the escape velocity for a body in space or for balancing the checkbook at home. Likewise, a database management program can be used for storing the properties of fluids, retrieving the elastic moduli of materials or keeping track of addresses and telephone numbers of friends.

Single-Purpose Programs In addition to the major categories of application software, programs are developed for solving specific problems. For example, a program written for solving the roots of a quadratic equation is also an application program. However, unlike programs in the general-purpose categories of equation solvers or spreadsheets, the quadratic equation solver has a single purpose, viz., finding the roots of a quadratic equation. Another example of a single-purpose application software is a set of programs developed for beam analysis in civil engineering. These programs are also known as *vertical market* or *niche* software. They have a smaller user base than general-purpose application software.

1.6 THE EVOLUTION OF COMPUTERS: A HISTORICAL PERSPECTIVE

Concepts of data processing can be traced back to the Egyptians and Chinese thousands of years ago. The *abacus* was one of the early devices used for arithmetic operations. This has since evolved over the ages into the present day personal computer as shown in Table 1.2.

1.6.1 The Age of Mechanical Calculators

In 1642, Blaise Pascal invented the adding machine, a mechanical calculator to assist his father in tax collection. The era of computation with mechanical devices had begun.

Joseph Jacquard's invention of the Jacquard head (circa 1801) set the stage for binary information processing. At any given point, the thread in a woven fabric can be in one of two positions: on the face of the fabric or on the back. The Jacquard head

TABLE 1.2 DEVELOPMENTS IN THE EVOLUTION OF THE PERSONAL COMPUTER

Date	Individual/organization	System
1100s	Egyptians and Chinese	Abacus
1642	Blaise Pascal	Adding machine
1801	Joseph Jacquard	Jacquard head and card
1832	Charles Babbage	Analytical engine
1885	Herman Hollerith	Punched-card adding machine
1944	Howard Aiken and IBM	Mark I
1946	Univ. of Pennsylvania	ENIAC
1965	Digital Equipment Corp.	PDP-8
1975	MITS	Altair 8800 Kit
1977	Jobs and Wozniak	Apple II (M-6502)
1981	IBM	Personal computer (I-8088)
1984	Apple	Macintosh (MC-68000)
1984	IBM	PC/AT (I-80286)
1988	NeXT	NeXT (MC-68030)
1988	Sun Microsystems	Sun 386i

was used on the weaving loom for raising and lowering the warp threads to form desired patterns. A hole in the card signified that the thread would appear on the face of the fabric, while a blank meant that the thread would be left down and appear on the back of the fabric. Jacquard's punched card is regarded as the inspiration for the punched computer card used in the three decades of the 1950s through the 1970s.

In 1832 Charles Babbage, an Englishman, started to build the "analytical engine" inspired by Jacquard's pattern cards. Babbage's contemporary Ada Augusta, Countess of Lovelace, is said to have remarked that the analytical engine wove algebraical patterns just as the Jacquard loom wove flowers and leaves. Incidentally, the U.S. Department of Defense has named the programming language Ada after the Countess. Although Babbage did not live to complete his work on the analytical engine, he developed the concepts of input-output, memory, arithmetic, and control, which form the basis for present-day computers. Around 1885, Herman Hollerith used a punched card to control the operations of a tabulating machine for use in the 1890 census.

In the late 1930s Howard Aiken of Harvard University and IBM Corporation began integrating the concepts developed by Charles Babbage and Hollerith's punched card for developing a calculator that would work without human intervention, an *automatic* calculator. The result was the Mark I in 1944.

1.6.2 The Electronic Calculator

The Mark I was quickly followed by the arrival of ENIAC (*E*lectronic *N*umerical *I*ntegrator *a*nd *C*omputer), which was put into operation at the University of Pennsylvania in February 1946. It was the world's first all-electronic, large-scale general-purpose digital computer, and it was built with 18,000 electronic (vacuum) tubes. It increased the speed of computation by approximately three orders of magnitude. Around the same time, John von Neumann proposed the concept of a **stored program computer** in which the instructions of a program would be stored in the computer's memory rather than being set by wires and switches. The age of the electronic computer had arrived.

This innovation was followed by the development of the transistor. The transistor, a small low-power amplifier, replaced the large, power-hungry vacuum tube, and this greatly reduced the size of the machines. With the advent of the integrated circuit in 1959 began the rapid trend toward miniaturization, which has been followed over the years by large-scale and very large scale integration technologies.

1.6.3 The Minicomputer

In the mid-sixties, Digital Equipment Corporation introduced the PDP-8 minicomputer. The price-performance ratio of this and similar computers made computers affordable for medium-sized organizations and universities. Of course, the PDP-8 was less powerful than today's microcomputer. The minicomputer line has evolved over time, and the distinction between minis and supermicrocomputers is gradually vanishing.

1.6.4 The Personal Computer

The era of the micro or personal computer was born in 1975 with the introduction, by MITS, of the ALTAIR 8800, a microprocessor-based do-it-yourself kit for the hobbyist. Despite its slow and cumbersome programming and the user's need to buy extra memory for storing programs, about 20,000 units were sold by the end of 1975. MITS was joined by two other companies, IMSAI and Processor Technology, in producing the first generation of personal computers.

The period 1976 to 1977 heralded the transition from do-it-yourself kits to ready-to-use systems, and the second-generation microcomputers emerged. With the Apple II, developed by Steve Wozniak and Steve Jobs, the personal computer revolution had begun. The Apple II, its derivatives, and other personal computers from companies such as Tandy, Vector Graphics, Commodore, and Atari were successful because they brought the power of the computer to the user's desk. One program that was largely instrumental in the success of the personal computer revolution was VisiCalc, the first spreadsheet program developed by Dan Bricklin and Bob Frankston. The rest is, of course, history.

IBM's introduction to the market of its IBM Personal Computer (IBM PC) in 1981 was a turning point in the annals of personal computing, indeed, of computing itself! This development led to the rapid proliferation of personal computers in the office environment.

The arrival of the Apple Macintosh in 1984 marked yet another milestone in the evolution of personal computing. The graphical interface using icons, derived from early work at Xerox Corporation and the Apple Lisa computer, greatly enhanced the human-computer interface, i.e., made it easier to communicate with the computer.

Recent Developments   The field of personal computing is continuously evolving with the advent of faster microprocessors, cheaper memory, and more powerful software. As of this writing the NeXT computer, from Steve Jobs's company of the same name, has optical disk drives that can be used to store Webster's dictionary and Roget's *Thesaurus* on a disk, and digital signal processing chips for superior sound quality. The computing power of Apple's MAC IIx, Compaq's 80386 machines, and Sun Microsystems' Sun 386i machine are rivalling those of more expensive and larger mainframes of yesteryears.

These welcome developments will enhance and aid the cause of complex scientific and engineering problem solving!

SUMMARY

A computer is essentially a device that accepts information, stores it, processes it, and communicates the results in a useful form. Software, a set of instructions or programs, is essential for harnessing the power of the computer. Computers can be classified as supercomputers, mainframes, minicomputers, and micro or personal computers.

We discussed the major components of personal computer hardware, viz., the CPU, primary and secondary memory, and input and output devices. Since the computer is a binary information processor, we introduced the basics of binary arithmetic and looked at the ASCII coding scheme for representing information in the computer. We also discussed the elements of boolean algebra used in logical computations.

We introduced the concept of systems and application software. Finally, we traced the origins of the present-day personal computer back to the Egyptians and Chinese to provide a historical perspective and appreciation for the growth of the industry.

CHAPTER 2

PROBLEM SOLVING IN SCIENCE AND ENGINEERING

In this chapter, we introduce the basic concepts associated with solving problems in science and engineering. We use an example from physics to develop a general framework for problem solving that will be adopted throughout the book.

2.1 WHAT IS PROBLEM SOLVING?

The word *problem* in *problem solving* can be used to describe a wide range of situations: trouble with starting the car, estimating the diameter of a pipe for satisfying certain fluid flow requirements, missing a flight, running out of paint near the end of a job, confronting "parity check 2" on a computer monitor, etc. The actual solution to each problem is obviously different. However, the underlying method of attacking a problem and solving it is essentially the same.

2.1.1 Solving the Car Problem

Let's consider the problem with starting the car. The first step is to stay calm (despite the fact that you might be late in reaching your destination) and analyze the situation, i.e., understand and identify the problem. In this case, the problem is easy to identify: The car doesn't start. The next step is to find out what could have gone wrong. Some of the questions that come to mind are:

1. Is it a cold day?
2. Is the battery weak?
3. Is there a short circuit in the electric system?
4. Is the starter malfunctioning?

5 Is there enough gas in the car?
6 Has there been a similar problem in the past?

The type, number, and sequence (order) of questions will vary with experience. This is why there are expert and not-so-expert auto mechanics. While answering each of these questions, you must carry out certain tests, some mentally, but mostly with your hands on the car. If the temperature is 60°F, then the first cause is ruled out. To test the battery, you turn on the lights. If they are bright, then it is not the battery. If you filled the gas tank the previous evening on the way home, then there is enough gas and you can rule out cause 5. If there has been no similar problem in the past, cause 6 can be ruled out too. (However, if you had a similar problem in the past, you could adopt the old solution strategy.) You are now left with causes 3 and 4. What you have been doing is developing a strategy for diagnosing the problem so that it can be solved. The solution strategy is called an *algorithm*.

To test the electric system and starter, you should not only have some knowledge of these systems but also have access to the appropriate tools. If not, you should seek the assistance of a knowledgeable person or find the appropriate tools, as the case may be. If you are capable and possess the necessary tools, you proceed with the diagnosis. If you find a short circuit in the electric system, you should identify the exact part and/or location and correct it. You should then try to start the car to check if the problem has been solved. If the car starts, the problem has been solved and you can be on your way. Or, if you are brave, you can turn the engine off and start it again to ensure that the problem has indeed been rectified. However, if you don't find a short circuit in the electric system, you check the starter (cause 4), and continue the solution process.

Methodology for Problem Solving Thus, the methodology for solving the problem includes the following steps: analysis and identification of the problem, evaluation of options and development of a solution strategy (i.e., the algorithm), testing of the algorithm, and solution of the problem. The same methodology can be adopted for solving other problems, including those in science and engineering.

2.1.2 Prime Number Algorithm

A positive integer is called **prime** if it has only trivial factors, i.e., no smaller positive integer other than 1 divides it evenly. For example, the first 10 primes, or prime numbers, are 2, 3, 5, 7, 11, 13, 17, 19, 23, and 29. They are not evenly divisible by any number greater than 1. How can we determine if a positive integer is a prime?

Analysis From the definition of a prime number it is clear that a number n (> 2) is a prime if n is not evenly divisible by any of the integers 2, 3, 4, 5, 6, ..., $n - 1$. This method of determining whether a number is a prime will become lengthy if n is large. However, we need to check only if n is divisible by any positive integer p (> 1) such that $p^2 \le n$. If it is, then n is not a prime number. Let's convert this strategy into a sequence of steps, i.e., an algorithm.

Algorithm

Step A If $n = 2$, then n is a prime and we can stop; otherwise, we proceed to step B.
Step B Let $p = 2$.
Step C Calculate n/p.
Step D If n/p is a whole number, then p is a factor of n, and so n is not a prime; otherwise, we go to step E.
Step E Check if $p^2 \le n$
Step F If $p^2 \le n$, then increment p by 1 and go to step C; otherwise, n is a prime and we can stop.

Testing the Algorithm Let's test the algorithm with $n = 25$.

Step A Since $n \ne 2$, we proceed to step B.
Step B Let $p = 2$.
Step C $n/p = 25/2 = 12.5$.
Step D $n/p = 12.5$ is not a whole number; so, we proceed to step E.
Step E Check if $p^2 \le n$: $2^2 \le 25$.
Step F Since $4 < 25$, $p = p + 1 = 2 + 1 = 3$, and we repeat action in step C.

Step C.1 $n/p = 25/3 = 8.33333$.
Step D.1 $n/p = 8.33333$ is not a whole number; so we proceed as in step E.
Step E.1 Check if $p^2 \le n$: $3^2 \le 25$.
Step F.1 Since $9 < 25$, $p = p + 1 = 3 + 1 = 4$, and we repeat action in step C.

Step C.2 $n/p = 25/4 = 6.25$.
Step D.2 $n/p = 6.25$ is not a whole number; so we proceed as in step E.
Step E.2 Check if $p^2 \le n$: $4^2 \le 25$.
Step F.2 Since $16 < 25$, $p = p + 1 = 4 + 1 = 5$, and we repeat action in step C.

Step C.3 $n/p = 25/5 = 5$.
Step D.3 $n/p = 5$ is a whole number; $p = 5$ is a factor of 25, and so 25 is not a prime number; we stop.

Another Example Let's show that 5 is a prime number.

Step A Since $n \ne 2$, we proceed to step B.
Step B Let $p = 2$.
Step C $n/p = 5/2 = 2.5$.
Step D $n/p = 2.5$ is not a whole number; so we proceed to step E.
Step E Check if $p^2 \le n$: $2^2 \le 5$.
Step F Since $4 < 25$, $p = p + 1 = 2 + 1 = 3$, and we repeat action in step C.

Step C.1 $n/p = 5/3 = 1.66666$.
Step D.1 $n/p = 1.66666$ is not a whole number; so we proceed as in step E.
Step E.1   Check if $p^2 \leq n$: $3^2 \leq 5$.
Step F.1 $9 > 5$, and we stop; since we could not find a factor p (> 1) of 5 such that $p^2 \leq 5$, we can conclude that 5 is a prime number.

From these two examples it is clear that the algorithm works correctly whether or not n is a prime. There are set sequences of computations (e.g., steps C through F) that are repeated several times, and computers are good at such repetitive computations. So the algorithm can be converted into a computer program and executed.

2.2 A FRAMEWORK FOR QUANTITATIVE PROBLEM SOLVING

Problem solving in science and engineering typically involves the major phases shown in Figure 2.1. This is similar to the methodology for solving the car problem discussed earlier.

2.2.1 Analysis of the Problem

The first step is the analysis of the problem: stating the problem clearly, writing down the appropriate equations and formulas, and identifying known (input) and unknown

FIGURE 2.1   ADTIRS: A problem-solving methodology.

Analyze the problem:
    State the problem clearly.
    Write down appropriate equations and formulas.
    Identify known and unknown variables.

Devise a solution strategy (algorithm):
    Has a similar problem been solved?
    Can any well-known techniques be used?
    Write down the sequence of steps for obtaining a solution.

Test the algorithm:
    Use sample data to test the algorithm.
    Has the problem been solved?

Interpret results:
    Are they in the expected range?
    Do they have a physical meaning?

Refine solution strategy:
    Is the algorithm correct?
    Is the algorithm generic?
    Are there any special cases?
    Generalize the algorithm.

Solve problems:
    Solve several sets of problems.
    Present results in desired formats (tables, graphs).

(output) variables. The equations or formulas constituting the mathematical model can come from textbooks or can be derived from the given information.

2.2.2 Devise a Solution Strategy

The next step is to devise a solution strategy, an algorithm, which is a step-by-step method for solving the mathematical model. If the problem has been analyzed carefully (step 1), the solution path may become apparent. Has a similar problem been solved? Are there any "canned" or well-known procedures that can be used? At times, the problem has to be simplified to a large extent before a solution becomes feasible. Then the complexities can be introduced one after another until the entire problem is solved. The result of this step is a sequence of steps that will lead to a solution of the problem.

2.2.3 Test the Algorithm

Testing the algorithm for correctness is a critical step in the problem-solving process. The algorithm is tested with sample sets of values to ensure its correctness. Whether the problem has been solved completely is also determined. Test values are also used to determine the scope of the algorithm, i.e., the range of problems (different combinations of input and output variables) that can be solved with the algorithm.

2.2.4 Interpretation of Results

The next step in the problem-solving process is interpretation of the results. This involves checking to see if the results obtained have physical significance and/or are within reasonable or expected ranges. This step will also bring out flaws in the algorithm that lead to incorrect results.

2.2.5 Refine Solution Strategy

Based on the results of the previous step, the algorithm is refined to make it generic. It is then validated with a wide range of test values.

2.2.6 Solve Problems

The refined algorithm is used to solve several sets of problems, and the solutions are presented in the form of plots, tables, and other formats.

The various phases in problem solving have been presented as separate steps for emphasizing the methodology. In practice, however, there is a great deal of overlap between the major phases. Moreover, problem solving is not only a creative process but also a process in which one learns from experience.

2.3 ILLUSTRATION OF PROBLEM-SOLVING METHODOLOGY

Consider the following example: A car starts from rest and accelerates uniformly to a speed of 20 ft/s in 10 s. Find its acceleration and the distance it traveled in this time. We will adopt the methodology shown in Figure 2.1 to solve this problem with the help of paper, pencil, and calculator.[1]

2.3.1 Analysis of the Problem

From the problem statement, it is clear that the equations for the uniform acceleration of a body will be applicable for solving the problem. The equations can be derived from first principles or taken from a basic physics textbook.

The appropriate equations are:

$$v_f = v_0 + at \tag{1}$$

$$s = v_a t \tag{2}$$

$$v_a = \frac{v_f + v_0}{2} \tag{3}$$

where v_0 = initial velocity, ft/s
v_f = final velocity, ft/s
v_a = average velocity, ft/s
s = distance traveled, ft
a = acceleration, ft/s^2
t = time taken, s

These three equations constitute the mathematical model for the solution of the problem.

The known and unknown variables are as follows:

Knowns: $v_0 = 0$ ft/s (from rest)
$v_f = 20$ ft/s
$t = 10$ s

Unknowns: v_a, ft/s
s, ft
a, ft/s^2

Total number of variables = 6.

---

[1] Although we believe that SI units should be used in science and engineering calculations, we cannot ignore the fact that British units are still being widely used. Therefore, we use both British and SI units in the text.

2.3.2 Devise a Solution Strategy

There are three equations and six variables of which three are known; consequently, an attempt can be made to solve the model. Note that Equation (3) is an intermediate equation and v_a is an intermediate variable. The solution strategy is:

From Equation (1): Calculate a.
From Equation (3): Calculate v_a.
From Equation (2): Calculate s.

2.3.3 Test the Algorithm

Assign values to the variables and solve the equations:

$$a = \frac{20 - 0}{10} = 2 \text{ ft/s}^2 \quad \text{From (1)}$$

$$v_a = \frac{20 + 0}{2} = 10 \text{ ft/s} \quad \text{From (3)}$$

$$s = (10 \text{ ft/s})(10 \text{ s}) = 100 \text{ ft} \quad \text{From (2)}$$

Since the three unknown variables have been evaluated, the problem has been solved.

2.3.4 Interpretation of Results

The values of acceleration and the distance traveled during the 10 s are well within the normal range associated with cars. Therefore, the results are reasonable and the algorithm stands validated.

2.3.5 Refine Solution Strategy

The next step is to find out if the model and algorithm are adequate for solving several types of problems. Consider the following example, a variation on the previous problem: A car starts from rest and accelerates at 12 ft/s^2 over a distance of 60 ft. What is its speed at that point, and how long does it take to reach this speed?
Total number of variables: 6.

Knowns: $v_0 = 0$ ft/s (from rest)
$a = 12$ ft/s^2
$s = 60$ ft

Unknowns: v_f, ft/s
t, s
v_a, ft/s

An examination of the three equations (1 through 3) shows that none of the three unknowns can be evaluated using the simple substitution strategy devised in Section 2.3.2.

There are two choices at this point: One is to solve the set of simultaneous equations through algebraic manipulation; the other alternative is to use (or derive) two other well-known equations involving uniformly accelerated motion of bodies. We will choose the latter route. The two additional equations are:

$$s = v_0 t + \tfrac{1}{2} a t^2 \tag{4}$$

$$v_f^2 = v_0^2 + 2as \tag{5}$$

With the addition of these two equations, it should be possible to solve the whole class of problems involving uniformly accelerated motion of bodies.[2]

2.3.6 Solve Problems

The present problem can be solved as follows:

$$v_f^2 = 0 + (2)(12 \text{ ft/s}^2)(60 \text{ ft}) = 1440 \text{ ft}^2/\text{s}^2 \quad \text{From (5)}$$

$$v_f = \sqrt{1440 \text{ ft}^2/\text{s}^2} = 37.95 \text{ ft/s}$$

$$t = \frac{37.95 - 0 \text{ ft/s}}{12 \text{ ft/s}^2} = 3.16 \text{ s}$$

All the unknowns have been evaluated and the values are in the normal range. Once again, the model and algorithm stand validated. Note that the values have been rounded off to two decimal places.

Units and Dimensional Analysis During the course of problem solving, it is important to keep track of the units associated with the variables. The numeric values by themselves have no meaning: Thus the value of v_f, 37.95, does not provide complete information about the velocity. The numeric value in conjunction with the unit of measurement provides a complete description of the variable. In addition, all calculations must be carried out in a consistent set of units. This is illustrated by the following example.

Another Example A car moving at 40 mi/h begins to slow down with a deceleration of 5 ft/s². How long does it take to travel 250 ft as it slows down? Determine the velocity at this point.

---

[2]It could be justifiably argued that these equations should have been used right at the beginning. Since the purpose is to introduce and illustrate the various steps involved in problem solving, we elected to not introduce the equations until now.

Solution An analysis of the problem reveals the following:

Knowns: $v_0 = 40$ mi/h
$\quad\quad\quad a = -5$ ft/s² (the minus sign denotes deceleration)
$\quad\quad\quad s = 250$ ft

Unknowns: t, s
$\quad\quad\quad\quad v_f$, ft/s

Since v_0 is specified in miles per hour, it should be converted to feet per second to maintain the proper balance of units in equation (5). We know that 1 mi = 5280 ft and 1 h = 3600 s. Therefore,

$$v_0 = (40 \text{ mi/h}) \times \frac{5280 \text{ ft}}{1 \text{ mi}} \times \frac{1 \text{ h}}{3600 \text{ s}}$$

$$= 58.67 \text{ ft/s}$$

From Equation (5), we have:

$$v_f^2 = (58.67 \text{ ft/s})^2 + (2)(-5 \text{ ft/s}^2)(250 \text{ ft})$$

$$= 3442.1689 - 2500 = 942.1689 \text{ ft}^2/\text{s}^2$$

i.e., $v_f = \sqrt{942.1689 \text{ ft}^2/\text{s}^2} = 30.7$ ft/s

From Equation (1), we have:

$$t = \frac{30.7 - 58.67 \text{ ft/s}}{-5 \text{ ft/s}^2} = 5.59 \text{ s}$$

Thus, using the same model and solution strategy, we solved a different problem. This modular or systematic approach to problem solving has several advantages. First, problem analysis lets you focus on the problem and minimizes your risk of getting "lost" in the solution process. Second, you can isolate the errors as they occur during each specific phase of the problem-solving process: model, algorithm, or arithmetic. Consequently, it is easier to correct them. And finally, as with anything structured, it minimizes the time taken to obtain a solution.

2.3.7 Role of Computers in Problem Solving

Among the many attributes of a computer is the ability to perform fast and repetitive computations; this means the computer can play a key role in the problem-solving process. For example, when a solution strategy, or algorithm, is developed for solving a problem, the sequence of instructions can be communicated to the computer; this step is known as *programming*, and the result of this activity is a computer *program*. Once the program is stored in the computer's memory, it can be executed as often as needed to solve a variety of problems.

From our perspective as end users, there are other means, besides programming, for solving problems on the computer. These include the use of application packages such as equation solvers, spreadsheets, and single-purpose programs. However, the computer cannot process ambiguous information: The program or set of instructions should be clear, precise, and specific. Even when the problem is solved by the computer, the interpretation of the results will be up to us. The computer can thus be an effective assistant in the problem-solving process.

2.4 COMMUNICATING WITH COMPUTERS: A HIERARCHY OF LANGUAGES

In everyday life, a language, e.g., English, is used to carry on a conversation which involves the exchange of information. If the two communicating parties can speak the same language, fast and meaningful transfer of information can occur. Otherwise, there will be the need for a third party—an interpreter or translator to assist in the information exchange process. If one is not available, the results are obvious—unclear, garbled communication. Similarly, if the computer were to play a role in problem solving, communication should be established between the computer and the human, i.e., there is a need for a common language.

2.4.1 Search for a Common Language

The computer is a binary information processor. Consequently, all instructions and information to the computer should be communicated in **machine language**, i.e., binary form—strings of 0s and 1s. As shown in Figure 2.2, machine language is at the lowest level in the hierarchy of computer languages. It was the only language available for programming on early computers. Since programming in machine language involves laying out a number of steps in binary notation for each logical process, it is prone to human error and requires a great deal of time and effort.

Assembler Language The **assembler** or **assembly** language is at the next higher level in the language hierarchy. Mnemonic symbols and words are used in place of binary patterns. For example, to perform an addition (in the arithmetic-logic unit, ALU) of two numbers stored in memory locations A and B, the sequence of instructions might be as follows:

LOAD A	Fetch value from A and Load ALU.
LOAD B	Fetch value from B and Load ALU.
ADD	Add the numbers in ALU.
STORE A	Store the contents of ALU in A.

An assembler, which is a program, recognizes the mnemonics or keywords (LOAD, ADD, STORE, and so on) in the set of instructions or source code and translates them

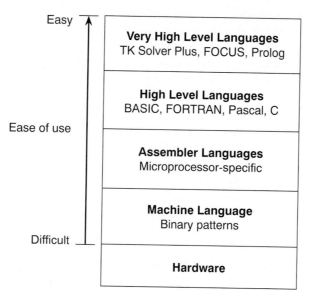

FIGURE 2.2 Hierarchy of languages for communication with computer: The higher in the hierarchy, the easier the language is to use.

into appropriate binary patterns or machine language code for execution by the microprocessor.

$$\text{Assembler source code} \longrightarrow \text{assembler} \longrightarrow \text{machine language}$$

The user is responsible for keeping track of the memory contents and also for specifying the individual operations. However, it is not as tedious as programming in machine language. Nevertheless, assembler language is very close to machine language in terms of logical content and program flow. An example of a distinguishing difference between the two languages is memory storage. Assembler introduces the use of named labels A and B to refer to storage locations in memory, as in the above sequence of instructions. Machine language requires direct or indirect reference to exact location in memory.

High-Level Languages The class of **high-level** programming languages relieves the programmer of having to provide a detailed set of instructions suited to the microprocessor for carrying out a computation. In the addition example, the programmer can simply state $A = A + B$ and let the program take care of the individual operations for carrying out the addition: fetching the contents of locations A and B, performing the addition and storing the ALU contents back in A. High-level programming languages greatly enhance the productivity of programmers.

Interpreters and Compilers The program written in a high-level language must, however, be converted into machine language code for execution by the computer. This is done by two broad categories of programs: **interpreters** and **compilers**. The interpreter translates a program, written in an interpreted language, one statement at a time into machine language code which is executed immediately. It then moves on to the next statement. So the entire process follows the interpret-execute cycle. If there is an error in the statement being processed, the interpreter will issue an appropriate message, and program execution may or may not continue depending on the error.

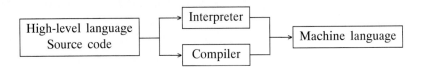

On the other hand, a compiler translates the entire program into machine language, which is then executed. Consequently, a compiled program runs much faster than a program in an interpreted language. The compiler spots syntax errors during the compilation process, which can be addressed before execution.

Common Programming Languages BASIC, FORTRAN, Pascal, and C are some of the commonly used programming languages in science and engineering applications. Of these, BASIC is the easiest to learn and use and is available on almost all personal computers. It is essentially an interpreted language, although compilers are presently available for BASIC.

FORTRAN, Pascal, and C are examples of compiled languages available on personal computers. FORTRAN is the oldest of the three languages; it has been used extensively in the development of several stand-alone packages for specific applications in various branches of engineering. Pascal and C are structured programming languages; the latter is gaining popularity, even on personal computers.

Just as every spoken language has its own syntax and grammar, each programming language has its own syntax. For example, the FOR-NEXT construct in BASIC does not exist per se in FORTRAN, which has the DO-CONTINUE construct for essentially the same task. High-level programming languages are also known as **procedural** languages because they are used by the programmer to specify the procedures or sequence of instructions necessary for processing the information.

Very High Level Languages Very high level languages are characterized by their extreme ease of use where the user specifies *what* needs to be accomplished and lets the software take care of *how* the task is performed. Recall that in a high-level language, the use specifies the procedure for obtaining the solution, though not at the byte level. This is the essential difference between high-level and very high level languages. For this reason, very high level languages are also known as **nonprocedural** or **declarative** languages, i.e., the user declares only the problem and not the solution strategy. Very high level languages also enable the users to communicate in the *natural* language of their discipline. Thus, TK Solver Plus, an equation-solving language, lets the user specify the problem—enter the equations and assign the known values—while it takes care of finding the solution.

Mathematical model ⟶ application software ⟶ execution ⟶ solution

Relative Roles of Computers and Humans Programming in a machine language requires a great deal of human effort and very little effort on the part of the computer. Conversely, using a very high level language is easier on the human while the computer has to do more work to interpret and carry out instructions (see Figure 2.2). This is analogous to the earlier example of two individuals with different native languages striking up a conversation. If the conversation is carried out in the native language of one of the individuals, it will be easier on that individual than on the other. Another example: Giving directions to an individual familiar with the streets in a city is different from giving directions to a newcomer to the city.

Over the course of evolution of computers, hardware speed and memory capacity have decreased in cost, while software programming costs have risen. Therefore, the trend is to acquire canned user-friendly software which needs very little expertise to use.

2.4.2 Problem Solving, Programming, and Application Software

Figure 2.3 shows the role of the computer in the problem-solving process. The first step, i.e., the development of the mathematical model, is a creative process and

FIGURE 2.3 Problem solving with computers.

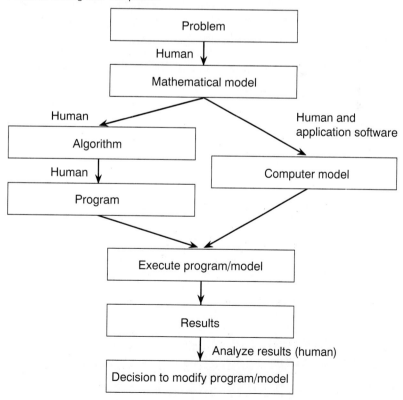

requires human involvement as shown in the figure. Once the model is developed, there are two routes: (1) programming and (2) application software. As the figure shows, the former route involves development of the algorithm, followed by its conversion into a program that can be subsequently executed to obtain results. This is a lengthy and time-consuming process and requires knowledge of a programming language. On the other hand, this is a very powerful route: Extremely complex problems can be tackled using this method.

When the latter course is chosen, the developed mathematical model can be directly incorporated in the application software, e.g., spreadsheet or equation solver, and used to solve problems. The choice of application software obviates the need to use a programming language for harnessing computer power in problem solving. This not only simplifies the problem-solving process but also reduces the extent of human involvement in certain phases. Typically, the time taken to solve a problem is also reduced.

SUMMARY

We introduced a methodology for quantitative problem solving in science and engineering. The methodology essentially involves the following phases: problem analysis, algorithm development and testing, interpretation of results, algorithm refinement, and problem solution. We used a few examples to illustrate this methodology. Advantages of this structured approach to problem solving were also highlighted.

To make effective use of computers in the problem-solving process, you should be able to communicate with the computer. This requires the use of a language. We discussed the hierarchy of computer programming languages. Relative roles of programming languages and application software in problem solving were also presented.

CHAPTER 3

FUNDAMENTALS OF COMPUTER PROGRAMMING

In this chapter, we introduce the major steps in the development of a program from an algorithm. They are coding, testing and debugging, documentation, and program maintenance. We illustrate these concepts by developing a program for solving the roots of a quadratic equation. Three additional examples—Ohm's law, sorting, and statistical computations—are used to illustrate other programming language concepts. Good programming practices are also covered.

3.1 STEPS IN PROGRAM DEVELOPMENT

A program is a set of instructions that directs the computer to perform specific tasks. As we saw earlier (Section 2.4), these instructions should be unambiguous and can be specified using any of the following languages: machine, assembler, high-level, or very high level. In the present discussion, however, we will use the term *program* to mean a set of instructions written in a high-level language such as BASIC, FORTRAN, Pascal, or C.

Figure 3.1 shows the various steps in the development of a program. Note the direct correspondence with the steps discussed in the problem-solving methodology shown in Figure 2.1. The first step, analysis of the problem, is followed by development of the algorithm. The algorithm is expressed in *pseudocode,* which is a set of English-like statements. The purpose of the pseudocode is twofold: First, it facilitates program writing, and second, it enables a nonprogrammer to understand the logic behind the program.

The next step is to convert the pseudocode into program code using appropriate language syntax. During this process, necessary comments and documentation should also be included. Documentation will enhance the readability of the program and is

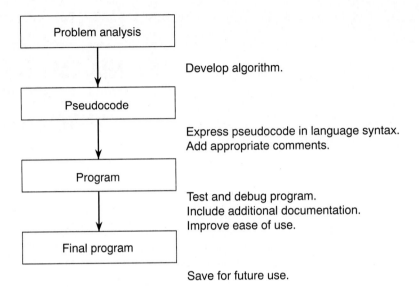

FIGURE 3.1 The various steps in program development.

almost essential if the program is to be used at a later date or by others. The program should be then tested and errors, or **bugs**, removed. Additional changes to and modifications of the program should be made to enhance the program's ease of use. The final step is to save the program for future use. This systematic approach will contribute to the development of good programming skills and practices. And the best way to learn a programming language is to practice using it.

3.2 ROOTS OF A QUADRATIC EQUATION

We will illustrate the programming process by developing a simple program for finding the roots of a quadratic equation: $4x^2 - 3x - 10 = 0$.

3.2.1 Developing the Algorithm

As you probably know, the roots x_1 and x_2 of a quadratic equation $ax^2 + bx + c = 0$ are given by the formulas:

$$x_1 = \frac{-b + \sqrt{b^2 - 4ac}}{2a} \tag{1}$$

$$x_2 = \frac{-b - \sqrt{b^2 - 4ac}}{2a} \tag{2}$$

So given the values of a, b, and c, we should be able to calculate the roots. In the present example, $a = 4$, $b = -3$, and $c = -10$. Using a calculator, here is how we would find the roots.

Inputs: $a = 4$
$b = -3$
$c = -10$

Calculations: $x_1 = \dfrac{-(-3) + \sqrt{(-3)^2 - 4(4)(-10)}}{2(4)}$

$= \dfrac{3 + \sqrt{9 + 160}}{8} = \dfrac{3 + 13}{8}$

$= 2$

$x_2 = \dfrac{-(-3) - \sqrt{(-3)^2 - 4(4)(-10)}}{2(4)}$

$= \dfrac{3 - \sqrt{9 + 160}}{8} = \dfrac{3 - 13}{8}$

$= -1.25$

Results: The roots of the quadratic equation are:

$x_1 = 2$

$x_2 = -1.25$

Figure 3.2 captures the sequence of steps we followed to compute the roots of the equation. This pseudocode is easily understandable and will serve as the basis for writing a program in a programming language. There are three major phases in the solution process: input, computation, and output. In the input phase, we obtain the values required for performing the calculations. In this case, they are the values of a, b, and c. In the next phase, we compute the values of the roots using the appropriate formulas. Finally, in the output phase, we print out the results of the computation, i.e., the roots. Thus, any program will contain statements for performing these three sets of activities. In addition, a program will typically have some initialization statements and termination statements as shown in Figure 3.2.

3.2.2 Translating the Algorithm into Program Code

Translation of the algorithm into program code is a simpler process than development of the algorithm. Because the free-form pseudocode should be specified without any

44 FUNDAMENTALS OF COMPUTER PROGRAMMING

	Algorithm	Phase/program
Step 0	Initiate the computation.	Initialization
Step 1	Ask for the values of a, b, and c.	Input
Step 2	Obtain the values of a, b, and c.	
Step 3	Compute x_1 using formula.	Computation
Step 3	Compute x_2 using formula.	
Step 4	Print the value of x_1.	Output
Step 5	Print the value of x_2.	
Step 6	End the computation.	Termination

FIGURE 3.2 An algorithm for computing the roots of a quadratic equation.

ambiguity in the programming language syntax, this task can be daunting for a novice programmer. Moreover, in a rush to develop program code, beginners often tend to overlook the importance of working out a sound algorithm and consequently end up with incorrect or poorly working programs. A systematic approach to the task of programming will minimize such risks and pitfalls. Such a systematic approach is also known as **structured** approach to programming.

Programming Language Concepts Every programming language has its own syntax. It is not our intention to cover the syntax of each in detail. Since our purpose is to teach basic programming concepts, we will limit our discussion to the fundamentals so that you can develop programs in the language of your choice. To simplify the learning process, we will use the syntax of the BASIC language in this discussion. Once the concepts are understood, it should be relatively easy to "pick up" the syntax of other languages. Figure 3.3 shows the first (and somewhat crude) attempt at developing a program in BASIC for the algorithm shown in Figure 3.2. Note the direct mapping from pseudocode to program code.

Statement Numbers Each program statement in the figure has a number associated with it—a characteristic of BASIC. This is similar to an address or an identifier for the statement. Typically, the program statements are executed sequentially starting with the first statement. However, the sequence can be altered during the course of program execution. The statement identifier assists in this transfer of control to a specific statement.

Comments Comment or remark statements are used to describe the program and to make it easy to read and understand. They are denoted by REM (for REMark) in the BASIC program shown in Figure 3.2. These statements are not executed by the computer. Good programming practice requires extensive documentation of the code. This is especially crucial in the real world of software development where several software engineers (programmers) work concurrently on a single program. Moreover, programmers move between projects, and it is necessary to maintain a guide to the program which takes the form here of comments and spaces between logical blocks of the program.

```
10    REM Program to Compute the Roots of a Quadratic Equation
20    REM Input Phase                                          ⎫
30    REM Obtain values of a, b, c                             ⎪
40    PRINT "Enter the value of a:"                            ⎪
50    INPUT A                                                  ⎬ Input
60    PRINT "Enter the value of b:"                            ⎪
70    INPUT B                                                  ⎪
80    PRINT "Enter the value of c:"                            ⎪
90    INPUT C                                                  ⎭
100   REM Computation Phase                                    ⎫
110   X1 = (-B+SQR(B^2-4*A*C))/(2*A)                           ⎬ Compute
120   X2 = (-B-SQR(B^2-4*A*C))/(2*A)                           ⎭
130   REM Output Phase                                         ⎫
140   PRINT "The first root of the equation is: ";X1           ⎬ Output
150   PRINT "The second root of the equation is: ";X2          ⎭
160   STOP                                                     ⎫
170   END                                                      ⎬ Terminate
```

FIGURE 3.3   BASIC quadratic equation solver: first attempt.

Variables Variables are used to represent values that can change. For example, *a*, *b*, and *c* are variables whose values can change when a different quadratic equation is solved. They represent specific memory locations or addresses in the computer's memory. Variable names must begin with a letter of the alphabet and can contain letters and numeric characters. The length of the variable name depends on the language. The names should be meaningful, when possible, e.g., ROOT, SUM. Certain words (names), specific to the language, are known as **reserved words** and cannot be used for variable names.

The three types of variables are **numeric**, **character**, and **logic**. Numeric variables represent numeric values and can be further classified into **integer** and **real** variables. The former are used to represent integer values, e.g., 45 or −23432, while the latter are used to represent real or floating-point numbers, e.g., 3.3234, −4.567, or 2.3×10^{-2}. Character variables are used to represent character strings such as "allow" and "available." In the present example, A, B, and C are the names of the three numeric variables.

Logic or boolean variables can take on one of two values: TRUE or FALSE. The specific value assigned to a logic variable depends on the condition being tested. For example, a logic variable, STATUS, is assigned TRUE if the condition being tested is "Is 5 greater than 3?"

Constants Constants are numeric values that do not change during a computation. For example, the value of π, 3.14159, is fixed and does not change during a computation. Numeric constants are of two types: real and integer. Thus 3.14159 is a real constant, while 1374 is an integer constant. Constants appearing several times in a program are normally saved in a variable (PI=3.14159) and later referenced by the variable name e.g., CIRCUM=PI*D. Note, however, that constants can also be strings or characters. Thus, "aluminum," "copper," and "zinc" are examples of string or character constants.

Declaration Statement A declaration statement is used to declare or define characteristics of variables. The declaration also allocates or reserves specific locations in the computer's memory for holding the values of the variables. A majority of BASIC dialects do not require declaration statements for simple variables. Some examples of declaration statements in FORTRAN are:

REAL A, B, C	A, B, C are real variables.
INTEGER NUM	NUM is an integer variable.
CHARACTER * 10 NAME	NAME is a string variable up to 10 characters long.

BASIC normally assumes variables are real, while FORTRAN treats variables beginning with the letters I through N as being integers.

Input Statement The input statement is used to obtain values for different variables in the program. Entered values are assigned to the respective variable names specified in the input statement. For example, in Figure 3.3, the first value that is entered in response to the prompt in statement 40 is assigned to the variable A, the second to the variable B, and so on. Another example of an input statement, in FORTRAN, is as follows:

READ * A, B, C	Read assigned values into A, B, and C.

Assignment Statement The assignment statement, a key construct of programming languages, is used to assign a value to a variable. The original value of the variable is replaced by the new value. For example, the assignment statement:

$$X = 45.25$$

assigns a value of 45.25 to the variable X. Its previous value, if any, is lost.

Computers store numbers into memory or read them from memory. The variable name X defines a memory location. The statement $X = 45.25$ should not be interpreted algebraically but should rather be read as "store the value 45.25 in the memory location X." Thus, if the above statement is followed by the statement:

$$X = X - 15.20$$

the value of X will be changed from 45.25 to 45.25 − 15.20 or 30.05. In everyday mathematics, the above statement is incorrect because the left-hand side value is not equal to the right-hand side value. However, the equal sign (=) in the assignment statement is not a true equal sign (as in mathematics) but is an assignment operator. So an assignment statement simply assigns the value of the right-hand-side expression to the variable on the left-hand side.

In Figure 3.3, the expressions on the right-hand side of statements 110 and 120 correspond to the formulas in Equations (1) and (2). Calculated values are assigned to variables X1 and X2 on the left-hand side of the assignment statements.

Arithmetic Operators and Precedence The five major arithmetic operations and their corresponding operators are: subtraction (−), addition (+), multiplication (*), division (/), and exponentiation (^ or ** in FORTRAN). The order in which the various operations in an expression are performed will affect the final value of that expression. For example, the statement:

$$A = 5 * 9 + 4/8$$

can assign several different values to A depending on the order in which the operations are performed. For example, if the operations are performed left to right, then A will be equal to 49/8 or 6.125, whereas if the order is multiplication, followed by division, and finally addition, the result will be 45 + 1/2 or 45.5. To avoid such confusion, every programming language has some conventions known as **rules of operator precedence**.

Typically, the operators are grouped into three major categories according to their *decreasing* order of precedence: (1) ^ or **; (2) / and *; and (3) + and −. Thus the exponentiation operator has the highest precedence and is applied first. Operators in the same group e.g., / and * in the second group, have the same precedence, and the operations are performed left to right in the order in which they appear in the expression.

Operator precedence can be altered by the addition of parentheses. The subexpressions in the leftmost and innermost parentheses are evaluated first using the precedence rules. Then any other subexpressions are evaluated. Finally, the resultant expression is evaluated left to right.

When these rules are applied to the earlier statement, A will be assigned a value of 45.5 as follows:

$$A = 5 * 9 + 4/8$$

Step 1 = 45 + 4/8
Step 2 = 45 + 0.5
Step 3 = 45.5

Another Example Consider the evaluation of the following statement:

$$A = 5 * (9 + 4/8)$$

Step 1 = 5 * (9 + 0.5)   Subexpression in parentheses
Step 2 = 5 * (9.5)
Step 3 = 47.5

In this example, the subexpression in the parentheses has been evaluated first followed by the remainder of the expression. Note that the same rules of precedence are applicable inside the parentheses.

Built-in Functions There are certain operations that are frequently performed in the course of problem solving. These include evaluating the square root of a number, finding the sine of an angle, or taking the logarithm of a number. Programming languages are equipped with such library or built-in functions that can be invoked when necessary. For example, the statement:

$$Y = SIN(X)$$

invokes the built-in SIN function on the argument X and assigns the result to the variable Y. Note that the function may require that the argument be specified in a particular unit. For instance, trigonometric functions such as the SIN function typically assumes that the argument is in radians. In the quadratic equation solver shown in Figure 3.3, the SQR function is used in statement 110 to find the square root of (B ^ 2 − 4 * A * C). The programming language manual contains the list of available library functions.

Output Statement The output statement is used to transmit computation results, i.e., values of different variables, error messages, and program status information. For example, in Figure 3.3, values of the roots are printed out in statements 140 and 150. Including such output statements for printing out values of intermediate variables in the program will greatly facilitate program **debugging**, the process of tracking and correcting errors in the program.

Termination Statement The termination statement is used to signify the end of the program. Since program statements are executed sequentially, the compiler or interpreter must know that the end of the program has been reached. END is commonly used as a program termination statement. The STOP statement in Figure 3.3 halts program execution, it does not terminate the execution. The STOP statement in BASIC allows you to execute only a certain portion of the code and is therefore a good debugging aid.

3.2.3 Executing the Program

The next step in the programming process is to execute the developed code, also referred to as the **source code.** How this step is executed depends on the programming language used. As we discussed in 2.4.1, if the source code is in a compiler language, e.g., FORTRAN, Pascal, or C, it has to be compiled prior to execution. During the compilation process, the compiler checks the syntax of the source code and flags errors with appropriate diagnostic messages. For example, if SIN is the correct syntax for the built-in trigonometric sine function and the program contains a statement Y = SINE(X), it will be detected by the compiler. It cannot, however, detect any flaws in the program logic. Once errors are corrected and the program is compiled without any errors, the resulting **object code** is ready to be executed.

If the source code is in an interpretive language such as BASIC, the interpreter checks the syntax during program execution one line at a time. If a syntax error is found in a particular statement, the execution is terminated with an appropriate message. When the error is corrected, the program can be executed again. The interpreter, like the compiler, will not detect flaws in the program logic. Flaws or errors in a program are referred to as bugs. The process of detecting and correcting bugs is known as debugging.

Program Tracing Let's trace the execution of the program in Figure 3.3 by "playing the computer." This means, we will execute the program step by step, just as the computer would, and record the actions. Such hand simulation is an effective means of testing the correctness of the program. It helps to understand the flow of computations and to detect flaws in the program logic. Hand simulation is also a simple debugging technique for small programs or small sections of a program.

Figure 3.4 shows a trace of the program when solving the equation $4x^2 − 3x − 10 = 0$. A part of each program statement is shown on the left and is followed by

3.2 ROOTS OF A QUADRATIC EQUATION

	Variable values					
Program statement	A	B	C	X1	X2	Action
REM Program to	?	?	?	?	?	Remark statement
REM Input phase	?	?	?	?	?	Remark statement
REM Obtain values	?	?	?	?	?	Remark statement
PRINT "Enter the	?	?	?	?	?	Print prompt for a
INPUT A	4	?	?	?	?	Assign 4 to A
PRINT "Enter the	4	?	?	?	?	Print prompt for b
INPUT B	4	−3	?	?	?	Assign −3 to B
PRINT "Enter the	4	−3	?	?	?	Print prompt for c
INPUT C	4	−3	−10	?	?	Assign −10 to C
REM Computation	4	−3	−10	?	?	Remark statement
X1 = (−B+SQR(B^2	4	−3	−10	2	?	Compute X1
X2 = (−B−SQR(B^2	4	−3	−10	2	−1.25	Compute X2
REM Output phase	4	−3	−10	2	−1.25	Remark statt.
PRINT "The first	4	−3	−10	2	−1.25	Print first root
PRINT "The second	4	−3	−10	2	−1.25	Print second root
STOP	4	−3	−10	2	−1.25	Halt program
END	4	−3	−10	2	−1.25	End program

FIGURE 3.4 Hand trace of program in Figure 3.3.

values of the variables after the execution of the statement. The action of each statement is also shown in the figure. To begin with, the values of the five variables are unknown and are indicated as such by the question marks. The three REM statements have no effect on the values of the variables. Then the program prints a prompt for the value of A. When the value of 4 is entered in response to the prompt, the variable A is assigned that value. Similarly, variables B and C are assigned −3 and −10, respectively.

Evaluation of Roots The evaluation of the expression for X1 in the program follows the rules of operator precedence discussed earlier and is carried out as follows:

$$X1 = (-B + SQR(B \wedge 2 - 4 * A * C))/(2 * A)$$

Starting from the left:

Step 1 $(B \wedge 2 - 4 * A * C) = ((-3)^2 - 4 * 4 * (-10)) = 169$
Step 2   SQR(169)   = 13
Step 3 $(-B + 13)$ $= (-(-3) + 13) = 16$
Step 4 $(2 * A)$ $= (2 * 4) = 8$
Step 5   16/8   = 2

Thus X1 is assigned the value of 2 as shown in Figure 3.4. In a similar manner, the expression for X2 is also evaluated and is assigned a value of −1.25.

Printing the Results The following REM statement is not executed, and the values of X1 and X2 are printed in the next two statements. The program halts at the STOP statement and is terminated by the END statement.

Thus, by playing the computer, we have executed the complete program and obtained the results. These results agree with those obtained with a calculator in 3.2.1. So the algorithm and the program are correct for solving the given quadratic equation. When the program is entered in the computer and executed or **run**, it should work fine. We can then save it on a secondary storage media, viz., hard or floppy disk for later use.

3.2.4 Improving and Generalizing the Program

Let's look at the two computation statements in Figure 3.3. They are:

110   X1 = (−B + SQR(B^2 − 4 * A * C))/(2 * A)
120   X2 = (−B − SQR(B^2 − 4 * A * C))/(2 * A)

There are several common subexpressions in the two statements, e.g., the discriminant (B^2 − 4 * A * C) and the denominator (2 * A). So instead of evaluating the common subexpressions in each statement, we can evaluate them once, assign the values to new variables, and use the variables to compute X1 and X2. In a simple example like the present one, this redundant computation may not appear to be wasteful, but bigger programs can be slowed down by such redundancies. It is also a good programming practice to avoid unnecessary computations. The program with these modifications is shown in Figure 3.5. Note the choice of the new variable names: They are mnemonic.

Solving Another Equation Let's use the program in Figure 3.5 to evaluate the roots of another quadratic equation: $2x^2 - 3x + 5 = 0$. To avoid repetition, we will discuss only the important steps in the execution of the program.

Based on the values of $A = 2$, $B = -3$, and $C = 5$, DISC is assigned a value of −31, and DENOM is assigned 4 in statements 110 and 120, respectively:

DISC = B^2 − 4 * A * C = (−3)² − 4 * 2 * 5 = 9 − 40 = −31
DENOM = 2 * A = 2 * 2 = 4

When the interpreter executes the next statement (130) to compute X1, it attempts to evaluate the expression SQR(DISC), i.e., SQR(−31); it encounters an error and terminates the program with the message "Illegal function call in 130." This is because an attempt has been made to evaluate the square root of a negative number which is an illegal operation in BASIC. Since the discriminant, DISC, is negative, we know that the roots of the equation are complex. So we realize now that the program in Figure 3.5 is designed to handle only real roots and fails, or **bombs out** (the more commonly used phrase), when the roots are imaginary! What we need is a way to control the computation flow in the program so that this problem does not arise, i.e., we should skip executing statements 130 and 140 when DISC is negative.

Controlling the Flow of Computation In general, statements in a program are executed one after another in sequence. The flow of computation is said to be

```
10      REM Program to Compute the Roots of a Quadratic Equation
20      REM Input Phase
30      REM Obtain values of a, b, c
40      PRINT "Enter the value of a:"
50      INPUT A
60      PRINT "Enter the value of b:"
70      INPUT B
80      PRINT "Enter the value of c:"
90      INPUT C
100     REM Computation Phase
110     DISC = B^2-4*A*C
120     DENOM = 2*A
130     X1 = (-B+SQR(DISC))/DENOM
140     X2 = (-B-SQR(DISC))/DENOM
150     REM Output Phase
160     PRINT "The first root of the equation is: ";X1
170     PRINT "The second root of the equation is: ";X2
180     STOP
190     END
```

} Input

} Compute

} Output

} Terminate

FIGURE 3.5   BASIC quadratic equation solver: modified version.

sequential as in the program in Figure 3.5. Starting with the first statement, each statement was executed one by one. However, there are times when we want to control the flow of computation, i.e., selectively execute specific segments of the program and in the desired order. Programming languages provide control statements for directing the flow of computation in the program.

Control or Decision-Making Statement   The control or decision-making statement is used to control the sequence and frequency of execution of parts of a program. Typically, the decision is based on a certain condition, e.g., the value of a particular variable. Relational operators such as equal to (=), less than (<), greater than (>), etc., are used for comparing the values of two expressions and arriving at a decision. The general syntax of a control or decision-making statement is as follows:

```
        IF <condition> THEN
            Execute statement 1
                Statement 2
                    ...
                Statement n
        [ELSE
            Execute statement a
                Statement b
                    ...
                Statement k]
        ENDIF
```

THEN clause

ELSE clause

The decision-making statement has two clauses: the THEN clause and the ELSE clause. If the condition being tested in <condition> is true, the first block of statements or the THEN clause is executed; otherwise, the second block of statements or the ELSE clause is executed. The ELSE clause is optional; hence the brackets around the ELSE clause. The various relational operators for testing the conditions are shown in Table 3.1. However, the exact syntax of the statement and the relational operator will vary from one language to another.

Figure 3.6 shows the pseudocode and the corresponding BASIC statement (125) for handling complex roots in the program. When DISC is negative, e.g., −31 in the

TABLE 3.1 RELATIONAL OPERATORS IN IF-THEN-ELSE STATEMENTS

Condition	Operator	Operator
Equal to	.EQ.	=
Not equal to	.NE.	<>
Less than	.LT.	<
Less than or equal to	.LE.	<=
Greater than	.GT.	>
Greater than or equal to	.GE.	>=
AND	.AND.	AND
OR	.OR.	OR

FIGURE 3.6 Controlling the flow of computation in the program. (*a*) Pseudocode. (*b*) Modified version of program in Figure 3.5.

```
Evaluate DISC (as in statement 110, Figure 3.5)
Begin Decision-Making
    IF DISC is negative, THEN print a message that roots are imaginary and stop
    Otherwise (ELSE)
        Go ahead with the evaluation of roots
End Decision-Making
```
(*a*)

```
10    REM Program to Compute the Roots of a Quadratic Equation
20    REM Input Phase
30    REM Obtain values of a, b, c
40    PRINT "Enter the value of a:"
50    INPUT A
60    PRINT "Enter the value of b:"
70    INPUT B
80    PRINT "Enter the value of c:"
90    INPUT C
100   REM Computation Phase
110   DISC = B^2-4*A*C
120   DENOM = 2*A
125   IF DISC<0 THEN PRINT "The roots are imaginary": GO TO 180
130   X1 = (-B+SQR(DISC))/DENOM
140   X2 = (-B-SQR(DISC))/DENOM
150   REM Output Phase
160   PRINT "The first root of the equation is: ";X1
170   PRINT "The second root of the equation is: ";X2
180   STOP
190   END
```
(*b*)

present equation, the condition IF DISC < 0 evaluates as true, and therefore the THEN clause is executed; a message is printed and control is transferred to statement 180 by the GO TO statement. The other statements, 130 through 170, are skipped. If the result is positive, e.g., 169 in the previous equation, the condition is not true; since there is no ELSE clause in 125, control is automatically transferred to the next statement, viz., 130, and the program proceeds with the evaluation of the roots. Thus, the IF-THEN-ELSE construct can be used effectively to sequence the flow of computation in the program.

The GO TO statement is a highly controversial statement among the computer science community. Inasmuch as it is a powerful construct, its overuse in the program to temporarily "fix" problems can be harmful because it can mask the program logic and render it incomprehensible for others and even for the programmer. Program code with extensive GO TO statements is referred to as "spaghetti code" — hard to figure out where computation starts and where it ends! One of the highlights of structured programming is the complete avoidance of GO TO statements and using other means or constructs to accomplish the same tasks.

Another Modification We know that when the discriminant, DISC, is zero, there is only one real root. The program shown in Figure 3.6 does not explicitly address this issue. It just computes the roots twice and prints them out. This is because there is no distinction made between the two values of DISC: zero and positive. The program merely traps the negative condition. The pseudocode for distinguishing between zero and positive values of DISC is shown in Figure 3.7.

When DISC is zero, the THEN clause is executed, the only real root is computed, and the appropriate message is printed. Control is transferred to the end of the IF-ENDIF block. If DISC is positive, as in the previous equation, the ELSE IF clause is executed: the two roots are computed and printed, and control is transferred to the end of the IF-ENDIF block. If neither of the above two conditions is true, the ELSE clause prints the complex roots statement and exits the block. Thus, each of the conditions in the IF-ENDIF block is tested in sequence, and when a condition is satisfied, the appropriate statements are executed and control transferred to the end of the block. So it is important to sequence the conditional tests properly to avoid logical flaws.

FIGURE 3.7 Pseudocode for distinguishing between zero and positive values of DISC.

```
IF DISC=0 THEN
    X1=-B/DENOM
    PRINT "There is only one real root"
    PRINT "The root is:";X1
ELSE IF DISC>0 THEN
    X1 = (-B + SQR(DISC))/DENOM
    X2 = (-B - SQR(DISC))/DENOM
    PRINT "The first root is:";X1
    PRINT "The second root is:";X2
ELSE
    PRINT "The roots are complex"
ENDIF
```

3.3 REPETITIVE COMPUTATIONS AND THE LOOP CONSTRUCT

In summary, using the quadratic equation example, we have seen how a program can be developed from an algorithm and then modified gradually to broaden its scope and remove its shortcomings. In the everyday world of software engineering, such a modular approach is adopted to enhance the program's capabilities.

We often find repetitive tasks cumbersome and boring. On the other hand, one of the major advantages of computers is that they excel in such repetitive tasks. Every programming language has a loop construct for performing repetitive computations, i.e., repeatedly executing a certain segment of the code. This repetitive process is also known as **iteration**. When the code is executed a specified number of times, it is known as **definite iteration.** When the number of executions depends on certain conditions, it is known as **indefinite iteration.** We will use a basic example from physics to illustrate the loop construct.

3.3.1 Ohm's Law

Ohm's law for electrical conductors is given by the relationship:

$$\text{Current} = \frac{\text{voltage}}{\text{resistance}} \tag{3}$$

For a given voltage, current increases as resistance decreases. Such an experiment can be easily carried out in the laboratory with the help of a battery and a variable resistor.

Programming Task   In an experiment, a 12-V battery is connected to a variable resistor. Write a program to compute the current in the circuit as the resistance is increased from 2 to 22 Ω in steps of 4 Ω.

This task is straightforward; the initial version of the pseudocode is shown in Figure 3.8. After the initialization in step 0, we obtain the values of the different variables in step 1. Then the first value of current CURR is computed and printed in step 3. Thereafter the process becomes repetitive: for every new value of resistance, the current should be computed and printed. This increment-compute-print cycle should be repeated until all values of resistance are exhausted. In the figure, a separate set of statements is shown for each of the cycles—a clear redundancy. The other problem with the code is that it can handle exactly six cycles of computations. If there are fewer values, it will not terminate properly, and if there are more than six sets, the program will terminate prematurely.

The Loop Construct   The loop construct is used for performing repetitive computations with a minimum number of program statements. The modified pseudocode in Figure 3.8 illustrates the concept of looping. The computations in step 2 of the pseudocode can be repeated as many times as needed until the final value of resistance, 22 Ω, is reached. The code is shorter, easier to modify, and easier to understand. The general syntax of the loop construct in BASIC is as follows:

Step 0	Initiate computation. Let the variables be VOLT, CURR, IRES, FRES, INCR; IRES, FRES, INCR correspond to the initial, final, and incremental steps of resistance.	
Step 1	Ask for the values of VOLT, IRES, FRES, INCR	[12, 2, 22, 4].
Step 2	Compute CURR = VOLT/IRES.	6
Step 3	Print the value of CURR.	
Step 4	Increment IRES = IRES + INCR.	6
Step 5	Compute CURR = VOLT/IRES.	2
Step 6	Print the value of CURR.	
Step 7	Increment IRES = IRES + INCR.	10
Step 8	Compute CURR = VOLT/IRES.	1.2
Step 9	Print the value of CURR.	
Step 10	Increment IRES = IRES + INCR.	14
Step 11	Compute CURR = VOLT/IRES.	0.85714
Step 12	Print the value of CURR.	
Step 13	Increment IRES = IRES + INCR.	18
Step 14	Compute CURR = VOLT/IRES.	0.66667
Step 15	Print the value of CURR.	
Step 16	Increment IRES = IRES + INCR.	22
Step 17	Compute CURR = VOLT/IRES.	0.54545
Step 18	Print the value of CURR.	
Step 19	Terminate computation.	

(a)

Step 0	Initiate computation. Let the variables be VOLT, CURR, IRES, FRES, INCR.
Step 1	Ask for the values of VOLT, IRES, FRES, INCR.

Step 2	Begin loop.
	    Compute CURR = VOLT/IRES.
	    PRINT CURR.
	    Increment IRES = IRES + INCR.
	    IF IRES > FRES exit loop.
	Repeat loop.

Step 3	Terminate computation.

(b)

FIGURE 3.8 Pseudocode for repetitive computations. (*a*) Initial version. (*b*) Modified version.

```
         ⎧ FOR <index> = <initial> TO <final> [STEP <size>]
         ⎪     Execute    Statement 1
         ⎪                Statement 2
  Loop  ⎨                Statement 3
         ⎪                   ...
         ⎪                Statement n
         ⎩ NEXT <index>
```

Or, in FORTRAN

```
DO <label> <index> = <initial>, <final>, [<size>]
    Execute    Statement 1
               Statement 2
               Statement 3
                  ...
               Statement n
<label>    CONTINUE
```
} Loop

The number of times the loop is executed is determined by the value of <index>. It is initially set to the value of <initial>, and its value is incremented at the end of the loop (in the NEXT <index> statement) by the value specified in <size>. When the value of <index> exceeds the value of <final>, the execution of the loop is terminated. Note that the STEP <size> part of the construct is optional. If it is not specified, <index> is incremented by 1. All the loop control values can be either positive or negative. In the FORTRAN DO LOOP, the <label> must be a valid label identity.

Figure 3.9 shows the implementation of the FOR-NEXT construct in the Ohm's law example and the trace of the execution of the loop. At the beginning of the loop, J is assigned the value of IRES, viz., 2, and CURR is computed in the next statement and printed. When the NEXT J statement is executed, J is incremented by the step size, viz., 4 (the value of INCR), and control is transferred to the FOR statement. Since the new value of J, viz., 6, is less than the value of FRES, viz., 22, execution proceeds and CURR is evaluated. This process is repeated six times, and at the end of the sixth cycle J is assigned a value of 26. When control is transferred to the FOR statement, J is greater than FRES and the loop is exited as shown in the figure. Thus the loop construct is a very powerful and useful feature in programming languages.

3.4 ARRAY DATA STRUCTURE AND APPLICATION OF LOOPS

One of the frequently encountered tasks in computing is the sorting of data in either increasing or decreasing order of magnitude. Whether the data pertains to scores in an examination, salaries in a corporation, or weights of individuals on a diet, comprehension of the information is enhanced if the data are presented in a meaningful manner. And sorting the data is an appropriate technique in all three cases.

Consider the points obtained by students on an examination: 35, 89, 23, 64, 79, 94, 55, 67, 81, and 43. The task is to sort these numbers in either descending or ascending order so that appropriate grades can be assigned based on preset criteria.

3.4.1 A Sorting Algorithm

Let's assume that the scores are to be sorted in increasing order of magnitude. A simple strategy is to start from the first position and compare the elements (or scores) in the first and second positions. If they are out of order, i.e., if the first element is larger than the second, we swap them. Then we compare the elements in the second and third

```
10      REM Illustration of the Loop Construct
20      INPUT "Enter the value of voltage:"; VOLT
30      INPUT "Enter the initial resistance value:"; IRES
40      INPUT "Enter the final resistance value:"; FRES
50      INPUT "Enter the incremental value:"; INCR
60      FOR J = IRES TO FRES STEP INCR
70          CURR = VOLT/J
80          PRINT "The current is:"; CURR
90      NEXT J
100     STOP
110     END
```

(a)

Loop cycle	Variable values						Action
	J	VOLT	IRES	FRES	INCR	CURR	
1	2	12	2	22	4	?	J = IRES
	2	12	2	22	4	6	Compute CURR.
	6	12	2	22	4	6	J = J + INCR
2	6	12	2	22	4	6	J < FRES
	6	12	2	22	4	2	Compute CURR.
	10	12	2	22	4	2	J = J + INCR
3	10	12	2	22	4	2	J < FRES
	10	12	2	22	4	1.2	Compute CURR.
	14	12	2	22	4	1.2	J = J + INCR
4	14	12	2	22	4	1.2	J < FRES
	14	12	2	22	4	0.85714	Compute CURR.
	18	12	2	22	4	0.85714	J = J + INCR
5	18	12	2	22	4	0.85714	J < FRES
	18	12	2	22	4	0.66667	Compute CURR.
	22	12	2	22	4	0.66667	J = J + INCR
6	22	12	2	22	4	0.66667	J = FRES
	22	12	2	22	4	0.54545	Compute CURR.
	26	12	2	22	4	0.54545	J = J + INCR
7	26	12	2	22	4	0.54545	J > FRES, exit loop

(b)

FIGURE 3.9 The loop construct in Ohm's law example. (a) BASIC program. (b) Execution trace of loop.

positions and swap them if they are out of order. We repeat this process of comparing adjacent elements until the end of the list of scores. At this time, the largest element should be in the last position. Since the largest element is being gradually "bubbled" through the list of scores, this sorting algorithm is known as the **bubble sort.**

Figure 3.10 shows the working of the algorithm on the scores. At the end of the first set of comparisons, also known as a "pass," 94 is in the last position. There have been nine comparisons during this pass, one less than the number of elements, N, viz., 10. Since the highest score is in its final position, only the first nine elements need

<u>35</u>	<u>89</u>	23	64	79	94	55	67	81	43
35	<u>89</u>	<u>23</u>	64	79	94	55	67	81	43
35	23	**89**	<u>64</u>	79	94	55	67	81	43
35	23	64	**89**	<u>79</u>	94	55	67	81	43
35	23	64	79	**89**	<u>94</u>	55	67	81	43
35	23	64	79	89	<u>94</u>	<u>55</u>	67	81	43
35	23	64	79	89	55	**94**	<u>67</u>	81	43
35	23	64	79	89	55	67	**94**	<u>81</u>	43
35	23	64	79	89	55	67	81	**94**	<u>43</u>
35	23	64	79	89	55	67	81	43	**94**

Pass = 1 (rows above)

23	35	64	79	55	67	81	43	**89**	94	After pass = 2
23	35	64	55	67	79	43	**81**	89	94	After pass = 3
23	35	55	64	67	43	**79**	81	89	94	After pass = 4
23	35	55	64	43	**67**	79	81	89	94	After pass = 5
23	35	55	43	**64**	67	79	81	89	94	After pass = 6
23	35	43	**55**	64	67	79	81	89	94	After pass = 7
23	<u>35</u>	<u>43</u>	55	64	67	79	81	89	94	After pass = 8
<u>23</u>	<u>35</u>	43	55	64	67	79	81	89	94	After pass = 9

FIGURE 3.10 The bubble sort algorithm: The highest test score has moved up to the last position after the first pass of the algorithm. *Notes:* (1) Elements being compared are underlined; when there is a swap, the bigger element is shown in boldface. (2) Only the first pass is shown in its entirety; the other positions represent states after the respective passes.

to be compared in the second pass, giving rise to $N - 2$ comparisons. So, for a given pass, the number of comparisons will be $N -$ pass number. And the total number of passes for sorting all the elements will be $N - 1$.

The figure also shows positions of the elements after each of the subsequent passes 2 through 9. In this example, it so happens that all the elements are in order after the seventh pass. There are no swaps in the last two passes. However, the current algorithm does not recognize the fact that when there are no swaps during a particular pass, all elements are in order and the sorting process can be terminated. It makes the maximum number of passes, viz., $N - 1 = 9$.

3.4.2 Implementing the Sorting Algorithm

We will follow our problem-solving methodology and move on to the next step, viz., implementation of the algorithm in a programming language.

Representing the Scores in the Computer The first thing we need is a scheme for representing the test scores in the program so that we can reference them for comparison and swapping purposes. Such a representation scheme is called a **data structure.** We will use the **array** data structure to represent the numbers.

Arrays An array is a set of data elements of the same type referred to by a single name. An individual element in the array can be referred to by its position, i.e., by its subscript. The type and number of elements in the array must be declared at the beginning of the program using a **dimension** statement. The declaration allocates a sequence of locations in the computer's memory for storing the data elements. The general syntax for array declaration is as follows:

DIMENSION <array name>(# elements)   (BASIC)
OR   <variable type><array name>(# elements)   (FORTRAN)

For example, the array called SCORE for storing the test scores could be declared as follows:

DIMENSION Score(10)
OR   REAL Score(10)
OR   DIM SCORE(10)

The number in the parentheses indicates the array size, i.e., the number of elements that can be stored in the array. The array can be pictured as a set of pigeon holes in a post office as shown below:

| SCORE(1) | SCORE(2) | SCORE(3) | SCORE(4) | ... | SCORE(10) |

The elements can be read into the array using input statements. The elements are referred to as SCORE(1), SCORE(2), SCORE(3), ..., SCORE(10). Let's assume that the test scores are stored in the array as shown below:

| 35 | 89 | 23 | 64 | 79 | 94 | 55 | 67 | 81 | 43 |
| SCORE(1) | SCORE(2) | SCORE(3) | SCORE(4) | | | ... | | | SCORE(10) |

Then, SCORE(5) refers to the fifth element in the array, viz., 79. Any of the elements can be accessed by specifying the array subscript. The statement Y = SCORE(8) would access the eighth element in the array, and assign the value 67 to the variable Y. Thus, the array is a powerful data structure for storing and retrieving data elements of the same type (numbers or strings). Instead of using 10 different variable names to refer to the 10 test scores, we have used a single array. This type of an array is known as a **one-dimensional array** because there is only one subscript. For storing the values of a two-dimensional matrix, a two-dimensional array can be declared and used. Once the data are stored in an array, they can be used as many times as needed in the program. This is another major advantage of arrays.

The Program Code Figure 3.11 shows the refined algorithm for sorting a set of numbers, while Figure 3.12 shows the code in BASIC. In Figure 3.11, a logic variable, FLAG, is used to see if an element swap occurs during a pass. If a swap occurs, it is set to .TRUE., else it remains .FALSE. After the initialization statements, we check to see if the number of elements is less than the declared array size and then proceed appropriately. Otherwise, an error will occur. This is yet another good programming practice.

Then we begin sorting. This should be carried out only if the number of passes is less than the number of elements and if a swap has occurred in the previous pass. This ensures the termination of the program as soon as all the elements are in their correct positions (an improvement over the previous algorithm). The WHILE-END WHILE loop is used for testing and carrying out the computations only if the conditions are satisfied. This is an example of indefinite iteration. In contrast, the number of comparisons in every pass is known (N – pass number): a case of definite iteration. So the FOR-NEXT loop is used for comparisons and swaps; if a swap occurs, the FLAG is set to .TRUE. When this loop is exited, it signifies the completion of a pass; the variable PASS is incremented by 1, and control is transferred back to the WHILE

FIGURE 3.11 A refined pseudocode for the bubble sort algorithm.

Initialize variables.
    Declare array SCORE with 100 elements.
    FLAG: Logic variable to keep track of element swaps; initialized to .TRUE.;
    changed to .FALSE. if no swap occurs.
    PASS: Number of passes, initialized to 1.
    N: Number of test scores.
    I, J: Temporary variables for loop control.

Read elements into array.
    Get the number of values, N.
    IF N > 0 and ≤ 100, go ahead with getting the numbers, else error message.
    FOR I = 1 to N
        Read the N elements into the array.
    NEXT I

Begin sorting.
    While PASS < N *and* FLAG = .TRUE.
        Execute the following:
            Set FLAG = .FALSE.
            FOR J = 1 to N – PASS
                IF SCORE (J + 1) < SCORE(J)
                    Swap the elements.
                    Set FLAG = .TRUE.
            NEXT J
            PASS = PASS + 1
    End while.
End sorting.

Print sorted elements.
    FOR I = 1 to N
        Print the array element SCORE(I).
    NEXT I

Terminate computations.

```
10    REM Sorting numbers in ascending order using Bubble Sort
20    REM Array size is set to 100.
30    REM N is number of elements, PASS is number of passes
40    REM If a swap occurs, FLAG is set to 1, else 0.
50    REM IF there is no swap during a pass, sorting is stopped.
60    DIM SCORE(100)
70    PASS=1: FLAG=1: TEMP=0
80    INPUT "Enter the number of values (0 < N <= 100):"; N
90    REM Test number of values and proceed appropriately
100   IF N <= 0 OR N > 100 THEN GOTO 80
110   FOR I = 1 to N
120      PRINT "Enter the test score:";
130      INPUT SCORE(I)
140   NEXT I
150   REM Sorting Phase
160   FOR I = 1 to N-1
170      FLAG = 0
180      FOR J = 1 to N-PASS
190         IF SCORE(J+1) < SCORE(J) THEN
200            TEMP = SCORE(J)
210            SCORE(J) = SCORE(J+1)
220            SCORE(J+1) = TEMP
230            FLAG = 1
240         END IF
250      NEXT J
260      PASS = PASS+1
270      IF FLAG=0 THEN 290
280   NEXT I
290   REM Printing routine
300   FOR I = 1 to N
310      PRINT SCORE(I)
320   NEXT I
330   STOP
340   END
```

FIGURE 3.12 A BASIC program for bubble sort. *Note:* The While loop in the pseudocode is represented with a For-next loop (150 to 280).

statement. Once all the elements are sorted, a condition denoted by FLAG = .FALSE., they are printed out.

Note the direct mapping between the pseudocode and the BASIC program. In the program, there are two loops, one within the other. This is called **nesting of the loops.** In nesting loops, care should be taken to ensure that the loop index of one loop does not cross the index of the other. Thus, the loop with J as the index variable is completely inside the loop with the I index. If this is not the case, the interpreter will be confused and the program will bomb out.

Implementing the Swapping of Elements Swapping two elements implies storing the value of SCORE(J) in SCORE(J + 1) and vice versa. Some programming languages have the SWAP statement, e.g., SWAP SCORE(J), SCORE(J + 1), which takes care of swapping the values. In Figure 3.12, we do not use the SWAP statement. Instead, we store the value of SCORE(J) in a temporary variable, TEMP, and after assigning the value of SCORE(J + 1) to SCORE(J), the value of TEMP is assigned to SCORE(J + 1). The sequence of statements is important. If the value of SCORE (J) is not stored temporarily, it will be overwritten by the value of SCORE(J + 1) during the swap.

Modular Approach   In developing the code, we have taken the modular approach: an input module to read in the values, a sorting module for the operations, and an output module to print out the results. This modular approach can be further enhanced by the use of subroutines, as we will see in the next example.

3.5 STATISTICAL COMPUTATIONS AND SUBROUTINES

Experimentation is at the heart of any scientific and engineering activity. And statistics deals with techniques for designing experiments, collecting and analyzing data, and drawing conclusions from the experimental data. The computer can be a valuable tool in statistical analysis. We will illustrate additional programming language concepts including subroutines by developing a program for some basic statistical computations.

3.5.1 A Program for Simple Statistics

Let's consider the following example. A new method of producing yarns has been developed in a laboratory. The properties of the yarn such as its tensile strength will be influenced by variability in the process. Table 3.2 shows the results of tensile tests carried out on a yarn produced by this new process. We need to assess the reliability of the process in producing yarn of uniform strength.

Need for Statistical Measures   An examination of the 15 data points in the table shows that the values range from 205 to 275 g. However, the numbers, as presented, are not adequate for assessing the process reliability in producing yarn of uniform strength. We can use some common statistical measures for a better interpretation of the data.

The **mean** is the most common location statistic. It is defined as the sum of the individual data points divided by the number of points. Mathematically, it is given by:

$$\bar{x} = \frac{1}{n} \sum_{i=1,n} x_i \qquad (4)$$

The **standard deviation** is a measure of dispersion of a set of data. Mathematically, it is the square root of the arithmetic mean of the squares of the deviation of each data point from the mean of the data set. It is given by:

TABLE 3.2   YARN STRENGTH RESULTS: BREAKING LOAD DATA, GRAMS

260	252	278	285	277
210	223	205	245	279
255	240	250	240	270

$$\sigma = \sqrt{\frac{\sum_{i=1,n}(x_i - \bar{x})^2}{n-1}} \tag{5}$$

The numerator in Equation (5) is often referred to as the "sum of squares."

The **variance** is another measure of the dispersion of a set of data and is given by the square of the standard deviation:

$$\text{Variance} = \sigma^2 \tag{6}$$

The **coefficient of variation** is a measure of variability of the data and is the ratio of the standard deviation to the mean. It is often expressed as a percentage and is given by:

$$\text{CV percent} = \frac{\sigma}{\bar{x}} 100 \tag{7}$$

Thus, these four measures of statistics will provide a reasonable estimate of the spread in the yarn strength data.

Programming Task The task is to develop a program for computing the four measures of statistics for the data in Table 3.2.

3.5.2 Developing the Algorithm

The programming task can be broken down into three distinct subtasks: (1) obtaining the individual data points; (2) computing the various statistics; and (3) printing the data points and the results. Each of these subtasks can be handled as individual modules and a code developed for each of the modules.

Obtaining the Data Points It is clear from an analysis of the task and the various formulas that the data points must be stored and used for repetitive computations: for computing the mean and again for the standard deviation. As in the previous example, an array data structure along with the loop construct will meet the needs. The pseudocode is shown in Figure 3.13.

Compute the Various Statistics Computing the mean requires summation of the individual data points stored in the array: The loop construct can be used. According to Equation (5), the computations of the deviations $(x_i - \bar{x})$ would also require the loop construct. Moreover, the value of $\bar{x}$ should be known for computing the mean deviations, and hence, the standard deviation. But, if we expand the squared term in Equation (5) and rearrange, we can perform all computations in a single pass. The new equation for a single-pass computation would be as follows:

$$\sigma = \sqrt{\frac{\sum_{i=1}^{n} x_i^2 - \frac{1}{n}\left(\sum_{i=1}^{n} x_i\right)^2}{n-1}} \tag{8}$$

1. Obtain the data points:
    Array: X(size) for storing data points
    Array: DEV(size) for storing $(x_i - \bar{x})^2$
    Variables: N = number of data points
        SUM = sum
        SUMSQ = sum of squares   } initialized to 0
        MEAN = mean
        SDEV = standard deviation
        VARI = variance
        CV = coefficient of variation
        I = temporary variable
    Obtain value of N.
    Use loop construct for getting individual data points.
    FOR I = 1 TO N
        ASK and STORE value in X(I).
    NEXT I
2. Compute various statistics:
    Compute the mean.
        Sum individual values using a loop construct.
        FOR I = 1 to N
            SUM = SUM + X(I)
        NEXT I
    MEAN = SUM/N
    Compute standard deviation.
        Compute deviation from mean $(x_i - \bar{x})$ and sum of squares of deviations $(x_i - \bar{x})^2$.
            FOR I = 1 to N
                DEV(I) = X(I) − MEAN
                SUMSQ = SUMSQ + DEV(I) * DEV(I)
            NEXT I
        STDEV = square root of (SUMSQ/(N − 1))
    Compute variance.
        VARI = STDEV^2
    Compute CV%.
        CV = STDEV/MEAN * 100
3. Print the values and results:
    Use loop construct to print the individual values.
    FOR I = 1 to N
        PRINT X(I)
    NEXT I
    PRINT MEAN, STDEV, VARI, CV

FIGURE 3.13 The pseudocode for statistical computations.

It is considered good practice to analyze problems and take such steps to minimize computations. In the present example, however, we will stick to the two-pass computation using Equation (5). Once the mean and standard deviation are computed, the variance and coefficient of variation can be computed easily. The pseudocode for this segment of the program is also shown in Figure 3.13.

Printing the Results The individual data points can be printed using a loop construct. However, if the values have to be printed as in the table, some **format** statements should be used for aligning the values. Examples of formatted PRINT statements in BASIC and FORTRAN, respectively, are as follows:

Print TAB(5) x TAB(15) y TAB(30) z   (BASIC)

Print 200, x, y, z
200   Format (T5, I10, T20, F5.2, T35, F8.2)   (FORTRAN)

3.5.3 Implementing the Algorithm

We will follow our programming methodology and move on to implementing the algorithm in a programming language. The structure of subprograms or subroutines lends itself well to the development of the code in separate modules.

Subprograms and Subroutines   Subprograms are central to programming. They can be thought of as subsections of a program which can be called upon by the main program to perform specific tasks. When the task is finished by the subprogram, control is transferred back to the main program which begins executing the next statement. The subprogram can be invoked several times in a main program with different arguments, resulting in a compact code. The subprograms can be written and debugged as separate entities by different programmers, thus speeding up the programming process.

Essentially, there are two kinds of subprograms: **function** and **subroutine**. The major difference is that the function subprogram returns only one value to the calling or main program. In contrast, several values can be returned by a subroutine subprogram. The typical structure of a program with subprograms is shown in Figure 3.14.

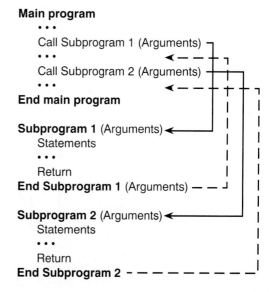

FIGURE 3.14   Structure of a program with subprograms. Note the transfer of control between the main program and the subprograms.

The transfer of control during the execution of the program is also shown. When the first CALL statement is executed, control is transferred to the first subprogram. The code in the subprogram is executed with the arguments passed by the main program. When the END or RETURN statement is reached, control is returned to the statement following the first CALL statement. Similarly, when the second CALL statement is executed, control is transferred to the second subprogram and so on. Moreover, one subprogram can call another subprogram.

In BASIC, the concept of subroutines is implemented in a different way. No arguments are passed to the subroutine when it is invoked nor are values passed back to the main program. The typical syntax is as follows:

>   Main program
>   Statements
>   ...
>   GOSUB <number>
>   Statements
>   ...
>   END
> <number> Statements
>   ...
>   RETURN

When the GOSUB <number> is executed, control is transferred to the statement with the <number> label. Execution continues with this and following statements until the RETURN statement is encountered at which time, control returns to statement following the GOSUB <number> statement. Although GOSUB in BASIC is not regarded as a "true" subroutine, it still has the advantage of writing one set of lines of source code to accomplish a certain task that will be used several times in different locations in the program.

The Program Code We will create a subroutine for computing the various statistics. From the pseudocode in Figure 3.13, it is clear that when the subroutine is invoked, the data points and the number of values should be passed to it. The subroutine, after the computations, will return the values of the different statistics. This illustrates another advantage of subroutines. Once a subroutine is developed, it can be called by any program at any time with the appropriate arguments. It is therefore a good practice to use subroutines, so that you can reuse the code in several different places and times.

Figure 3.15 shows the program code in BASIC. Although we could have created separate subroutines for each of the three subtasks, we have created only one for the computation of the various parameters. After the initialization of the arrays and variables, we obtain the data points from the user and store them in the array X. In statement 120 we invoke the subroutine. Note that the arguments for the subroutine are not specified in BASIC—they are, in other languages such as FORTRAN. Control

```
10    REM Program to compute statistical parameters
20    REM Illustration of the subroutine concept
30    REM Declare variables
40    DIM X(100), DEV(100)
50    SUM = 0: SUMSQ = 0
60    REM Obtain the data points and store them in X array
70    INPUT "Enter the number of data points:"; N
80    FOR I = 1 to N
90        INPUT "Enter value:" X(I)
100   NEXT I
110   REM Use a subroutine for statistical computations
120   GOSUB 230
130   REM Control returns here after execution of subroutine
140   REM Output the various values
150   FOR I = 1 to N
160       PRINT X(I)
170   NEXT I
180   PRINT "The mean is:"; MEAN
190   PRINT "The standard deviation is:" STDEV
200   PRINT "The variance is:"; VARI
210   PRINT "The coefficient of variation is:"; CV
220   END
230   REM Subroutine for statistical measures
240   FOR I = 1 to N
250       SUM = SUM + X(I)
260   NEXT I
270   MEAN = SUM/N
280   REM Compute standard deviation
290   FOR I = 1 to N
300       DEV(I) = X(I)-MEAN
310       SUMSQ = SUMSQ + DEV(I) * DEV(I)
320   NEXT I
330   STDEV = SQR((SUMSQ)/(N-1))
340   VARI = STDEV^2
350   CV = STDEV/MEAN * 100
360   RETURN
```

FIGURE 3.15 A BASIC program for statistical computations.

is transferred to the subroutine beginning at statement 230. The subroutine uses the data points stored in array X to compute the different parameters. After this, the RETURN statement returns control to the main program. The individual values and calculated parameters are printed and the program stops.

Thus, using several examples beginning with the quadratic equation, we have illustrated a programming methodology that, when followed, will result in the development of a program with relatively few errors in a short period of time.

3.6 GOOD PROGRAMMING PRACTICES

Programming is a skill that is perfected with practice. What distinguishes an expert from a novice programmer is not just the ability to create programs faster but the thoroughness with which good programming practices and standards are followed. We will discuss a few major tips that will help you improve your programming skills and develop good programs.

A good program is typically characterized by its efficiency, correctness, and robustness. A well-designed program should be easy to understand, modify, and use.

3.6.1 A Modular Approach to Programming

A structured approach to the programming process is essential for the development of programs that are easier to understand, debug, and maintain. Problem solving involves hierarchically breaking down a big problem into smaller ones and solving the smaller problems individually to eventually obtain the solution for the big problem (Section 2.1). Likewise, any programming activity should also be broken down into smaller and smaller tasks and code developed for each of the subtasks or **modules**. As we saw in the earlier example, each of these modules may be subroutines or subprograms. This approach to programming is known as **modular programming.**

As in problem solving, there are several advantages to this modular approach. Programs will be compact and easier to modify and debug; several programmers can work simultaneously on a single task. Once developed, the modules can be used in other programs.

Good Algorithm Design A correct, efficient, and detailed algorithm is the key to a good program. You should spend time and effort in developing the algorithm and in testing it with sample values. Though this process does not yield tangible program code, it can greatly simplify and speed up the subsequent coding process. Moreover, a well-designed algorithm will also speed up program execution.

Declare and Initialize Variables Since variables, i.e., names, are associated with specific memory locations in the computer, you cannot be sure of the contents of those locations. Consequently, it is a good practice not only to declare the type of variable (real, integer, character) but also to initialize the variable to a desired value, e.g., zero. Use mnemonic variable names whenever possible. For example, CURR is a good choice for *current* while RESIS is suitable for *resistance,* and so on. Make sure you declare the type and dimensions of array variables at the beginning of the program. A common pitfall to avoid is accessing elements in an array that are out of the declared size.

Document the Code A big and complex program with no documentation is of very little value either to you (when you try to use it at a later date) or to others. You should develop the habit of adequately documenting the program logic and flow. Any special conditions should be clearly indicated.

Programming Errors Programs may contain two types of errors: syntax and semantic or logic. Syntax errors arise from improper use of the language: misspelled words, e.g., FIR instead of FOR, REEL instead of REAL; improper arguments, e.g., character value for the SIN function; and so on. Such errors frequently occur during the learning stages and are easy to correct — interpreters and compilers usually provide helpful diagnostic messages.

Semantic errors occur when the program doesn't perform the computation the way you had intended it to. They are harder to detect. During development of a program, it is a good practice to use output statements (such as PRINT) to print out the values of variables during the intermediate stages of the computation. This will help you isolate errors, making debugging easier. Another valuable tool is "playing the computer" and hand tracing the program.

Defensive Programming You should test the program with several sets of test data to identify conditions under which the program is likely to bomb, and you should take defensive measures to guard the user against them. For example, in the quadratic equation solver in Figure 3.3, if the value of A is 0, a "division of zero by zero" error will occur in statement 110. As you know, 0/0 is indeterminate and the program will crash. A simple control statement could have been included to bypass the computations when A is assigned 0. An appropriate error message should also be printed out. If the scope of the program is restricted to a certain range of values, the user should be warned of the limits. These defensive tactics will lead to the development of robust code.

SUMMARY

We discussed the basic concepts of computer programming and developed a methodology for creating programs for solving science and engineering problems. We used several examples to illustrate programming language concepts and modular programming techniques. These concepts are applicable to all programming languages. However, exact syntax will vary from one language to another. The pseudocode presented for the various problems clearly spells out the solution strategy, viz., the algorithm, and helps in the development of the program code. Finally, we covered some programming skills that will aid in the development of efficient, easy-to-read, and easy-to-use computer programs.

CHAPTER 4

SPREADSHEETS FOR PROBLEM SOLVING

In this chapter, we will introduce the electronic spreadsheet, one of the most popular software tools on personal computers. We will use an example from physics to illustrate the basics of a spreadsheet and its role in quantitative problem solving. We will follow the problem-solving methodology discussed in Chapter 2 and develop a structured approach to the design and use of spreadsheet templates. Finally, we will present a brief history of the evolution of spreadsheets.

4.1 TRAJECTORY OF A PROJECTILE

Let's consider the trajectory of a ball thrown in the air. The trajectory of the ball can be altered by varying the angle and/or speed with which it is thrown. Ignoring air resistance, the movement of the ball is governed by the mathematical relationships shown in Figure 4.1.

Example Consider the following problem: A ball is thrown with a velocity of 20 m/s and at an angle of 30° to the horizontal. Assuming acceleration due to gravity is 9.8 m/s^2 and ignoring air resistance, calculate the maximum height and distance traveled by the ball.

The solution obtained with the help of a calculator and paper and pencil is shown in Figure 4.2. This is a simple straightforward computation. However, to examine the effect of changing the throwing angle on the dependent parameters (range, height, and time), the calculations have to be repeated all over. Here's where the spreadsheet comes in handy.

Figure 4.3 shows the spreadsheet representation or model for the ball trajectory problem. Such a model is known as a **template** in spreadsheet terminology. Once you define the mathematical relationships, the program lets you simply change the values of the desired variables, and the unknowns will be automatically computed. You spend

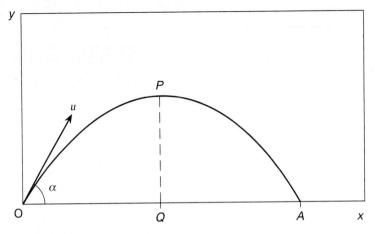

α = angle of departure, degrees
u = initial velocity, m/s
R = range, OA, m
h = maximum height, PQ, m
t = time of flight, s

$$h = \frac{u^2 \sin^2 \alpha}{2g} \quad (1)$$

$$R = \frac{u^2 \sin 2\alpha}{g} \quad (2)$$

$$t = \frac{2u \sin \alpha}{g} \quad (3)$$

FIGURE 4.1 Projectile trajectory: mathematical relationships.

your time efficiently and creatively on exploring various facets of the problem, not just plugging and chugging. As shown in Figure 4.4, you can display the results of your analysis in the form of graphs and tables too. You can save the template, modify it by adding new relationships or deleting existing ones, and use it at a later date.

4.2 WHAT IS A SPREADSHEET?

Traditionally, the term *spreadsheet* refers to an oversized accounting ledger in which the various entries are related to each other. The checkbook is an everyday example of a spreadsheet in which the remaining balance is related to the deposits made and checks written. Likewise, an electronic spreadsheet is a computational framework that allows you to define relationships between various entries, enter data, and utilize these relationships and data to solve problems (see Figure 4.5). Once you create a template of relationships, you can examine various alternatives and perform "what-if" analysis quickly and effortlessly. Thus, the spreadsheet can serve as an efficient prototyping tool in engineering design. Indeed, with an assortment of built-in functions, instantaneous error diagnostics, and help facility, the spreadsheet provides a very powerful and easy-to-learn-and-use interactive problem-solving environment for the scientist and engineer.

Inputs:

α — angle of departure = 30°
u — initial velocity = 20 m/sec
g — acceleration due to gravity = 9.8 m/sec²

Unknowns:

R — range: OA (m)
h — maximum height: PQ (m)
t — time of flight (sec)

Calculations:

$$h = \frac{u^2 \sin^2 \alpha}{2g} = \frac{20^2 \times \sin^2(30)}{2 \times 9.8} = 5.10204 \, m \quad [\text{from (1)}]$$

$$R = \frac{u^2 \sin 2\alpha}{g} = \frac{20^2 \times \sin(2 \times 30)}{9.8} = 35.34797 \, m \quad [\text{from (2)}]$$

$$t = \frac{2u \sin \alpha}{g} = \frac{2 \times 20 \times \sin(30)}{9.8} = 2.040816 \, sec \quad [\text{from (3)}]$$

FIGURE 4.2 Paper-pencil-calculator solution for the Projectile problem.

```
C4: 45                                                    READY

           A              B           C        D      E      F
  1              Projectile Motion
  2       ---------------------------------------
  3   Angle of departure    [deg]         30
  4   Initial velocity      [m/sec]       20
  5   Gravity               [m/sec^2]     9.8
  6   Maximum height        [m]           5.102040
  7   Range                 [m]          35.34797
  8   Time                  [sec]         2.040816
  9
 10
```

FIGURE 4.3 A spreadsheet template for Projectile trajectory.

Path of a Projectile

Angle of departure	[deg]	5	15	25	35	45	55	65	75	85
Initial velocity	[m/sec]	22	22	22	22	22	22	22	22	22
Gravity	[m/sec^2]	9.8	9.8	9.8	9.8	9.8	9.8	9.8	9.8	9.8
Maximum height	[m]	0.187577	1.654	4.410	8.124	12.347	16.570	20.283	23.040	24.506
Range	[m]	8.576093	24.694	37.833	46.409	49.388	46.409	37.833	24.694	8.576
Time	[sec]	0.391311	1.162	1.897	2.575	3.175	3.678	4.069	4.337	4.473

(a)

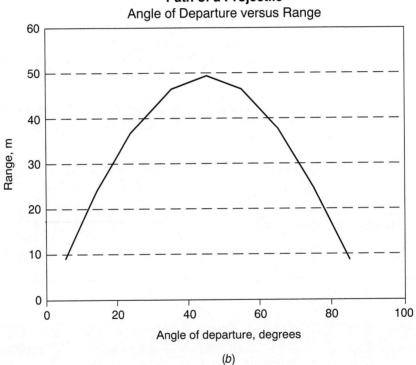

(b)

FIGURE 4.4 Effect of Angle of departure on Projectile trajectory. (a) Tabular representation of results. (b) Graphical output.

4.2.1 Spreadsheet Basics

Common to all spreadsheet programs is the underlying structure. A spreadsheet is essentially a two-dimensional array, or matrix, of entries known as **cells**. The cell contents, referred to as **cell entries,** can be numbers, formulas or text. Just as an element in a matrix is referred to by its row and column number, i.e., C_{ij}, each cell

> A spreadsheet lets you do the following:
> - Declare variable names (in cells).
> - Define relationships between variables (in cells).
> - Assign values to variables (in cells).
> - Calculate unknown variables based on defined relationships.
> - Perform "what if" analysis by examining various alternatives.
> - Present results in tables and/or graphs.
> - Save, modify and reuse the template of relationships.

FIGURE 4.5   A spreadsheet features capsule.

in the spreadsheet has a unique **cell address.** The address specifies the coordinates of the cell in the matrix. Typically, the horizontal coordinate (column) is specified by letters and the vertical coordinate (row) by numbers. Thus, the address D5 in Figure 4.6 refers to the cell at the intersection of the fourth column and fifth row of the array. In certain programs, numbers are used to designate both rows and columns. The cell address is then given by R#C#, where the R and C stand for Row and Column, respectively, and # specifies the corresponding number. Thus, cell D5 would be addressed as R5C4.

The number of rows and columns in the spreadsheet depends on the particular program or product. For example, Microsoft Excel has 256 columns labeled A through Z, AA through AZ, ... , IA . . . IV, and 16,384 rows numbered 1 through 16,384. Given the large number of cells in the entire spreadsheet (e.g., 256 × 16,384 = 4,194,304 cells), it is possible to view only a certain segment of these cells on the display screen at any given time. The display screen becomes a window through which the desired segment of the spreadsheet is viewed.

4.2.2 The Spreadsheet Window

Let's look a little closer at the spreadsheet window displayed in Figure 4.6. Cell D5 appears in reverse video on the screen, and the highlighted bar, or rectangle, is called the **cell pointer.** The pointer is also referred to as a **cursor**, and instead of a rectangular bar, it can be a blinking underline or a cross. The cell pointer marks the active cell whose address is displayed on the status line (first line) of the **Control Panel.** Though the control panel may be displayed at the top of the screen or at the bottom (depending on the particular spreadsheet program), its role is the same: to provide status information and to assist in "navigating" through the spreadsheet.

The cell contents are also displayed on the status line. The **mode indicator** displays the current mode of operation. When the spreadsheet is ready to accept data or commands, the indicator reads READY, signaling its status. When the command menu is invoked, it displays MENU as shown in Figure 4.7.

The second line of the control panel is the **command** or **entry line.** When we enter or edit information in a cell, it is displayed on this line; as shown in Figure 4.7,

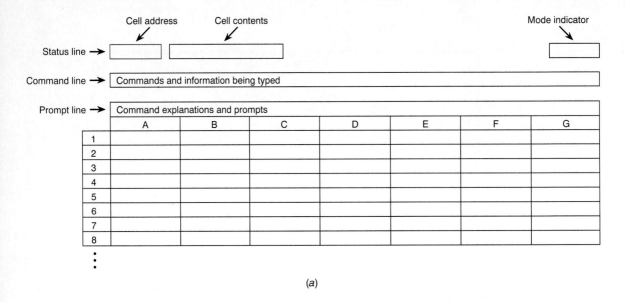

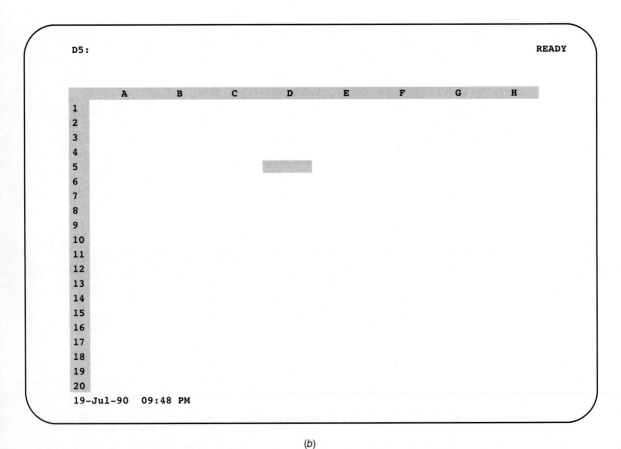

FIGURE 4.6 Spreadsheet screen display. (*a*) Display screen explained. (*b*) Lotus 1-2-3 screen.

```
D4: 20                                                                      MENU
Worksheet  Range  Copy  Move  File  Print  Graph  Data  System  Quit
Global, Insert, Delete, Column, Erase, Titles, Window, Status, Page
           A              B           C         D         E         F
 1                             Path of a Projectile
 2     ------------------------------------------------------------------
 3     Angle of departure    [deg]         5        15        25        35
 4     Initial velocity      [m/sec]      20        20        20        20
 5     Gravity               [m/sec^2]   9.8       9.8       9.8       9.8
 6     Maximum height        [m]      0.155092  1.367689  3.646560  6.716717
 7     Range                 [m]      7.090823  20.41645  31.27737  38.36243
 8     Time                  [sec]    0.355896  1.056866  1.725695  2.342043
 9
10
```

FIGURE 4.7 Command menu invoked by typing /; mode indicator reads MENU.

commands are also displayed on this line. The command explanation or message is displayed on the third line of the control panel and is also known as the **prompt line.** The prompt for the next appropriate action is displayed on this line. Typically, the date and time of day are also displayed on the screen (Figure 4.6).

4.2.3 Moving around the Spreadsheet

When the spreadsheet is loaded initially, the first 20 rows and 8 columns are displayed as shown in Figure 4.6. The initial position of the pointer, cell A1, is called the **home** position. We can move the pointer around the worksheet with the help of the **Arrow** keys, keys on the **Numeric** keypad, the **GoTo Function** key or the **Scroll bars.**

The Arrow Keys The pointer can be moved from the current cell to the adjacent cell (left, right, above, or below) using the Arrow keys. For example, if we press the Down Arrow key thrice from the home position, the pointer is moved to the fourth row (A4); then, pressing the Right Arrow key twice positions the pointer in cell C4. Note that the status line displays the current location of the pointer and the contents of the cell. When we move the pointer down past row 20, the rows at the top disappear from the window, and the new rows (and row numbers) appear in the window. Similarly, if we move the pointer in the horizontal direction, new columns will appear in the window and the old ones will scroll out of view. Thus, we can bring any part of the spreadsheet into the display window for working.

The Numeric Keypad Using the Arrow keys to move from one part of the spreadsheet to another (e.g., from cell A4 to CD234) can be rather tedious. This is where the keys on the numeric keypad come in handy. The PgUp (page-up) and PgDn (page-down) keys enable us to scroll the spreadsheet vertically page by page, i.e., to move 20 rows at a time. To scroll horizontally by a window width to the right, we can press the Control key and Right Arrow key simultaneously. The combination of the Control key and Left Arrow key scrolls the sheet horizontally to the left. In some programs, the Tab and Back Tab keys can be used for scrolling horizontally. The home key moves the pointer to the home position, i.e., the upper-left corner of the spreadsheet.

The GoTo Function Key   The Function keys (F1, F2, ..., F10) on personal computers are utilized by the spreadsheet program to execute some frequently used commands. The keyboard overlay supplied with the program (see Figure 4.8) denotes the specific key bindings. One of these keys is the **GoTo** key which allows you to move the pointer to a specific cell in the worksheet. When the GoTo key is pressed, the program prompts you for a cell address. When you specify the address (e.g., BC145 in Figure 4.9) and press the Enter key, the pointer is moved to that cell and a new portion of the spreadsheet is displayed in the window.

The pointer movement commands for three major spreadsheet programs are shown in Quick Reference Table 4.1.

4.3 A STRUCTURED APPROACH TO PROBLEM SOLVING WITH SPREADSHEETS

The use of spreadsheets in problem solving follows the methodology discussed in Chapter 2 and shown in Figure 2.1. The first step in problem solving is the analysis

FIGURE 4.8   Function key overlays.

(a) 1-2-3

F1 Help	F2 Edit
F3 Name	F4 ABS
F5 GoTo	F6 Window
F7 Query	F8 Table
F9 Calc	F10 Graph

(b) EXCEL

F1 Help	F2 Edit
F3 Paste name	F4 ABS/REL
F5 GoTo	F6 Next pane
F7 Formula find next	F8 Extend
F9 Calculate all	F10 Menu
F11 New chart	F12 Save as

(c) Quattro

F1 Help	F2 Edit
F3 Name	F4 ABS
F5 GoTo	F6 Window
F7 Query	F8 Macro
F9 Calc	F10 Graph

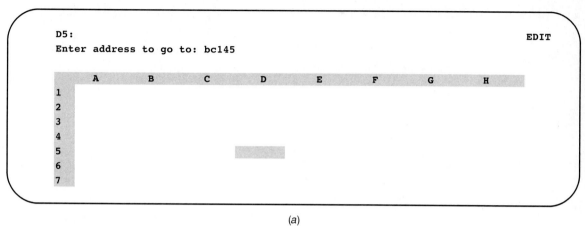

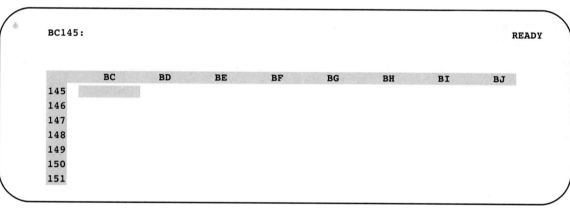

FIGURE 4.9 The GoTo command. (*a*) Prompt for cell address in control panel. (*b*) Cursor moved to cell BC145.

TABLE 4.1 QUICK REFERENCE TABLE FOR CURSOR MOVEMENT COMMANDS

Pointer/cursor movement	1-2-3	Quattro	Excel
Cell to cell	↑↓→←	↑↓←→	↑↓←→
Page to page (horizontal)	Ctrl→; Ctrl←	Ctrl→; Ctrl←	Scroll Lock Ctrl→; Ctrl←
Page to page (vertical)	PgDn; PgUp	PgDn; PgUp	PgDn; PgUp
To a specific cell	F5	F5	F5

of the problem and the development of the mathematical model (see Figure 2.3). Then the model is incorporated in the spreadsheet framework. This step is analogous to programming in a conventional language with some notable differences: It is simpler, easier, and faster. Once this is done, the spreadsheet template or model is tested and debugged. The template is then documented and saved for later use.

4.3.1 Developing the Spreadsheet Template

Figure 4.10 shows the major steps in the development of a spreadsheet template from a mathematical model. As in programming, a structured approach to the development of a spreadsheet template will result in one that is easy to understand, use, and modify.

FIGURE 4.10 Steps in the development and use of a spreadsheet template in problem solving.

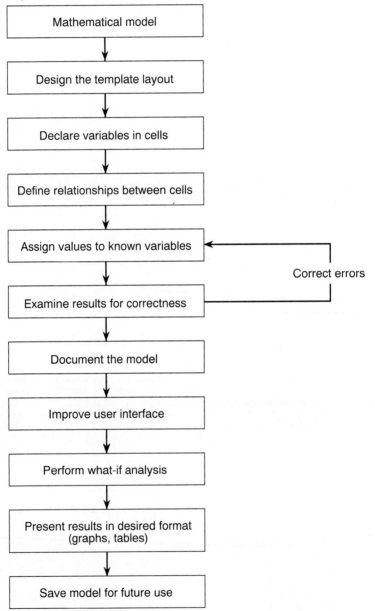

Design the Template Layout A typical layout for a template is shown in Figure 4.11. Just as any letter or document is identified by the author, the date of origination, and subject title, the top section of the template should be used for similar identification. This should be followed by a description of the scope of the template: its intended use and how it works. Any assumptions, restrictions or limitations on the applicability of the template should be noted in the following section.

The next section constitutes the heart of the template, the mathematical or cell relationships. When defining the relationships, attention should be paid to the flow of computation. Any specialized sequence of instructions, known as **macros** in spreadsheet terminology, for menus or computations should be isolated in a separate section either at the bottom of the spreadsheet or at the extreme right. In some programs such as Excel, separate macro sheets are available for storing macros. There are several advantages to adopting this modular structure for the design of a template. The template will be easy to understand and use. Moreover, errors can be traced to a certain portion of the template, making debugging easier.

Declare Variables in Cells The next step is to declare the variables in cells (see Figure 4.10). Unlike programming languages, the major advantage of a spreadsheet is that there are no real restrictions on variable names or descriptions. The cell widths can be expanded to the desired number of characters. However, it is a good practice to keep the variable names and descriptions succinct. Another useful practice is to arrange the variables in some meaningful fashion, e.g., inputs followed by outputs.

Define Relationships between Cells Once variables have been associated with cells, the relationships between the cells should be defined using the various mathematical operators and special functions available in the spreadsheet. The formulas should not contain important factors or numbers that are likely to vary during the

FIGURE 4.11 A typical layout of a spreadsheet template.

Identification
    Developer
    Revision date
    Title

Scope
    Purpose of the template
    How it works: identifying inputs and outputs
    Literature references for formulas and constants (if any)

Restrictions
    Assumptions underlying the model
    Limitations on range of values

Mathematical relationships
    Cells, formulas, values

Macros
    Macros for menus, computations

course of problem solving: there should be references only to the cells that contain these values. If such numbers and factors are hidden in the formulas, the template will be harder to modify or even to understand. Debugging the template will also be more difficult. For example, when interest rate is used in computing the monthly payments on a loan, the address of the cell containing the value of the interest rate should be used instead of the value itself.

Defining the cell relationships is the most important step in the template development process. In doing this, it is necessary to rearrange the equation so that only the unknown variable appears on the left-hand side of the equation. The expression on the right-hand side of the rearranged equation should be entered in the cell for the unknown variable. Care should be taken to avoid mutual references to cells in formulas. This problem of **circular references** is discussed in detail in Sections 5.1.2 and 6.2.2. Moreover, if a cell with no value is referenced in a formula, it is assumed to have a value of zero—a major and annoying drawback of spreadsheets.

Assign Values to Variables   Testing the template with sample data is the next step in the development process (Figure 4.10). Recall that this is similar to testing a program by "playing the computer." The formulas in the cells must be printed and checked. The solution path must be traced to ensure the proper flow of computations; results obtained must be confirmed independently.

There are three main sources of errors in spreadsheet templates; they are: (1) incorrect cell references in formulas generally caused by mistyping cell addresses; (2) lack of understanding of operation sequencing, i.e., how the expressions in cells will be evaluated; and (3) semantic or logic errors. Consequently, the results you see on the screen may be different from what you expect. Errors in the template should be traced to the source(s), corrected, and the test-modify-test cycle repeated until all the bugs are eliminated (more on this in Section 5.1.2).

Document the Template   Once the template is validated, it should be documented with appropriate comments. Documentation should cover the template's purpose and scope, how the template works, and other pertinent information associated with its identification and limitations.

Improve the User Interface   If the template is to be used extensively, especially by others, it is a good idea to improve the user interface. This can be accomplished by adding helpful hints on what variables need to be specified, formatting values for improving the template appearance, protecting cell formulas from being accidentally overwritten or erased, etc. Where appropriate, menus should be created for displaying the various options specific to the template. At the end of this stage, the template is ready for future use.

4.3.2 Using the Template: What-If Analysis

The spreadsheet is a very powerful analytical tool. The time spent in initially developing a template may be slightly greater than that spent using a calculator for solving the problem. However, once developed, various alternatives can be evaluated with very little time and effort. The ability to perform what-if analysis is one major advantage of spreadsheets over traditional paper-and-pencil methods. Once the various alternatives have been evaluated, results can be displayed in the form of tables or graphs for presentation. The template can then be saved for future use.

Should You Use a Spreadsheet Program? The time spent in the development of the template should be regarded as an investment with a long-term payoff. Therefore, you should carefully weigh the pros and cons before developing a spreadsheet template for the problem at hand. Some issues to bear in mind are: Will it be useful at a later date and/or for others? Is there a need to perform what-if analysis? Should the results be presented in the form of tables and/or graphs? Are there any features in the spreadsheet, such as built-in functions, that will simplify the solution process? Does the solution require extensive procedural computation? Once you are convinced that the spreadsheet framework is appropriate, you follow the structured methodology discussed here.

4.4 THE PROJECTILE EXAMPLE

We will illustrate the development methodology by creating a template for analyzing the trajectory of the ball discussed earlier in Section 4.1.

4.4.1 Creating a Spreadsheet Template

As shown in Figure 4.10, the first step is to design the layout of the template. This is shown in Figure 4.12. Note that we have used the format discussed in Figure 4.11.

Declare Variables in Cells In this template, we will use the same variable descriptions used in Figure 4.3. Let's load the spreadsheet program and with the pointer in the home position, type the first variable description <u>Angle of departure</u>.* As it is being typed, the letters appear on the second line of the control panel as shown in Figure 4.13a. It does not appear in the cell, yet. The mode indicator changes from READY to LABEL as soon as the first character is typed denoting that text information is being entered. We can use the Backspace key to correct errors while typing. When we finish typing the variable description, we press the Enter or Return key. The typed information is "entered" into the cell, and the contents are also displayed next to the cell address in the control panel as shown in Figure 4.13b. The mode indicator is back to READY.

Variable Names and Descriptions Variable names in conventional programming languages are used to store the values of variables; values are assigned to variable names. In contrast, the variable description in a spreadsheet is used as a descriptor for the variable. The value of the variable itself is stored in a neighboring cell. For example, the value of the descriptor "Angle of departure" will be stored in cell B1. When referring to a variable in a formula, the address of the cell containing the value

---

*All entries that we type at the computer are underlined in the text.

84 SPREADSHEETS FOR PROBLEM SOLVING

Identification
    Developer: Sundaresan Jayaraman
    Revision date: May 1, 1990
    Title: The Projectile Trajectory

Scope
    Purpose: This template has been developed for analyzing the trajectory of projectiles. It can also be used to generate tables and plots.
    Working: Given the values of angle of departure, initial velocity, and acceleration due to gravity, the other unknown variables are calculated using the standard formulas.
    Formulas: Obtained from any physics textbook.

Restrictions
    Assumptions: Air resistance to the projectile is ignored.
                        Angle of departure should be specified in degrees; it is converted to radians in the formula.
    Limitations: Angle of departure: 0 to 90° range

Mathematical relationships
    Angle of departure   [Given]
    Initial velocity   [Given]
    Gravity   [Given]
    Maximum height   [Compute using Equation (1) in Figure 4.1]
    Range   [Compute using Equation (2) in Figure 4.1]
    Time   [Compute using Equation (3) in Figure 4.1]

FIGURE 4.12   Layout of the projectile trajectory template. *Note:* No mutual cell references in formulas; no computational problems are likely to be encountered.

---

should be used and not the address of the variable description. Thus, we would use B1 (and not A1) when referring to the value of angle of departure for computing the other parameters.

Aligning Text in Cells   Note the apostrophe (') added by the program in front of the label "Angle of departure" in the control panel (Figure 4.13*b*). By default, labels are aligned to the left edge of the cell and the apostrophe prefix denotes left-justification. Other prefixes include the caret (^) for centering labels in the cell and double quotes (") for aligning the labels to the right edge of the cell.

Cell Width and Walking Labels   The pointer in Figure 4.13*b* highlights only the first nine characters we entered. The word *departure* appears in the adjacent cell. Therefore, when the entered label or text is longer than the width of the cell (nine, in the present example), it spills over to the adjacent cell or cells. This feature is known as "walking labels." There is an exception, however. If the adjacent cell contains information (labels, values, or formulas), the label cannot walk and the display is restricted to the width of the cell. However, as we will see later, the width of the cell can be changed to display more than nine characters.

Entering Information and Moving the Pointer   Let's move the pointer to the second row and type <u>Initial velocity</u>; this time, instead of pressing the Enter key, we press the Down Arrow key. The information is entered into the cell and the cursor is moved to the next row. So we can enter information and move the cursor in a single

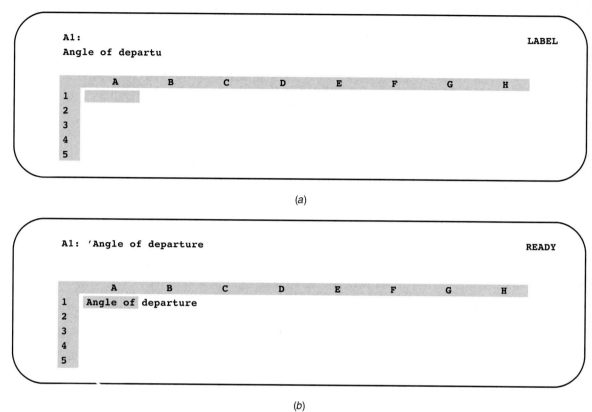

FIGURE 4.13 A LABEL entry in a cell. (a) Mode indicator displays LABEL when the first character is typed. (b) Mode indicator back to READY when label is entered; cell contents are displayed in the control panel.

step with the help of the Arrow keys. In a similar manner, we enter the other variable names or descriptions shown in Figure 4.14.

Define Relationships Between Cells Having declared the variables in the problem, the next step is to define the mathematical relationships between them. Let's enter the first relationship [Equation (1) in Figure 4.1] for calculating the maximum height attained by the projectile. We move the cursor to B4, the cell adjacent to Maximum height. We want the calculated value to be displayed in this cell. This value will depend on the known values of the variables entered in the cells above B4. For example, the angle of departure will be entered in cell B1, the initial velocity in cell B2, and so on (for reference, see the template layout in Figure 4.12).

With the pointer in cell B4, we type +B2^2*@SIN(B1*@PI/180)^2/(2*B3) (Figure 4.15a). The mode indicator reads VALUE, and when we press the Enter key, the formula is entered and ERR appears in the cell as shown in Figure 4.15b. Let's compare Equation (1) and its equivalent expression just entered.

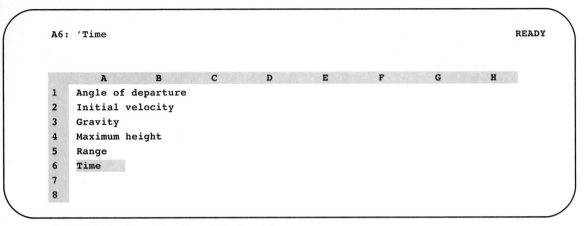

FIGURE 4.14 Template with variable descriptions.

Equation   Spreadsheet entry

$h = \dfrac{u^2 \sin^2 \alpha}{2g}$ +B2^2*@SIN(B1*@PI/180)^2/(2*B3)

B1, B2, and B3, in the spreadsheet entry, are the cell addresses where the values of the three variables α, u, and g, respectively, will be stored. The first + in the entry signifies a VALUE entry (i.e., an entry that will yield a numeric value) to the spreadsheet. The spreadsheet judges the type of entry (LABEL or VALUE) based on the first character typed in a cell. Recall that if it begins with a letter of the alphabet, it is treated as a LABEL entry. So, by typing a + before "B2," we are indicating to the program that what is being entered should be treated as a VALUE entry.

Arithmetic Operators and Precedence   The following operators are used to specify the major arithmetic operations:

Operator	Operation	Precedence
^	Exponentiation (raising to a power)	Highest
* /	Multiplication and division	↓
+ −	Addition and subtraction	Lowest

As in conventional programming languages, the arithmetic operations are performed from left to right according to their order of precedence. The exponentiation operation has the highest priority and is performed first. Then multiplication and division, and finally the addition and subtraction operations are performed. Parentheses ( ) can be used to group operations. The subexpression in the left-most and innermost parentheses are evaluated first, using the precedence rules. Then any other subexpressions are evaluated. Finally, the resultant expression is evaluated left to right.

Built-in Functions   A host of trigonometric, hyperbolic, statistical, financial, date, and other special functions is built into a spreadsheet program. Such a built-in function is specified in a formula with the @ prefix, and its arguments are enclosed in paren-

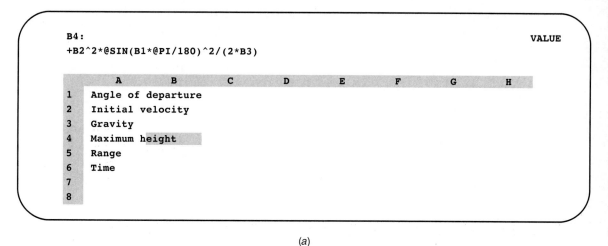

FIGURE 4.15 Entering a formula in cell B4. (*a*) Mode indicator reads VALUE. (*b*) ERR message in cell B4 (Division by zero); value of a blank cell referenced in a formula is 0.

theses. In the present example, the @SIN stands for the trigonometric *sine* function whose argument is the value in cell B1. Since the arguments for trigonometric functions are required to be expressed in radians, the value in cell B1 (assumed to be in degrees) is converted into radians using the conversion factor @PI/180. The built-in function @PI represents π and is approximately equal to 3.1415926. Built-in functions are case-insensitive, i.e., they can be specified in either upper or lowercase or in any combination.

Evaluation of the Expression **+B2^2*@SIN(B1*@PI/180)^2/(2*B3)** Let's examine how this expression for maximum height (in cell B4) is evaluated in the spreadsheet. According to the rules of precedence, the expression (B1*@PI/180) is evaluated first followed by the expressions @SIN(B1*@PI/180) and (2*B3),

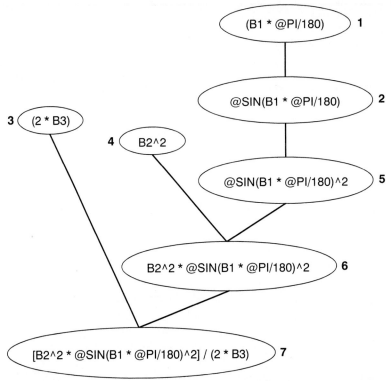

FIGURE 4.16 Evaluation tree for the formula in cell B4. The subexpressions are evaluated in the numbered sequence.

respectively. Then B2^2 and @SIN(B1*@PI/180)^2 are evaluated in sequence. This is followed by the multiplication and division operations, respectively. The entire evaluation can be represented in the form of a tree structure as shown in Figure 4.16.

The ERRor Message Let's analyze the reasons for the ERR (for error) message in cell B4. Since no values have been entered in cells B1, B2, and B3, the cells referenced in the formula, the program treats them as having 0s. Therefore, when the last operation in Figure 4.16 is performed, both the numerator and the denominator are zero, and the result of 0/0 is indeterminate. Hence the error message. Remember this caveat when dealing with cell values: *If a blank cell is referenced in a formula, the spreadsheet assigns it a value of zero.*

Entering Other Formulas Let's move the cursor to B5 and enter the formula for Range: +b2^2*@sin(2*b1*@pi/180)/b3. Then, the formula for Time is entered in cell B6: 2*B2*@sin(b1*@pi/180)/B3. Note that the cell addresses are also case-insensitive. The spreadsheet program will typically display the address in uppercase in the control panel. Observe the similarities between Equations (2) and (3) in Figure 4.1 and the cell entries for B5 and B6. The resulting screen is shown in Figure 4.17. Note that the variable description for maximum height in cell A4 has been truncated because its neighboring cell, B4, now contains an expression.

```
B6: 2*B2*@SIN(B1*@PI/180)/B3                                            READY


        A         B         C         D         E         F         G         H
1   Angle of departure
2   Initial velocity
3   Gravity
4   Maximum h      ERR
5   Range          ERR
6   Time           ERR
7
```

FIGURE 4.17 Template with all formulas entered; the reason for the error messages: The empty cells B1, B2, and B3 are treated as having 0s.

Assign Values to Known Variables   Now that the various mathematical relationships have been defined, the next step in the problem-solving process is to assign values to known variables. We move the pointer to B1, enter 30 and press the Down Arrow key. Likewise 20 is entered in cell B2 and 9.8 in B3. As soon as the third value is assigned, the program instantaneously computes the unknown values and displays them as shown in Figure 4.18. The ball will travel 35.34797 m in 2.040816 s, and it will attain a maximum height of 5.10204 m during its journey. We have now created our first spreadsheet template and solved a problem!

Check the Correctness of Results   The solution in Figure 4.18 matches the one obtained by the paper-and-pencil method shown in Figure 4.2. This means the relationships between the cells have been defined correctly. If, for instance, we had left the parentheses out of the denominator in the formula in cell B4, the calculated value for maximum height would have been different and incorrect.

FIGURE 4.18 Template with solution of projectile problem (angle = 30°, velocity = 20 m/s).

```
B3: 9.8                                                                 READY


        A         B         C         D         E         F         G         H
1   Angle of      30
2   Initial v     20
3   Gravity       9.8
4   Maximum h 5.102040
5   Range     35.34797
6   Time       2.040816
7
```

```
B1: 50                                                          READY


        A         B         C         D         E         F         G         H
1    Angle of           50
2    Initial v          20
3    Gravity            9.8
4    Maximum h   11.97600
5    Range      40.19623
6    Time        3.126712
7
```

FIGURE 4.19   What if angle = 50°? New solution.

4.4.2 Using the Template

Once a template is created, it is very easy to solve several problems using the defined relationships. We will use the projectile template to illustrate this feature.

What-If Analysis   Let's say we want to find out how high and how far the ball travels when it is thrown at an angle of 50°. We move the pointer to B1, type 50, and press the Enter key. The new values are displayed instantaneously as shown in Figure 4.19. This feature of changing variable values and recalculating other variables is known as *what-if analysis*; this powerful feature has contributed to the enormous popularity of spreadsheets as a powerful problem-solving tool for financial analysts and engineers.

Playing on Mars   What if the ball is thrown on Mars? As you probably know, the acceleration due to gravity on Mars is one-fifth of that on earth. There are several ways to enter this new value in cell B3. We can enter 1.96 and overwrite the current value of 9.8. Alternately, we can type 9.8/5 and press the Enter key. The spreadsheet will perform the division and display 1.96 in the cell. However, let's use a third alternative, the EDIT feature, and change the value.

Editing an Entry   We move the cursor to B3. Observe the keyboard template supplied with the program (recall Figure 4.8). One of the Function keys (e.g., F2 in 1-2-3) is the Edit key, and pressing it places us in the edit mode; the mode indicator reads EDIT. A blinking cue appears at the end of the cell contents, 9.8. In the edit mode, the Left and Right Arrow keys can be used to position the cue in the entry. The Function keys HOME and END move the cue to the beginning and end of the cell entry, respectively. Since we want to divide the current entry by 5, and the cue is correctly positioned at the end of the entry, we type /5 so that the cell entry reads 9.8/5 (Figure 4.20a). When we press the Enter key, the expression is evaluated and displayed as 1.96. At the same time, the other variables are evaluated as shown in Figure 4.20b. The ball does travel a much longer distance on Mars!

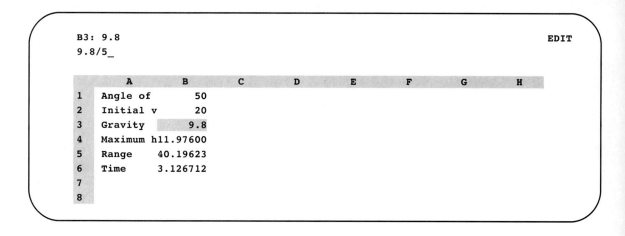

FIGURE 4.20 Using the Edit feature. (a) Entering the Edit mode with the Edit key. (b) Playing on Mars! New solution.

We will now examine some spreadsheet commands that will help us utilize the tool's other capabilities.

4.5 THE COMMAND STRUCTURE

The spreadsheet has a set of well-structured commands that are invoked from the ready mode generally by pressing the Slash (/) key. These commands enable us to perform a whole host of operations including modifying the appearance of the template, printing portions of the spreadsheet, saving and retrieving templates, creating, displaying and saving graphs, etc.

```
A1: 'Angle of departure                                              MENU
Worksheet  Range  Copy  Move  File  Print  Graph  Data  System  Quit
Global, Insert, Delete, Column, Erase, Titles, Window, Status, Page
          A         B         C         D         E         F         G         H
    1  Angle of       50
    2  Initial v      20
    3  Gravity      1.96
    4  Maximum h59.88000
    5  Range    200.9811
    6  Time     15.63356
    7
    8
```

FIGURE 4.21 The command menu invoked by typing /; mode indicator reads MENU.

4.5.1 Invoking a Command*

With the pointer in cell A1, we press the / key. A set of commands appears in the control panel, and the mode indicator reads MENU as shown in Figure 4.21. The pointer highlights Worksheet on the first line, and a set of subcommands appears on the next line. The **command menu,** as it is generally called, is similar to a restaurant menu where the entrees are grouped under different categories. Selecting a command from the menu is similar to picking out an entree from the menu. For example, a typical dinner menu may have four major food categories, i.e., Salads, Chicken, Meat, and Seafood (see Table 4.2). To order lobster, we would have to turn to the Seafood section of the menu first and then select the item from the list of entrees in that category.

Coming back to the spreadsheet display, the subcommands or items under the Worksheet command are listed on the second line, e.g., Global, Insert, Delete, etc. To select a subcommand, you have to choose the top level command first and then go down the menu. This type of command structure is also referred to as a "tree structure."

Moving around the Command Menu   We press the Right Arrow key. The cursor highlights the next command, Range. Its subcommands (e.g., Format, Label, and Erase) appear on the next line as shown in Figure 4.22. When we press the Right Arrow key again, the cursor highlights Copy, and an one-line explanation of the command appears in place of the usual subcommands (Figure 4.23). So, when a command is highlighted, either its subcommands or its explanation is displayed on the next line. Thus, the command menu and submenus assist you in selecting the command.

*While the specific command sequences and menus will vary from program to program, the underlying logic for executing the commands remains essentially the same across all programs. To enhance the usefulness of the text, the specific command sequences for Lotus 1-2-3, Quattro, and Excel are given in quick reference tables. As of this writing, Quattro Pro, an enhanced version of Quattro, has been released. Though Quattro Pro's command structure differs from that of Quattro, we have chosen to include the commands for Quattro since Quattro Pro's commands are similar to that of Excel.

TABLE 4.2 ENTREES IN A TYPICAL DINNER MENU

Salads	Meat
Spinach salad	Beef stroganoff
Tossed salad	Corned beef
Tuna salad	Meat loaf
Chef salad	Swiss steak
Chicken	**Seafood**
Stir-fried chicken breast	Broiled lobster
Chicken kebabs	Fish steaks
Chicken Kiev	Fish kebabs
Stuffed chicken breasts	Scalloped oysters
Chicken Marengo	Shrimp teriyaki

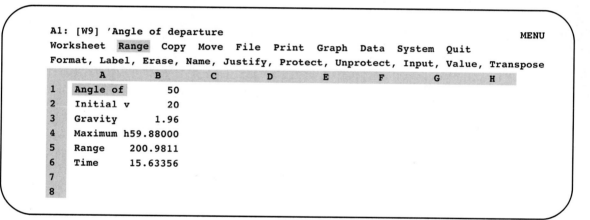

FIGURE 4.22   The Range command submenu.

4.5.2 Increasing the Cell Width

Let's compare the screen displays shown in Figures 4.3 and 4.20. In Figure 4.20, the variable description "Angle of departure" has been shortened to "Angle of." (Note that the number of characters displayed corresponds with the width of cell A1, i.e., 9 characters.) This is because the adjacent cell, B1, now contains a value (50), thus preventing the label in A1 from walking over to B1. However, the label itself has not been truncated (see the first line in the control panel of Figure 4.21), only its display in the cell has been curtailed. We will display the label in full and improve the appearance of the template with the help of some spreadsheet commands. To display the label in full, we have to increase the width of column A to, for example, 20 characters.

We move the pointer to highlight Worksheet in the command menu (Figure 4.21). Since the width of column A needs to be increased, we have to invoke the Column command under Worksheet. At first, we select the Worksheet command by pressing the Enter key. The options under Worksheet are now displayed as shown in Figure 4.24.

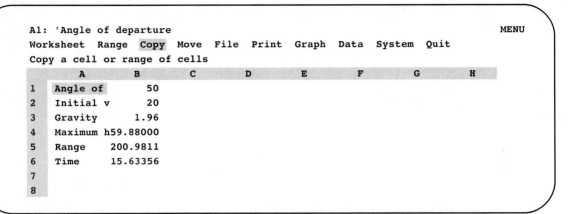

FIGURE 4.23 The Copy command: Note the command explanation in the control panel.

FIGURE 4.24 The Worksheet command submenu.

The Global subcommand affects the entire spreadsheet. We move the cursor to highlight Column subcommand and press the Enter key. The prompt line of the control panel reads as shown in Figure 4.25. We then select the Set-width option by pressing the Enter key. The current width (9) is displayed, and we are prompted to enter a new value between 1 and 240 (see Figure 4.26a). We type 20 (the required column width) and press the Enter key. The column is widened and all the variable descriptions are displayed in full as shown in Figure 4.26b.

The sequence of commands for changing cell width is shown in Quick Reference Table 4.3.

4.5.3 Inserting Rows and Columns

Continuing the spreadsheet development process, let's document the model and improve its user interface. We will add the identification information (from Figure 4.12)

```
A1: 'Angle of departure                                              MENU
Set-Width  Reset-Width  Hide  Display
Set width of current column
         A          B          C          D          E          F          G          H
1    Angle of       50
2    Initial v      20
3    Gravity        1.96
4    Maximum h59.88000
5    Range       200.9811
6    Time         15.63356
7
8
```

FIGURE 4.25   The Column command options.

TABLE 4.3   QUICK REFERENCE TABLE FOR CHANGING THE CELL WIDTH

Task	1-2-3	Quattro	Excel
Set cell width to # characters	/Worksheet Column Set width <#>	/Column Width <#>	/Format Column width <#>
Reset cell width to global width	/Worksheet Column Reset width	/Column Reset width	/Format Column width Standard width

and associate units with the different variables. For brevity, we will omit the scope and restrictions sections shown in the figure. However, they should be included in the template to ensure completeness.

In the screen display shown in Figure 4.26, there is no empty row at the top of the template nor is there an empty column after the variable description for the units. The Insert command in the spreadsheet can be used to insert rows and columns in the template. The new rows are inserted above the current cursor location while the new columns are inserted to the left of the cursor. The sequence for invoking the Insert command for the three major spreadsheet programs is shown in Quick Reference Table 4.4.

Using the Insert Row Command   Since we want the new rows to be inserted at the top of the template, we move the pointer to the home position and invoke the command menu by pressing /. The Insert command appears under the Worksheet command. So we press the Enter key to display the Worksheet submenu. We select the Insert option from the submenu. The options shown in Figure 4.27a appear.

We need to select the Row option. There are two ways of doing this. We can move the cursor to Row and press the Enter key. Alternately, we can select the Row option

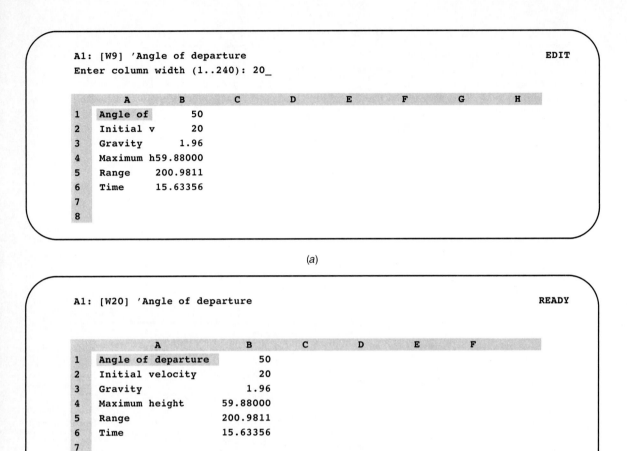

FIGURE 4.26 Using the Column-width command. (a) Prompt for column width. (b) The first column is widened; variable descriptions appear in full.

by pressing the first character of the option (R, in this case). Note that the cursor need not highlight the desired option. Thus, there are two ways of selecting a command option: by moving the cursor to the option and pressing the Enter key or by pressing the first character of the desired option. As you become more proficient with the spreadsheet command structure, you will gravitate toward the latter method.

When we press R, a new prompt for the range of rows to be inserted is displayed along with the cell address (Figure 4.27b). If we press the Enter key, a single row will be inserted above the current row. To insert multiple rows, we can use the Arrow key (Up or Down) to specify the number of rows to be inserted. Since we need six rows, we press the Down Arrow key till the pointer is in the sixth row and the cell address reads A6 in the control panel (see Figure 4.28a). When we press the Enter key, six rows are inserted at the top as shown in Figure 4.28b.

TABLE 4.4 QUICK REFERENCE TABLE FOR INSERT COMMAND

Task	1-2-3	Quattro	Excel
Inserting row(s)	/Worksheet Insert Row <..>	/Row Insert <..>	Select Row(s) /Edit Insert
Inserting column(s)	/Worksheet Insert Column <..>	/Column Insert <..>	Select Column(s) /Edit Insert

Note: <..> denotes a range.

```
A1: [W20] 'Angle of departure                                         MENU
Column  Row
Insert one or more blank columns to the left of the cell pointer
         A              B           C        D        E        F
  1  Angle of departure       50
  2  Initial velocity         20
  3  Gravity                  1.96
  4  Maximum height          59.88000
  5  Range                  200.9811
  6  Time                    15.63356
  7
  8
```

(a)

```
A1: [W20] 'Angle of departure                                        POINT
Enter row insert range: A1..A1_

         A              B           C        D        E        F
  1  Angle of departure       50
  2  Initial velocity         20
  3  Gravity                  1.96
  4  Maximum height          59.88000
  5  Range                  200.9811
  6  Time                    15.63356
  7
  8
```

(b)

FIGURE 4.27   Invoking the Insert command. (*a*) The Insert command options. (*b*) Prompt for Insert Row command; mode indicator reads POINT.

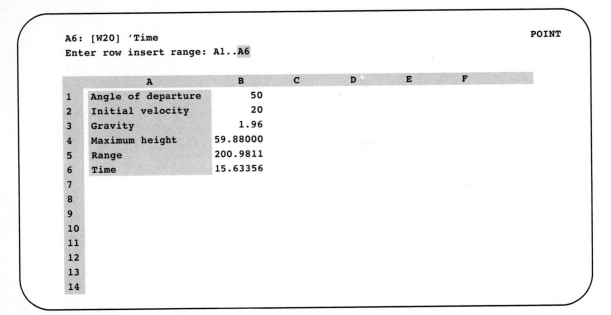

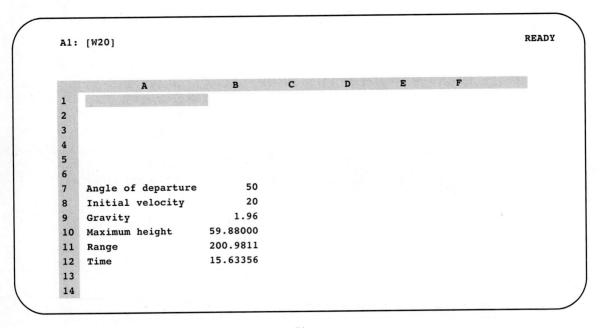

FIGURE 4.28 Using the Insert Row command. (*a*) Since the cursor is in the sixth row, the address changes in the control panel. (*b*) Six new rows are inserted.

```
B10: +B8^2*@SIN(B7*@PI/180)^2/(2*B9)                                    READY


              A                 B           C         D         E         F
   1
   2
   3
   4
   5
   6
   7   Angle of departure       50
   8   Initial velocity         20
   9   Gravity                 1.96
  10   Maximum height      59.88000
  11   Range               200.9811
  12   Time                15.63356
  13
  14
```

FIGURE 4.29 Control panel with formula for maximum height. Cell references have been changed automatically.

Formulas Not Affected Since six new rows have been inserted, you would expect the formulas for maximum height, range, and time to be "messed up." Let's move the cursor to the value for maximum height (cell B10) and compare the formula in the control panel (Figure 4.29) to the one in Figure 4.15. The cell addresses have changed; for example, B1 is now B7, B2 is B8, and B3 is B9.

The cell addresses in the formula for maximum height in Figure 4.29, viz., B7, B8, and B9, are *relative* to the cell in which the formula is contained, viz., B10. This is the same as the original relationship defined in B4 containing references to B1, B2, and B3 (Figure 4.15). So the spreadsheet preserves the implied positional relationship between the cells and automatically modifies the formula to account for new locations of cells referenced in the formula. Likewise, the formulas for range and time are automatically modified by the program. We can verify this by moving the cursor to those cells.

To enter the identification information shown in Figure 4.12, we move the cursor to the home position and type <u>Identification</u> and press the Down Arrow key. Since labels can walk, it is a good practice to enter text information in a single cell. Thus, the next line of text, "Developer: Sundaresan Jayaraman," is entered in A2. If it is inserted in two or more adjacent cells in a row, the insertion of columns (at a later date) will split up the information. In a similar manner, the remainder of the information is entered in the following rows. However, the identification information is not shown in the figures here.

Repeating a Character Just as we would draw a line under the title (on a paper) to enhance the appearance of a document, we will use the Repeat Character feature

```
A6: [W20] \-                                                    READY


          A              B           C         D         E         F
 1
 2
 3
 4
 5    Path of a Projectile
 6    ---------------------- ---------
 7    Angle of departure      50
 8    Initial velocity        20
 9    Gravity                  1.96
10    Maximum height          59.88000
11    Range                  200.9811
12    Time                    15.63356
13
14
```

FIGURE 4.30 The Repeat Character command: Cell filled with hyphens.

to draw a line of hyphens (-) to separate the identification information from the variables. The Repeat Character feature can be invoked by pressing a special key such as the backslash key (\) or apostrophe (') followed by the character to be repeated all across the cell. With the cursor in cell A6, the \ key is pressed; the mode indicator reads LABEL. The repeat character (- in our example) is typed followed by the Enter key. The hyphen is repeated all across the cell as shown in Figure 4.30. We move the cursor to the neighboring cell (B6) and fill it with another series of hyphens.

Using the Insert Column Command The variable values in the template are not very useful unless their associated units are known. To further improve the template's usefulness, we will enter the units of measurement next to the variable descriptions. For this, we need to insert a column between the variables and their values.

With the cursor in any cell of the second column, we type / to display the command menu. As before, we select the Worksheet command, followed by the Insert command, and then the Column option (see Quick Reference Table 4.4). The address for the new column location is displayed in the control panel as shown in Figure 4.31a. Since we want only one column, we press the Enter key. An empty column appears to the left of the cursor as shown in Figure 4.31b. The variable values now appear in the third column.

When we move the cursor to the value for maximum height, we find that the formula has changed, i.e., the references to cells B7, B8, and B9 have been automatically changed to C7, C8, and C9. To reemphasize, when rows or columns are inserted, the spreadsheet program automatically changes the cell references to maintain the

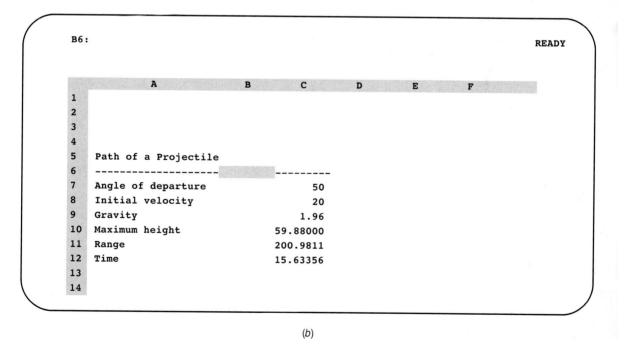

FIGURE 4.31 Using the Insert Column command. (*a*) Prompt in Insert Column command. (*b*) A new column is inserted.

relationships defined between the cells during the creation of the template. The program does the housekeeping and relieves you of the chore of changing individual cell references every time you insert rows or columns. This is a major strength of spreadsheets.

We move the cursor to the inserted empty column and enter the units for the different variables shown in Figure 4.32. Using the Repeat Character feature, we fill in B6 with a series of hyphens and the screen looks like the one in Figure 4.32.

Thus, using the different features in the spreadsheet program, we have enhanced the appearance of the template and improved its usefulness.

4.5.4 Escaping from Trouble

The Escape key (ESC) acts like a safety valve and lets you cancel a command or abort an action in progress. For example, if you invoke the command menu by mistake and want to get out of the command mode, you can press the ESC key. The menu will disappear and the mode indicator will be back to READY. The ESC key also comes in handy when you want to pop back to a higher-level menu from a submenu. Let's illustrate this with a simple example.

We type /W to display the submenu for Worksheet. Then, we realize that the Print command is not one of the options in the submenu. So we press the ESC key. The top-level menu will be displayed, and we can select any of its commands, including the Print command. Suppose we change our mind and decide not to execute any command. We then press the ESC key and exit the command menu.

FIGURE 4.32 Complete projectile template with units of measurement.

```
B12: '[sec]                                                           READY


              A              B          C         D         E        F
    1
    2
    3
    4
    5    Path of a Projectile
    6    ------------------------------------
    7    Angle of departure  [deg]        50
    8    Initial velocity    [m/sec]      20
    9    Gravity             [m/sec^2]    1.96
   10    Maximum height      [m]          59.88000
   11    Range               [m]          200.9811
   12    Time                [sec]        15.63356
   13
   14
```

Cancelling an Entry If we press the ESC key while entering or editing data (before pressing the Enter or Arrow keys), the action will be interrupted and the original contents of the cell (if any) will be restored. Suppose, we are entering a numeric value in a cell and suddenly realize that the cell already contains a formula. If we press the ESC key (before pressing the Enter key), the numbers we typed will disappear from the control panel and the original contents will remain in the cell. Thus, the ESC key gives you the flexibility and power to "escape" from your actions.

4.5.5 Printing the Template

One of the major features of a spreadsheet program is the ability to print hard copies for record-keeping and reporting purposes. Depending on the particular spreadsheet program, the subcommand names may vary. However, they all have a common set of functions. The output can be sent either to a data file (File option) or to a printer (Printer option). The area of spreadsheet to be printed can be specified using the Range subcommand. The paper can be advanced a line at a time (Line subcommand) or a page at a time (Page subcommand). After selecting the range and formatting options, printing can be initiated with the Go subcommand.

Using the Print Command Let's print a hard copy of the current screen display. To find out the exact sequence for the print command, we can use the **Help** facility in the spreadsheet. Referring to the keyboard overlay in Figure 4.8, we press the F1 key. The Help screen is displayed on top of the current sheet with instructions on how to use this facility (Figure 4.33a). When we select the desired topic, i.e., printing the worksheet, appropriate Help information is displayed on the screen (Figure 4.33b). To exit the Help facility, we can press the ESC key. The Help information is context-sensitive; this means if we invoke the facility in the midst of a command, only the information that is pertinent to the context will be displayed.

We move the cursor to A5 and invoke the command menu by pressing /. To select the Print command from the command menu (Figure 4.21), we type P, the first character in the Print command. The submenu is displayed as shown in Figure 4.34a.

Specifying the Range From the Print command submenu, we select the Printer option to direct output to the printer. From the menu in Figure 4.34b, we select the Range subcommand. A new prompt appears in the control panel for the desired range. A **range** in a spreadsheet refers to a cell or a set of contiguous cells in rows and columns (more on this in Section 5.2.2). Since we want to print the three columns and twelve rows, the print range is A5 through C12 (or R5C1 through R12C3). We press . — this causes the pointer to be "anchored" in A5, and the display in the control panel changes to A5..A5. We press the Right Arrow key; the neighboring cell is also highlighted and the address changes in the control panel, for example, from A5..A5 to A5..B5. To include the third column in the print range, we press the Right Arrow key again. The third cell is also highlighted, and the cell address changes in the control panel. This process is known as "painting" the range.

We press the Down Arrow key; the second row of cells is painted. Likewise, we move the cursor till the entire range (A5..C12) is painted as shown in Figure 4.35a. To signal the end of range, we press the Enter key. The original display is restored.

```
B12: '[sec]                                                          HELP


READY Mode

The mode indicator, READY, in the upper-right corner of the screen means you
can select a command or type a cell entry.  The first key you press
determines your action:

Formula or Number:   Type a digit (0..9) or one of the characters
                     +, -, ., (, @, #, or $.
Label:               Type any character except those that begin a
                     formula or number.  Start with a label-prefix character to
                     create a label of a particular type: ' for left-aligned,
                     " for right-aligned, ^ for centered, or \ for repeating.
Command:             Type /.
Special Function:    Press a special key.


To learn more about this Help facility, press [END], then [RETURN].
Cell Entries           Mode Indicators
Help Index             How to Use Help
```

(a)

```
B12: '[sec]                                                          HELP


/Print commands -- Sends data to the printer or to a file

Select   o   Printer to send data directly to the printer or

         o   File to create a print file so you can print later or
             use the file with another program.

With either choice, 1-2-3 next displays a menu of print settings and
operations.  During a single /Print command, you can print several different
ranges (or the same range several times), using various print options.

Use the /Worksheet Global Default Printer command to establish default print
settings.  These settings specify standard printing procedures, such as page-
length, margins, and a printer setup string.

While using /Print, you can override the default settings using /Print
Options.
Next Step--Print Menu              Ranges          /Print Options
Help Index
```

(b)

FIGURE 4.33 Using the Help facility. (a) The Help menu. (b) Help for the print command.

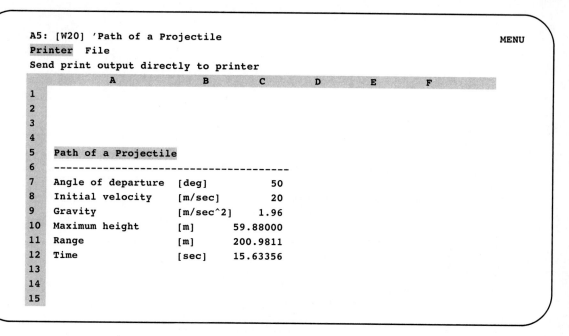

(a)

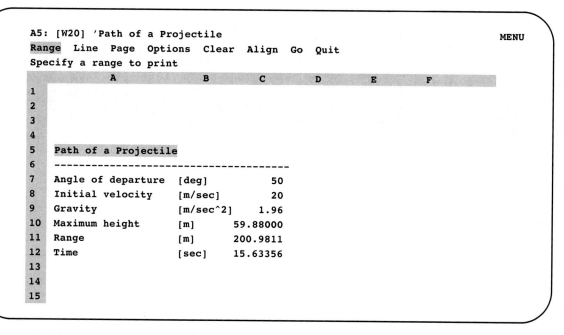

(b)

FIGURE 4.34 Invoking the Print command. (*a*) The Print command menu. (*b*) The Printer option submenu.

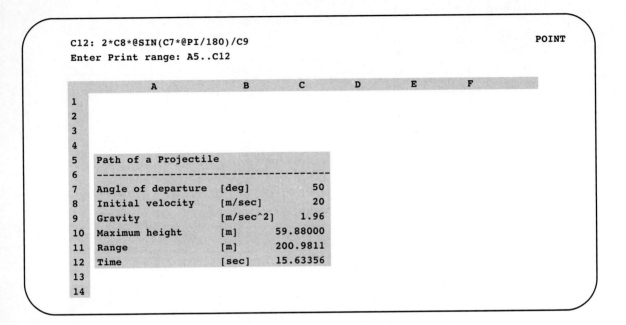

FIGURE 4.35 Using the Printer option. (a) Painting the range to be printed. (b) Printer output.

Aligning and Printing We set the print parameters to format from the top of the page by selecting the Align option (Figure 4.34b). We initiate printing by selecting Go from the menu. The specified range is printed (see Figure 4.35b) and the print menu reappears. We exit the print menu by selecting the Quit option. The sequence for invoking the Print command for the three major spreadsheet programs is shown in Quick Reference Table 4.5.

4.5.6 Saving the Template

Yet another frequently used command in the spreadsheet is the Save command. This command enables you to save your template for subsequent use and modification. When working with a spreadsheet, it is advisable to save your work every 15 to 20 min to guard against loss of work caused by power or other hardware failures.

TABLE 4.5 QUICK REFERENCE TABLE FOR PRINT COMMAND

Task	1-2-3	Quattro	Excel
Printing a file	/Print Printer Range <..> Align Go	/Print Block <..> Adjust Go	/File Print
Creating a print file	/Print File Range <..> Align Go	/Print Destination Disk <..> Adjust Go	F12 Options Text <name>

Note: <..> denotes a range.

The File Save Command Since it is a file operation, the Save subcommand typically appears under the File command. We invoke the command menu and select the File option. The File command menu is displayed as shown in Figure 4.36a. We type <u>S</u> to select the Save option. The prompt for the filename appears as shown in Figure 4.36b. Typically, the filename should not exceed eight characters. The spreadsheet program appends an extension to the specified name. We type <u>PROJ1</u> and press the Enter key to save the file under the name PROJ1. Note that we have typed the entire pathname so that the file is stored in the desired directory. The file is written and the prompt disappears. The spreadsheet is ready for further commands. The saved file can be subsequently retrieved using the file Retrieve option of the file command. The sequences for invoking the Save and Retrieve commands for the three major spreadsheet programs are shown in Quick Reference Table 4.6.

4.5.7 Exiting the Spreadsheet

If you have been following us on the computer, as we hope you have been, you deserve a break. The Quit command lets you leave the spreadsheet program and return to the operating system. However, before invoking this command, you should save your work. Otherwise, it will be lost. In fact, most programs will seek your confirmation before exiting when you invoke the Quit command.

TABLE 4.6 QUICK REFERENCE TABLE FOR FILE COMMAND

Task	1-2-3	Quattro	Excel
Saving a file	/File Save <name>	/File Save <name>	/File Save <name>
Retrieving a file	/File Retrieve <name>	/File Retrieve <name>	/File Open <name>

```
A5: [W20]  'Path of a Projectile                                    MENU
Retrieve  Save  Combine  Xtract  Erase  List  Import  Directory
Erase the current worksheet and display the selected worksheet
           A              B          C         D        E       F
 1
 2
 3
 4
 5    Path of a Projectile
 6    ---------------------------------------
 7    Angle of departure    [deg]        50
 8    Initial velocity      [m/sec]      20
 9    Gravity               [m/sec^2]    1.96
10    Maximum height        [m]          59.88000
11    Range                 [m]         200.9811
12    Time                  [sec]        15.63356
13
14
15
```

(a)

```
A5: [W20]  'Path of a Projectile                                    EDIT
Enter save file name: C:\DEMO\PROJ1_
           A              B          C         D        E       F
 1
 2
 3
 4
 5    Path of a Projectile
 6    ---------------------------------------
 7    Angle of departure    [deg]        50
 8    Initial velocity      [m/sec]      20
 9    Gravity               [m/sec^2]    1.96
10    Maximum height        [m]          59.88000
11    Range                 [m]         200.9811
12    Time                  [sec]        15.63356
13
14
15
```

(b)

FIGURE 4.36 Using the File command. (a) The File command menu. (b) Prompt for filename; the complete pathname should be specified.

The sequence for invoking the Quit command for the three major spreadsheet programs is shown in Quick Reference Table 4.7.

Using the Quit Command We press / and choose the Quit command. The confirmation prompt appears in the control panel as shown in Figure 4.37. If we select the No option, the Quit command is cancelled, and the spreadsheet returns to ready mode. Since we want to end the session, we select the Yes option. The spreadsheet disappears from the screen.

4.6 A TOUCH OF HISTORY

No other category of software has influenced the field of personal computers more than spreadsheets. VisiCalc, the first electronic spreadsheet developed by Dan Bricklin and Bob Frankston of Software Arts in 1979, was predicted to be "the software tail that wags (and sells) the personal computer dog." VisiCalc certainly lived up to this prediction and went on to spawn several generations of spreadsheets. Multiplan from Microsoft was another popular spreadsheet on the first-generation personal computers.

TABLE 4.7 QUICK REFERENCE TABLE FOR THE QUIT COMMAND

Task	1-2-3	Quattro	Excel
Exiting the program	/Quit Yes	/Quit	/File Exit

FIGURE 4.37 The Quit command.

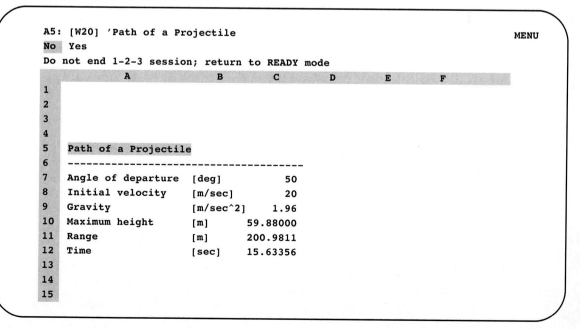

Since its introduction in 1982, Lotus Development Corporation's 1-2-3 has become the "industry-standard" spreadsheet for the MS-DOS environment. This program has undergone several revisions, and Version 3.1 was released as of this writing. Among the host of other spreadsheet programs available in the market, Quattro Pro from Borland International, VP Planner from Paperback Software, and Excel from Microsoft are popular.

SUMMARY

We introduced the basics of an electronic spreadsheet and developed a structured methodology for using spreadsheets in quantitative problem solving. We applied this methodology for solving problems related to projectile trajectory. Major spreadsheet commands were illustrated with suitable examples. The origins of spreadsheets were also briefly discussed.

As you can see, learning to use the spreadsheet for solving problems is extremely simple and requires no knowledge of programming languages. The methodology discussed for creating templates of mathematical relationships is applicable for problems in various fields of science and engineering; this process very closely resembles the classical paper-and-pencil method for solving problems, minus its tedium.

In the next chapter, the other features and capabilities of spreadsheets will be explored with suitable examples. Finally, several application examples will be presented as case studies to further illustrate the power of the spreadsheet as a quantitative problem-solving tool.

CHAPTER 5

SPREADSHEET COMMANDS AND FEATURES

In this chapter, we continue the exploration of other features in spreadsheets useful in quantitative problem solving. We develop a template for solving a simple problem in thermodynamics and look at commands for copying cells, naming ranges, and creating graphs and tables of data. By the end of this chapter, you should have gained a good appreciation for the power of a spreadsheet as a problem-solving tool and be fairly conversant with its major features.

5.1 AN EXAMPLE FROM THERMODYNAMICS

Thermodynamics is the study of the relationship between heat, work, and the stored energy contained by the matter under consideration. In doing so, the behavior of matter is described in terms of its state variables, pressure, volume, temperature, internal energy, and entropy. The field of thermodynamics rests on two fundamental laws, known as the first and second laws of thermodynamics. Thermodynamics is widely used in all branches of science and engineering for tasks ranging from the design of giant power plants to understanding the molecular motion in polymers.

Example Consider the following example: An ideal gas expands reversibly and adiabatically from an initial volume (V_1) of 1 L (at temperature $T_1 = 450$ K) to a final volume (V_2) of 5 L. Assume further that the heat capacities are $C_p = 5$ and $C_v = 3$ cal/(g·mol·K), respectively.

    a Estimate the final temperature (T_2), the work done (W), and the enthalpy change (ΔH) of the gas.

    b Show how the values of T_2, W, and ΔH change as the final volume (V_2) changes from 1.5 L to 5 L, in steps of 0.5 L.

5.1.1 Problem Analysis and Template Layout

Following our problem-solving methodology discussed in Chapter 2, the first step is to develop the mathematical model. Figure 5.1 shows the mathematical relationships describing the behavior of the system. The equations can be derived from first principles or obtained from any physics or thermodynamics textbook. From the data given in the problem, it is clear that the unknowns T_2, W, and ΔH can be calculated using the equations shown in the figure.

Then we need to solve the problem several times for each value of V_2. This repetitive computation will require some features analogous to the loop construct in programming languages (Section 3.3).

Our strategy is to first develop the template and obtain one single solution after working out the bugs. Once this is done, we can use the Copy command in spreadsheets to replicate the formulas and carry out the required number of computations. Then we can use the Graph or Chart feature to produce the desired graphical representation of the solution. Note that this is similar to the modular approach to problem solving discussed in Section 3.6.

Designing the Template Layout Figure 5.2 shows the layout for the template. Let's look a little closer at the mathematical relationships section. Following the model

FIGURE 5.1 Adiabatic expansion of an ideal gas: mathematical relationships.

From first principles in thermodynamics, we know that if an ideal gas undergoes an adiabatic change from P_1, V_1, T_1 to P_2, V_2, T_2, then:

$$P_1 V_1^{\gamma} = P_2 V_2^{\gamma} \tag{1}$$

where γ = ratio of heat capacities = C_p/C_v. \hfill (2)

From the ideal-gas law, we know that:

$$P_1 V_1 = nRT_1 \quad \text{and} \quad P_2 V_2 = nRT_2 \tag{3}$$

From relationships (1) and (3), we have:

$$\frac{T_1}{T_2} = \left(\frac{V_2}{V_1}\right)^{\gamma-1}$$

or

$$T_2 = T_1 \left(\frac{V_1}{V_2}\right)^{\gamma-1} \tag{4}$$

The relationships for work done (W) and change in enthalpy (ΔH) are as follows:

$$W = \text{change in internal energy} = -C_v(T_2 - T_1) \quad \text{cal/(g·mol)} \tag{5}$$

$$\Delta H = C_p(T_2 - T_1) \quad \text{cal/(g·mol)} \tag{6}$$

Identification
  Developer: Sundaresan Jayaraman
  Revision date: July 1, 1990
  Title: A Thermodynamics Example

Scope
  Purpose: This template has been developed for analyzing the behavior of an ideal gas undergoing a reversible, adiabatic compression or expansion. It can also be used to generate tables and plots.
  Working: Given the values of C_p, C_v, T_1, V_1, and V_2, the other unknown variables are calculated using the standard formulas.
  Formulas: From any thermodynamics or physics textbook.

Restrictions
  Assumptions: Process is adiabatic compression or expansion. C_p and C_v are constant.
  Limitations: Applicable for ideal gas.

Mathematical relationships

C_p (cal/..)	[Given]	V_1 (liter)	[Given]
C_v (cal/..)	[Given]	T_1 (K)	[Given]
Gamma [Equation (2) in Figure 5.1]			
V_2	T_2	W	ΔH
(liter)	(K)	[cal/(g·mol)]	[cal/(g·mol)]
[Given]	[Equation 4]	[Equation 5]	[Equation 6]
[Given]	[Equation 4]	[Equation 5]	[Equation 6]
[Given]	[Equation 4]	[Equation 5]	[Equation 6]

FIGURE 5.2 Layout of the thermodynamics template. *Note:* Computations are flowing top-down and left-right. Also, all equation numbers refer to equations in Figure 5.1.

layout in Figure 4.11, we have the known variables C_p, C_v, V_1, and T_1 at the top. Based on the values of C_p and C_v, γ is calculated using Equation (2) in Figure 5.1. Since we need to find the values of T_2, W, and ΔH, for different values of V_2, these variables are laid out in the form of a table in the figure. The flow of computation in the template will be top-down and left-right. So, it is good practice to devote some time to designing the template layout.

5.1.2 Creating the Template

In science and engineering, there are some variable names which, despite being short, are easily recognized and widely used. The variables in this example, viz., V_1, V_2, T_1, T_2, C_p, and C_v belong to this class. Consequently, we will use these variable names as variable descriptors in the template. To reiterate, the variable descriptor or label in a spreadsheet is not the same as the variable name in a programming language. In a computer program, the variable name is used to refer to its value, while in a spreadsheet the cell address is used to refer to its contents.

Figure 5.3a shows the template with the defined relationships. We have used the formula display command to display the formulas in the cells instead of their values. The cell formula display commands for the three major spreadsheets are shown in

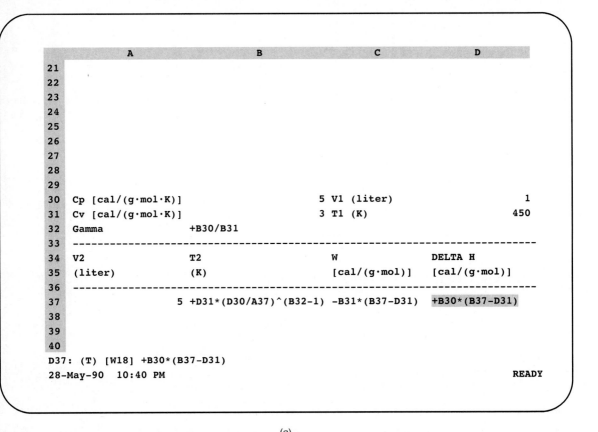

(a)

FIGURE 5.3 Thermodynamics template: initial version. (a) Formula displays in cells. (b) The correct solution.

Quick Reference Table 5.1.[1] Also, the columns have been specifically widened so that all the formulas may be seen. Figure 5.3b shows the results of the computation. The temperature drops to 153.8978 K, and the positive value of W indicates that in the expansion process work is done by the system, which is correct. So the cell relationships in the template are defined correctly.

Error Detection and Correction However, for the same set of given values, Figure 5.4a shows a negative value for the work done, W. Indeed, this is puzzling, and the question comes up, "What could have gone wrong?" To debug the template, we widen the columns and use the formula display command to reveal the underlying

---

[1]While the specific command sequences and menus will vary from program to program, the underlying logic for executing the commands remains essentially the same across all programs. To enhance the usefulness of the text, the specific command sequences for Lotus 1-2-3, Quattro and Excel are given in quick reference tables. As of this writing, Quattro Pro, an enhanced version of Quattro, has been released. Though Quattro Pro's command structure differs from that of Quattro, we have chosen to include the commands for Quattro since Quattro Pro's commands are similar to that of Excel.

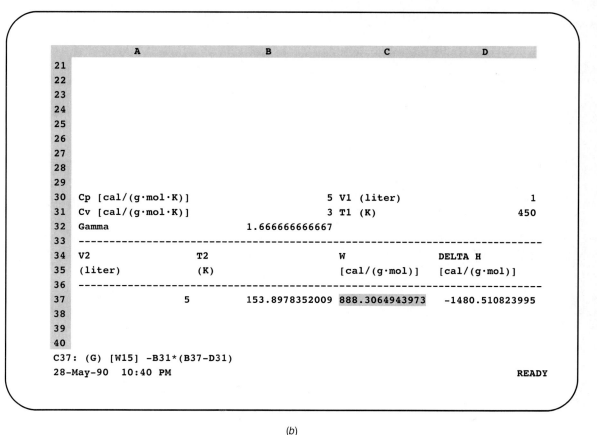

(b)

FIGURE 5.3 (continued).

formulas in the cells. This is shown in Figure 5.4b. If the template is large, it is better to print out the template and use the hard copy to debug it. Recall also that when the cursor highlights the cell containing a formula, the formula is displayed in the control panel next to the cell address (see bottom of Figure 5.4b).

Types of Errors Just as there are syntactic and semantic errors in programming, there are three main types of errors that occur during the use of spreadsheets. They are syntactical or cell reference errors, improper computational flow errors, and semantic or logic errors. Cell reference errors are the most common and are the easiest to detect while semantic or logic errors are the hardest to spot. When entering formulas, it is important to keep track of the precedence of the various operators to avoid logical errors.

Certain types of computational flow errors are flagged by the program. These include the so-called **circular reference** conditions. A circular reference occurs in a spreadsheet when the formula in a cell contains a reference to itself (a **direct** circularity) or the formulas in two cells reference each other (an **indirect** circularity). This means

TABLE 5.1 QUICK REFERENCE TABLE FOR FORMULA DISPLAY COMMAND

Task	1-2-3	Quattro	Excel
Display formula in cells	/Range Format Text <..>	/Block Display Text	/Options Display Formulas OK

Note: <..> denotes a range of cells.

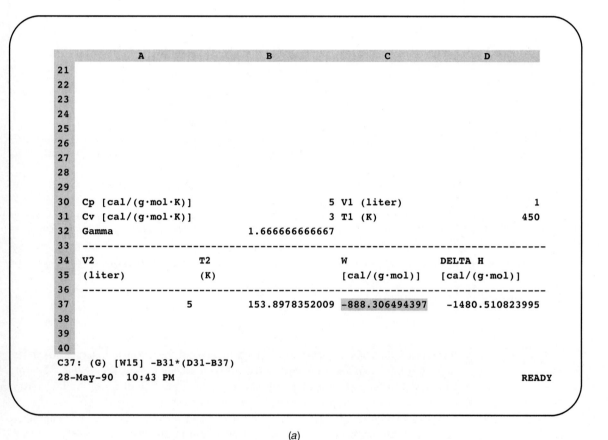

(a)

FIGURE 5.4 Problem in the thermodynamics template. (*a*) The incorrect solution. (*b*) Formula displays in cells. *Note:* The formula in cell C37 should be –B31*(B37 – D31).

that when a direct circularity condition is present, the value being computed in the cell depends on the value of the cell; in the case of indirect circularity, the value of a cell depends on the value of another cell whose value itself depends on this cell. Whenever a circular reference occurs, the spreadsheet displays a CIRC indicator on the screen or an appropriate message in the alert box.

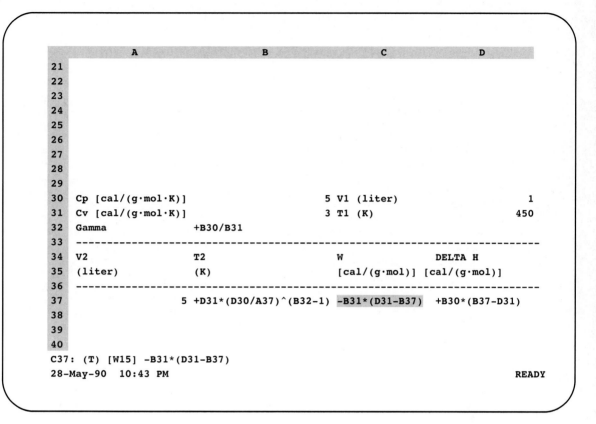

(b)

FIGURE 5.4 (continued).

For example, if the formula in cell B5 in a certain spreadsheet reads +B5*B4-B2, then the value in B5 depends on the value in B5 creating a direct circularity condition. This kind of circularity cannot be resolved by the spreadsheet because each time a new value is computed for B5, the value of B5 changes and the process could go on indefinitely. Therefore, it typically returns a value of 0 in the cell. Care should be taken to avoid such situations.

The indirect circularity condition, where two cells reference each other in the respective formulas, can be resolved using the **Iteration** feature. This feature causes the spreadsheet to be recalculated a certain number of times that can be specified, anywhere from 2 to 32,765, depending on the spreadsheet. An example of resolving an indirect circularity using the iteration feature is discussed in detail in the case study in Section 6.2.

It is extremely important to understand the order (or sequence) in which a spreadsheet evaluates the formulas in the various cells so that you can avoid errors caused by improper computational flow (see Capsule 5.1).

CAPSULE 5.1

ORDER AND MODE OF RECALCULATION

Orders of recalculation

The formulas in spreadsheet cells can be recalculated in three different orders. In the **natural** or default order, the cells referenced in a formula are calculated prior to the evaluation of the formula itself. For example, if the formula in B5 contains a reference to cell D44, then D44 is evaluated prior to the evaluation of B5. In the **columnwise** order, calculation commences at the top of the first column (A) and proceeds to the bottom of the column and continues on to the second (B), third (C) columns, and so on. In the **rowwise** order, calculation commences at the beginning of the first row and proceeds to the end of the row and continues on to the second row, third row, and so on.

Modes of recalculation

There are two modes of recalculation: **automatic** and **manual**. In the **automatic** mode, which is also the default, the cells in the spreadsheet are recalculated every time the contents of any cell are changed. This can be pretty annoying in the case of very large spreadsheets involving extensive calculations. In such situations, it is better to select the manual mode of recalculation. When the **manual** mode is selected, the spreadsheet is recalculated only when the CALC key (see keyboard overlay in Figure 4.8) is pressed. The commands for selecting the appropriate order and mode of recalculation in the three major spreadsheets are shown in Quick Reference Table 5.2.

TABLE 5.2 QUICK REFERENCE TABLE FOR RECALCULATION COMMAND

Task	1-2-3	Quattro	Excel
Selecting recalculation order	/Worksheet 　Global 　　Recalculation 　　　Natural 　　　Rowwise 　　　Columnwise	/Default 　Recalculation 　　Order 　　　Natural 　　　Rowwise 　　　Columnwise	Natural order cannot be changed
Selecting recalculation method	/Worksheet 　Global 　　Recalculation 　　　Automatic 　　　Manual	/Default 　Recalculation 　　Mode 　　　Automatic 　　　Manual	/Options 　Calculations 　　Automatic 　　Automatic (except tables) 　　Manual

Continuing the Error Search We begin at the top of the spreadsheet (home position) and look for the first cell containing a formula (cell B32 in Figure 5.4b); we check to see if the cells referenced in the formula are the correct ones.* A typical error is to refer to the cell containing the variable description or label instead of the variable value. For example, in B32, if instead of B30/B31, we had typed A30/B31, the result would have been zero since A30 corresponds to the variable description and spreadsheets typically assign a value of zero to a label. Since there is no problem in this cell, we move on to the formula in B37 and find that there are no problems.

When we come to C37, we find that the negative sign is present at the beginning of the expression, but the expression in the parentheses is the culprit. It should have

---

*It can be argued justifiably that one should go to the error first, i.e., cell C37, and backtrack from there. Our objective here is to introduce the concept of debugging and present a general methodology. Therefore, we begin at the top of the spreadsheet and look for the first formula.

been B37 − D31 and not D31 − B37, since the temperature difference is given by $T_2 - T_1$ and not $T_1 - T_2$. We can enter the edit mode using the Edit key and correct the error. When this is done, the correct value is displayed. To complete the process of debugging, we look at the formula in D37 and find it to be correct. So, even during the initial phase of the template development process, it is important to debug it (see Figure 4.10) through a structured process, such as the one just completed.

5.2 REPETITIVE COMPUTATIONS AND THE COPY COMMAND

One of the major features of a spreadsheet program is its ability to copy the contents of a cell or range of cells to a different location on the spreadsheet. This feature comes in particularly handy when values or formulas have to be used in several different places or for repetitive evaluation of a template when one or more of the variables change. This is the case with part (*b*) of the present example, where the value of V_2 changes from 1.5 L to 5 L in steps of 0.5 L. Note, however, that the mathematical relationships remain the same for all the steps.

5.2.1 Generalizing the Template: Looking for Relationships

Since the value of V_2 changes from 1.5 to 5 L, we might be tempted to enter each of the values (1.5, 2, 2.5, ..., 5) in the first column under V2 in Figure 5.3, starting with cell A37. What will happen if V_2 changes from 5 to 10 L, instead of from 1.5 to 5 L? Then we would have to reenter all the values; not only is this time-consuming but it is also not taking full advantage of the power of spreadsheets.

Looking for a Pattern (Formula) In our example, each of the cells following A37 is 0.5 greater than the preceding one (see Figure 5.5*a*). So the relationship between the *i*th cell and the (*i* + 1)st cell is given by

$$\text{Value in } (i+1)\text{st cell} = \text{value in } i\text{th cell} + 0.5$$

and this is shown in Figure 5.5*b*. Thus, the formula in cell A44 has the same relationship to cell A43, as the formula in cell A40 to cell A39. This is commonly referred to as **relative addressing** in spreadsheet programs. When we use this relationship, if the starting value of V_2 is 4 instead of 1.5, all we need to do is to change the first entry. All the other dependent cell values will automatically change (see Figure 5.5*c*), provided they contain the formulas shown in Figure 5.5*b*.

Therefore, it is good practice to look for relationships between the variables (and their values) and use them to generalize the template; not only is this elegant, it will also enhance the usefulness of the template.

5.2.2 The Copy Command

We can see from Figure 5.5*b* that the formulas in the various cells under the descriptor V2 are similar. So, if we create the first formula, it can be easily replicated in the other

120 SPREADSHEET COMMANDS AND FEATURES

```
            A                    B                  C                    D
27
28
29
30   Cp [cal/(g·mol·K)]                       5 V1 (liter)                     1
31   Cv [cal/(g·mol·K)]                       3 T1 (K)                       450
32   Gamma                 1.666666666667
33   ------------------------------------------------------------------------
34   V2                    T2                  W                    DELTA H
35   (liter)               (K)                 [cal/(g·mol)]        [cal/(g·mol)]
36   ------------------------------------------------------------------------
37                       1.5        343.414272766      319.757181702        -532.92863617
38                       2
39                       2.5
40                       3
41                       3.5
42                       4
43                       4.5
44                       5
45
46
A44: [W19] +A43+0.5
28-May-90  10:48 PM                                                         READY
```

(a)

FIGURE 5.5 Using relationships instead of values alone. (a) Values of V_2 entered in column A. (b) Values replaced by cell relationships. (c) With cell relationships, when the first value under V2 changes, the succeeding values change.

cells. The **Copy** command in the spreadsheet allows us to copy the contents of one or more cells (whether labels, values, or formulas) to any other location on the spreadsheet.

At first, we change the value in A37 to 1.5. Then we enter the formula +A37+.5* in cell A38; the result 2 is displayed in the cell. When we use the Copy command, we need to specify what (information) needs to be copied and where it should be copied to, i.e., the **source range** and the **target range,** respectively.

The **source range** is the set of one or more cells that needs to be copied, while the **target range** is the location at which the source range contents need to be copied to. In the present example, cell A38 is the source range, while cells A39 through A44 make up the target range. Note that the term *range* can refer to a single cell or a set

---

*All entries that we type at the computer are underlined in the text.

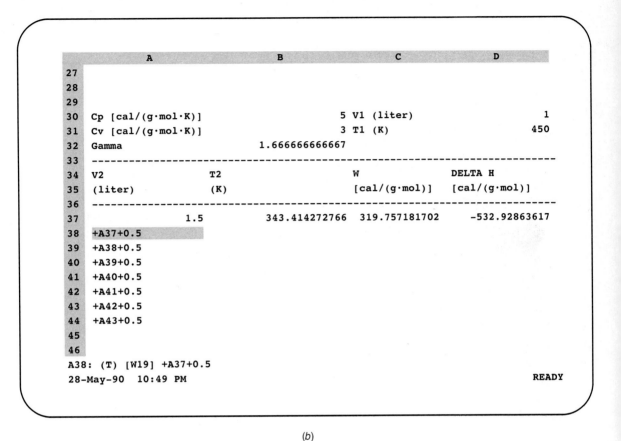

FIGURE 5.5 (continued).

of contiguous cells in rows and columns. Thus, the range A30..D37 includes all the cells from A30 through D37, i.e., cells A30 through A37, B30 through B37, C30 through C37, and D30 through D37.

Using the Copy Command With the cursor in A38, we invoke the command menu and select the Copy command (/C). Figure 5.6a shows the prompt for the source range. Since only one cell (A38) is the source, we press the Enter key; the prompt for the target range appears as shown in Figure 5.6b. We move the cursor to A39, the first cell in the target range, and press . to anchor the pointer. Then we press the Down Arrow key and move the cursor to A44. Recall that this task of specifying the range is similar to the one for printing a template (Figure 4.35a). The address in the prompt changes with the cursor movement. When we press the Enter key, the formula in cell A38 is copied to the other cells and the computed values are displayed in the cells. Thus, using the Copy command, the formula from one cell can be copied to a range of cells.

```
              A                    B                 C                  D
27
28
29
30  Cp [cal/(g·mol·K)]                    5 V1 (liter)                    1
31  Cv [cal/(g·mol·K)]                    3 T1 (K)                      450
32  Gamma               1.666666666667
33  -----------------------------------------------------------------------
34  V2                  T2                W                  DELTA H
35  (liter)             (K)               [cal/(g·mol)]      [cal/(g·mol)]
36  -----------------------------------------------------------------------
37         4                 178.5826183464 814.2521449607   -1357.086908268
38              4.5
39                5
40              5.5
41                6
42              6.5
43                7
44              7.5
45
46
A37: (G) [W19] 4
28-May-90  10:49 PM                                                  READY
```

(c)

FIGURE 5.5 (continued).

While specifying the source and target ranges in the Copy command (or any other command involving a range of cells), you can enter the cell addresses in response to the prompts instead of moving the cursor and painting the ranges. This is faster, especially if you know the cell addresses. The Copy command for the three major spreadsheets is shown in Quick Reference Table 5.3.

Absolute Cell Addresses The next step is to copy the formulas for T_2, W, and ΔH (in cells B37, C37, and D37, respectively) down the respective columns so that they can be evaluated for each of the values of V_2 in the first column. Before we do that, let's look at one of the common errors encountered in the use of the copy command. Consider the formula for T_2 in cell B37 (Figure 5.3a):

$$+D31 * (D30/A37)\wedge(B32 - 1)$$

If this is copied into cell B38 (as we want to), then the formula will become:

$$+D32 * (D31/A38)\wedge(B33 - 1)$$

```
Source block of cells : A38..A38_

          A                    B                  C                   D
27
28
29
30   Cp [cal/(g·mol·K)]                    5 V1 (liter)                          1
31   Cv [cal/(g·mol·K)]                    3 T1 (K)                            450
32   Gamma                  1.666666666667
33   --------------------------------------------------------------------------------
34   V2                    T2                    W                   DELTA H
35   (liter)               (K)                   [cal/(g·mol)]       [cal/(g·mol)]
36   --------------------------------------------------------------------------------
37                         1.5            343.414272766    319.757181702       -532.92863617
38                         2
39
40
41
42
43
44
45
46
A38: (G) [W19] +A37+0.5
28-May-90  10:54 PM                                                                POINT
```

(a)

FIGURE 5.6 Using the Copy command. (a) Highlighting the source cell. (b) Specifying the target cells.

TABLE 5.3 QUICK REFERENCE TABLE FOR COPY COMMAND

Task	1-2-3	Quattro	Excel
Copying a cell	/Copy <enter> <..> <enter>	/Block Copy <enter> <..> <enter>	Select cell /Edit Copy <..> Paste
Copying a range of cells	/Copy <..> <enter> <..> <enter>	/Block Copy <..> <enter> <..> <enter>	Select range of cells /Edit Copy <..> Paste

Note: <..> denotes a range of cells.

This is because the Copy command retains the relative positions of the cells specified in the original formula. For example, since the difference between D37 and D31 is six positions in the source formula, when it is copied into D38, D31 will be changed

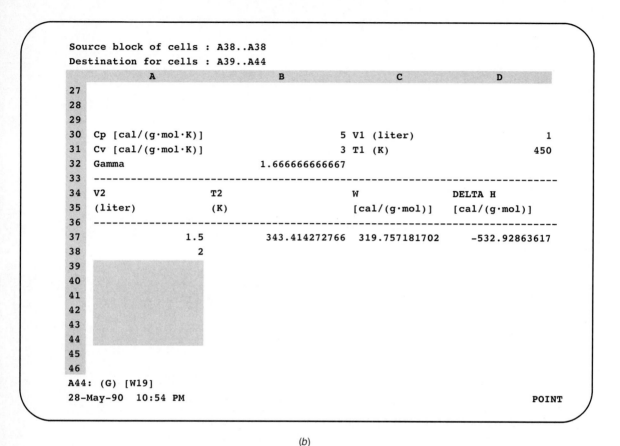

(b)

FIGURE 5.6 (continued).

to D32 to maintain the difference of 6. Likewise, D30 will become D31, A37 will change to A38, and B32 will become B33. However, from Figure 5.3a, we know that the values of V_1, T_1, and γ in D30, D31, and B32, respectively, should be used in the formula for computing T_2 for every value of V_2.

This means, no matter where the formula for T_2 is entered in the spreadsheet, it should make *absolute* references to the cells D30, D31, and B32. This form of making specific references to cells in formulas is known as **absolute addressing.** When an absolute cell address is specified in a formula, the Copy command retains the address in the new location. Note, however, that the reference to the variable V_2 (in A37) should be relative, since its value changes for each computation.

Specifying Absolute Cell Addresses Many spreadsheet programs use a dollar ($) sign as a prefix to the row and column of an absolute address. For example, the address D31 becomes $D$31 when it is specified as an absolute cell address. It is possible to specify either the row or the column as absolute and let the other be relative. For example, the address $D31 fixes the column to be D, while the row remains relative.

Conversely, D$31 implies that the column is relative while the row remains absolute (fixed) at 31.

Let's edit the formula in cell B37. We press the Edit key (F2 in 1-2-3). With the cue in front of D31 in the formula, we press the ABS key (see function key overlay in Figure 4.8). A $ sign appears in front of D and 31 so that it reads $D$31. Likewise, we change D30 to $D$30 and B32 to $B$32. Since the cell address A37 should be relative, it is not altered. The new formula looks as follows:

$$+\$D\$31 * (\$D\$30/A37)^{\wedge}(\$B\$32 - 1).$$

Note that the ABS key is a toggle switch. This means that if an address is relative, it will be changed to absolute and vice versa.

In a similar manner, we edit the formula in C37 from $-B31 * (B37 - D31)$ to $-\$B\$31 * (B37 - \$D\$31)$ and the formula in D37 from $+B30 * (B37 - D31)$ to $+\$B\$30 * (B37 - \$D\$31)$.

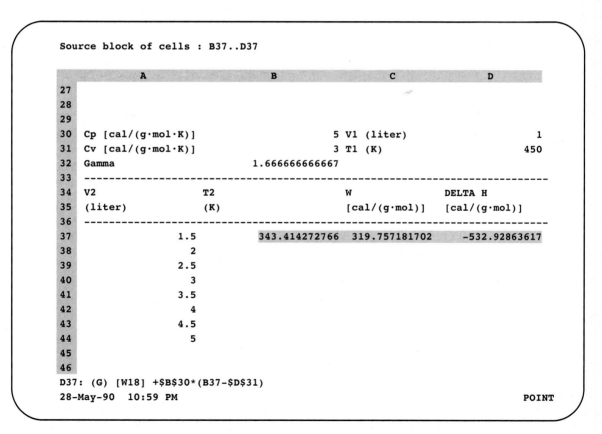

(a)

FIGURE 5.7 Copying a range of cells. (a) Highlighting the source range. (b) Specifying the target range.

```
Source block of cells : B37..D37
Destination for cells : B38..B44
                 A                    B              C              D
27
28
29
30  Cp [cal/(g·mol·K)]               5 V1 (liter)                   1
31  Cv [cal/(g·mol·K)]               3 T1 (K)                     450
32  Gamma              1.666666666667
33  -----------------------------------------------------------------
34  V2                 T2             W              DELTA H
35  (liter)            (K)            [cal/(g·mol)]  [cal/(g·mol)]
36  -----------------------------------------------------------------
37                     1.5            343.414272766  319.757181702  -532.92863617
38                     2
39                     2.5
40                     3
41                     3.5
42                     4
43                     4.5
44                     5
45
46
B44: [W23]
28-May-90  10:59 PM                                              POINT
```

(b)

FIGURE 5.7 (continued).

Copying a Range of Cells We can now copy the formulas in cells B37 through D37 to the desired number of rows in one step. We invoke the Copy command (/C) and highlight the source range as shown in Figure 5.7a. The target range is cells B38 through D44. We anchor the pointer in B38 and use the Down Arrow key to move the cursor to B44, the last row in the target range (Figure 5.7b). When we press the Enter key, all the formulas are copied and calculations are performed (Figure 5.8).

Note that while indicating the target range, we had to specify only the first and last cells (B38 and B44, respectively) in the first column of the target range (B38..D44), and the program copied the formulas into all the cells in the range. Thus, when copying a rectangular range of cells, it is necessary to specify only the first and last cells of a row (or column) in the target range — the program will take care of copying all the entries from the source range.

In Figure 5.8, as the final volume V_2 increases, the final temperature decreases, the work done increases, and the enthalpy changes. Thus, using a step-by-step approach, we have built a template for solving problems related to thermodynamics of an ideal gas.

```
            A                    B                    C                    D
27
28
29
30   Cp [cal/(g·mol·K)]                         5 V1 (liter)                          1
31   Cv [cal/(g·mol·K)]                         3 T1 (K)                            450
32   Gamma              1.666666666667
33   ---------------------------------------------------------------------------
34   V2                 T2                   W                    DELTA H
35   (liter)            (K)                  [cal/(g·mol)]        [cal/(g·mol)]
36   ---------------------------------------------------------------------------
37                      1.5          343.414272766  319.757181702   -532.92863617
38                      2            283.4822362263 499.553291321   -832.5888188683
39                      2.5          244.2975854935 617.1072435194  -1028.512072532
40                      3            216.3374355461 700.9876933617  -1168.312822269
41                      3.5          195.2092910491 764.3721268526  -1273.953544754
42                      4            178.5826183464 814.2521449607  -1357.086908268
43                      4.5          165.0963624447 854.7109126658  -1424.518187776
44                      5            153.8978352009 888.3064943973  -1480.510823995
45
46
D37: (G) [W18] +$B$30*(B37-$D$31)
28-May-90   11:06 PM                                                        READY
```

FIGURE 5.8 The thermodynamics template: final solution. Note that the values are displayed in the general format, with a varying number of digits after the decimal place.

What-If Analysis How will the values change if the heat capacities of the gas change to $C_p = 3$ and $C_v = 5$ cal/(g·mol·K), respectively, and the initial temperature is 400 K?

To solve this problem, all we have to do is change the values of the three variables. Since the program is in the automatic recalculation mode, the default mode, the output values will automatically change when each of the new input values is entered. Figure 5.9 shows the final result. To reiterate, once a template has been developed, it can be used to solve a wide variety of problems. We save this template under the name thermo1.

5.3 IMPROVING THE TEMPLATE APPEARANCE: THE RANGE FORMAT COMMAND

In Figure 5.8, all calculated values are displayed with several digits after the decimal place. Some values are not properly aligned in the columns; the labels V2, T2, W, and DELTA H are left-justified by default, and they are not above the numeric values. In

```
             A                    B                  C                    D
27
28
29
30    Cp [cal/(g·mol·K)]                      3  V1 (liter)                        1
31    Cv [cal/(g·mol·K)]                      5  T1 (K)                          400
32    Gamma                                 0.6
33    ----------------------------------------------------------------------
34    V2                  T2                   W                    DELTA H
35    (liter)             (K)                  [cal/(g·mol)]        [cal/(g·mol)]
36    ----------------------------------------------------------------------
37                       1.5         470.4316090099   -352.158045049        211.2948270296
38                         2         527.8031643092   -639.015821546        383.4094929275
39                       2.5         577.0799623629   -885.399811814        531.2398870887
40                         3         620.7382295661  -1103.69114783         662.2146886984
41                       3.5         660.2177695796  -1301.0888479          780.6533087388
42                         4         696.4404506369  -1482.20225318         889.3213519107
43                       4.5         730.0372102718  -1650.18605136         990.1116308155
44                         5         761.4615754864  -1807.30787743        1084.384726459
45
46
D31: [W18] 400
28-May-90   11:07 PM                                                           READY
```

FIGURE 5.9 The thermodynamics template: Solution to another problem.

short, the appearance of the results is not quite what we would want. We can use the format commands in spreadsheets to enhance the appearance of the template.

5.3.1 The Range or Block Command

Every spreadsheet program has a Range or Block command that enables you to execute a set of commands on a range or block of cells. These include formatting the display of labels and values, erasing the range, naming the range, and protecting the range from changes. These commands can also be used for deleting a named range, unprotecting a protected range, and transposing the range, i.e., copying columns to rows or vice versa. The range is usually specified by the addresses of the top-left and bottom-right cells, separated by ellipsis (..), e.g., A1..D5. The typical actions under the Range or Block command are shown in Figure 5.10.

The Label-Prefix or Label-Align Command The Label-Prefix or Align command is used to align the label entry in the cell. Recall from Section 4.4.1 that a label entry can be left-justified (' prefix, the default), centered (^), or right-justified ("). Since

```
                                                            Copy a block

              A                    B                  C               D
    27
    28                                                         Block
    29                                                         Copy
    30  Cp (Cal/(g·mole)(K)                  3 V1 (liter)      Move
    31  Cv (Cal/(g·mole)(K)                  5 T1 (K)          Erase
    32  Gamma                              0.6                 Display Format
    33  ------------------------------------------------------ Label Align
    34  V2                   T2                  W             Fill
    35  (liter)              (K)                 [cal/(g·mol)] Search/Replace
    36  ------------------------------------------------------ Reformat
    37              1.5        470.4316090099  -352.158045049  Advanced
    38              2          527.8031643092  -639.015821546
    39              2.5        577.0799623629  -885.399811814
    40              3          620.7382295661 -1103.69114783     662.2146886984
    41              3.5        660.2177695796 -1301.0888479      780.6533087388
    42              4          696.4404506369 -1482.20225318     889.3213519107
    43              4.5        730.0372102718 -1650.18605136     990.1116308155
    44              5          761.4615754864 -1807.30787743    1084.384726459
    45
    46
    D31: [W18] 400
    28-May-90   11:08 PM                                                  MENU
```

FIGURE 5.10 The Range or Block command options.

we want the labels in cells A34..D35 to be centered, we select the Label-Align option (L) from the menu shown in Figure 5.10. From the next-level menu, we select the Center option. The system then prompts us for the range of cells (Figure 5.11a), and we specify A34..D35 (Figure 5.11b). The labels are immediately centered in the specified cells as shown in Figure 5.11c.

Formatting Numeric Values A spreadsheet program provides a great deal of flexibility in displaying numeric values in the template. The major numeric formatting options are shown in Table 5.4. The **General format** is the default format. In the **Scientific format,** also known as the **E format,** the number is displayed with one digit to the left of the decimal place and the remaining digits to the right. As shown in the table, 2345.6786 is displayed as 2.3456786E + 03. The sign and magnitude of the exponent, +03, indicate the direction and number of places to which the decimal point should be moved to determine the number. Thus, 2.3456E − 09 stands for 0.0000000023456. The Scientific format is extremely useful for representing very large and very small numbers.

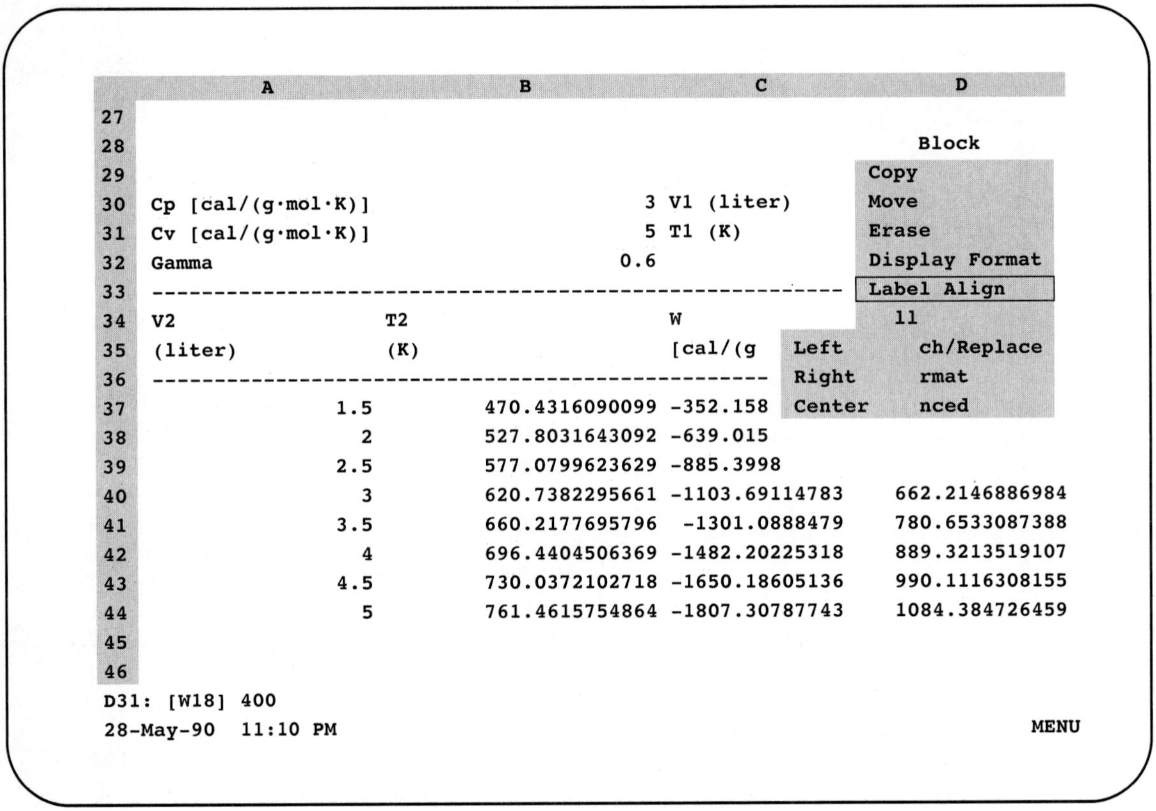

```
              A                       B                C              D
        27
        28                                                         Block
        29                                                         Copy
        30   Cp [cal/(g·mol·K)]              3  V1 (liter)         Move
        31   Cv [cal/(g·mol·K)]              5  T1 (K)             Erase
        32   Gamma                         0.6                     Display Format
        33   ----------------------------------------------------- Label Align
        34   V2                 T2                  W              11
        35   (liter)            (K)               [cal/(g  Left    ch/Replace
        36   -----------------------------------------------Right  rmat
        37                     1.5       470.4316090099 -352.158 Center nced
        38                       2       527.8031643092 -639.015
        39                     2.5       577.0799623629 -885.3998
        40                       3       620.7382295661 -1103.69114783   662.2146886984
        41                     3.5       660.2177695796 -1301.0888479    780.6533087388
        42                       4       696.4404506369 -1482.20225318   889.3213519107
        43                     4.5       730.0372102718 -1650.18605136   990.1116308155
        44                       5       761.4615754864 -1807.30787743  1084.384726459
        45
        46
        D31: [W18] 400
        28-May-90  11:10 PM                                            MENU
```

(a)

FIGURE 5.11 Using the Range Label-Align or Prefix command. (a) Selecting the Label-Align command. (b) Prompt for range of labels to be aligned. (c) The labels are centered.

Using the Numeric Formatting Feature Let's use the Numeric Formatting feature to display the values of T2, W, and DELTA H with four digits after the decimal place in the range B37..D44. To do this, we need to first specify the range and then select the Fixed option of the Format command and set the number of digits, viz., 4.

With the cursor in cell B37, we invoke the command menu and select the Range or Block option. The submenu appears on the screen and is shown in Figure 5.10. We select the Display Format option followed by the Fixed option. In response to the prompt for the number of digits, we specify 4. The program prompts us for the range as shown in Figure 5.12a. We type B37..D44 and press the Enter key. The range is formatted, and the screen looks as shown in Figure 5.12b. The Range Format commands for the three major spreadsheets are shown in Quick Reference Table 5.5.

Naming Ranges or Blocks of Cells Another useful feature of spreadsheet programs is the ability to name individual cells or groups of cells. Once a cell is named,

```
Block to be modified : A34..D35

           A                    B                   C                      D
    27
    28
    29
    30  Cp [cal/(g·mol·K)]                     3 V1 (liter)                   1
    31  Cv [cal/(g·mol·K)]                     5 T1 (K)                     400
    32  Gamma                                0.6
    33  -----------------------------------------------------------------------
    34  V2                     T2                   W                     DELTA H
    35  (liter)                (K)              [cal/(g·mol)]          [cal/(g·mol)]
    36  -----------------------------------------------------------------------
    37             1.5       470.4316090099    -352.158045049         211.2948270296
    38             2         527.8031643092    -639.015821546         383.4094929275
    39             2.5       577.0799623629    -885.399811814         531.2398870887
    40             3         620.7382295661   -1103.69114783          662.2146886984
    41             3.5       660.2177695796   -1301.0888479           780.6533087388
    42             4         696.4404506369   -1482.20225318          889.3213519107
    43             4.5       730.0372102718   -1650.18605136          990.1116308155
    44             5         761.4615754864   -1807.30787743         1084.384726459
    45
    46
    D35: (G) [W18] '(cal/(g mole))
    29-May-90   09:56 AM                                                    POINT
```

(b)

FIGURE 5.11 (continued).

the name can be used in place of the cell address in formulas. Likewise, the range or block name can be specified in commands. In large spreadsheets, it is easier to jump to a named range using the GoTo command.

Let's use the Range or Block Name command and associate the name VOL2 with the range of cells in A37..A44. With the cursor in A37, we invoke the command (/RNC in 1-2-3) and specify the name VOL2 in response to the prompt (Figure 5.13a). We then anchor the pointer in A37 and paint the range of cells up to A44 and terminate the command with the Enter key (5.13b). The name VOL2 is now associated with the range. Note that VOL2 does not imply a variable name as in a programming language. Moreover, the range name should not correspond to a cell address; thus V2 would be an invalid name since V2 is a spreadsheet cell address. This naming feature comes in handy when specifying ranges in the Graph command (see Section 5.4.2). The commands for naming ranges in the three major spreadsheets are shown in Quick Reference Table 5.6.

```
                A                    B                    C                  D
27
28
29
30   Cp [cal/(g·mol·K)]                             3  V1 (liter)                     1
31   Cv [cal/(g·mol·K)]                             5  T1 (K)                       400
32   Gamma                                        0.6
33   ------------------------------------------------------------------------------
34        V2                         T2                   W               DELTA H
35      (liter)                     (K)              [cal/(g·mol)]      [cal/(g·mol)]
36   ------------------------------------------------------------------------------
37           1.5             470.4316090099       -352.158045049       211.2948270296
38           2               527.8031643092       -639.015821546       383.4094929275
39           2.5             577.0799623629       -885.399811814       531.2398870887
40           3               620.7382295661      -1103.69114783        662.2146886984
41           3.5             660.2177695796      -1301.0888479         780.6533087388
42           4               696.4404506369      -1482.20225318        889.3213519107
43           4.5             730.0372102718      -1650.18605136        990.1116308155
44           5               761.4615754864      -1807.30787743       1084.384726459
45
46
A34: (G) [W19] ^V2
29-May-90   09:57 AM                                                          READY
```

(c)

FIGURE 5.11 (continued).

5.4 GENERATING TABLES AND GRAPHS

The presentation of the results generated during problem solving is almost as important as the solution process itself. Scientists and engineers typically present the problem data and results in tabular form or display it graphically for two main reasons. First, it helps to organize the information in a form that can be easily understood by others. Second, this organization assists in spotting trends in the information or establishing relationships between the variables.

Typically, in tables the independent-variable values are entered in the first column followed by the dependent-variables data in the subsequent columns.

The major graphical formats used by scientists and engineers are bar graphs (also known as bar charts), *x-y* plots, and pie charts. The *x-y* plot is the most popular of the three formats in scientific and/or engineering applications.

We will now use some spreadsheet features to display the results from the thermodynamics problem in a tabular form and also produce graphical outputs of the relationships.

TABLE 5.4 **NUMERIC FORMATTING OPTIONS**
(Value entered in cell is 2345.67860)

Option	Description	Example
General	Values displayed as entered or calculated; zeros after decimal point are suppressed.	2345.6786
Fixed	Values displayed with specified number of digits after the decimal place; 3 in this example; note the rounding up of the last digit.	2345.679
Scientific	Values displayed in scientific notation, i.e., in terms of powers of 10. Useful for very small or large numbers.	$2.345678E+03$
Percent	Values multiplied by 100 and displayed with a % sign. This format is counterintuitive: opposite of what you would normally expect. Two decimal places are displayed.	234567.86%
Currency	Values displayed as dollars and cents with a $ prefix, commas in the appropriate places and two digits after the decimal place. Negative values typically displayed in parentheses.	$2,345.68
Date	Displays date in selected format: e.g., mm/dd/yy dd/mm/yy	11/21/90 22/11/90

TABLE 5.5 **QUICK REFERENCE TABLE FOR RANGE FORMAT COMMAND**

Task	1-2-3	Quattro	Excel
Select numeric format for values	/Range Format Fixed Currency Scientific General Percent Date <..>	/Block Display Fixed Currency Scientific General Percent Date <..>	/Format Number Select format from menu <..>

Note: <..> denotes a range of cells.

5.4.1 Table of Results

In designing the layout of the template, we took care to ensure that the calculated values will be displayed in the form of a table on the screen that could be easily viewed and printed for analysis. As seen in Figure 5.8, the values of the independent variable V_2 are in the first column, and the values of dependent variables are in neighboring columns. Therefore, we can simply use the Print command (Section 4.5.5) and produce the table of results (see Figure 5.14). Otherwise, we would have to move the rows and columns to the desired locations and then print the results.

The Move Command The **Move command** is used for moving the contents of a cell or range of cells to a specified location in the spreadsheet. Unlike the Copy

134 SPREADSHEET COMMANDS AND FEATURES

```
Block to be modified : B37..D44

            A                B                   C                  D
27
28
29
30    Cp [cal/(g·mol·K)]                 3  V1 (liter)                          1
31    Cv [cal/(g·mol·K)]                 5  T1 (K)                            400
32    Gamma                            0.6
33    ------------------------------------------------------------------------
34           V2                T2                  W                 DELTA H
35         (liter)             (K)             [cal/(g·mol)]        [cal/(g·mol)]
36    ------------------------------------------------------------------------
37           1.5         470.4316090099   -352.158045049         211.2948270296
38            2          527.8031643092   -639.015821546         383.4094929275
39           2.5         577.0799623629   -885.399811814         531.2398870887
40            3          620.7382295661  -1103.69114783          662.2146886984
41           3.5         660.2177695796  -1301.0888479           780.6533087388
42            4          696.4404506369  -1482.20225318          889.3213519107
43           4.5         730.0372102718  -1650.18605136          990.1116308155
44            5          761.4615754864  -1807.30787743         1084.384726459
45
46
B37: (G) [W23] +$D$31*($D$30/A37)^($B$32-1)
29-May-90   10:01 AM                                                        EDIT
```

(a)

FIGURE 5.12   Using the Range or Block Format command. (*a*) Prompt for range to be formatted. (*b*) The formatted display.

command, the original cell or range of cells is left blank. As in the Copy command, absolute cell references in formulas are retained, while relative cell addresses are adjusted only if the referred cells themselves have been moved to a different location. The Move command for the three major spreadsheets is shown in Quick Reference Table 5.7.

If the cells in the target location specified in the Move command contain any information (labels, values, formulas), they are overwritten by the new information. Therefore, care should be taken when using the Move command. However, if the target location is protected (using the Range or Block Protect option of the spreadsheet program), the contents cannot be replaced and the Move command will be terminated with an appropriate error message.

Using the Move Command   Let's move the formula for T_2 in cell B37 to the location B45 using the Move command. With the cursor in B37, we type /M (Move command in 1-2-3) and press the Enter key to specify the source range (Figure 5.15*a*).

```
           A                   B                  C                  D
27
28
29
30  Cp [cal/(g·mol·K)]                      3 V1 (liter)                        1
31  Cv [cal/(g·mol·K)]                      5 T1 (K)                          400
32  Gamma                                 0.6
33  ----------------------------------------------------------------------------
34         V2                  T2                 W               DELTA H
35       (liter)               (K)          [cal/(g·mol)]      [cal/(g·mol)]
36  ----------------------------------------------------------------------------
37         1.5              470.4316          -352.1580           211.2948
38          2               527.8032          -639.0158           383.4095
39         2.5              577.0800          -885.3998           531.2399
40          3               620.7382         -1103.6911           662.2147
41         3.5              660.2178         -1301.0888           780.6533
42          4               696.4405         -1482.2023           889.3214
43         4.5              730.0372         -1650.1861           990.1116
44          5               761.4616         -1807.3079          1084.3847
45
46
B37: (F4) [W23] +$D$31*($D$30/A37)^($B$32-1)
29-May-90  10:02 AM                                                    READY
```

(b)

FIGURE 5.12 (continued).

In response to the prompt for the target range, we move the cursor to B45, the target cell, and press the Enter key. The formula in B45 reads as shown in the control panel at the bottom of Figure 5.15*b*. Note that since none of the cells referenced in the formula in B37 (the new B45) has changed locations, the references are retained. However, the formulas in C37 and D37 have changed to reflect the new location of T_2, viz., B45 (bottom of Figure 5.15*c*). The cell B37 remains blank.

Protecting a Range of Cells In the real world, a template is typically created by an expert scientist or engineer and used by others in the organization. For instance, in the academic world, professors often create templates that the students play with during laboratory sessions. Under those circumstances, it is important that the underlying cell relationships (formulas) are not accidentally altered or deleted by the user. The spreadsheet program has a range- or block-protect feature that enables us to guard against such disasters. The protected cells can also be "unprotected" at a subsequent time.

```
Enter Name to Create/Modify : VOL2

         A                    B                  C                    D
27
28
29
30   Cp [cal/(g·mol·K)]                       3 V1 (liter)              1
31   Cv [cal/(g·mol·K)]                       5 T1 (K)                400
32   Gamma                                 0.6
33   ------------------------------------------------------------------------
34          V2                 T2                 W               DELTA H
35        (liter)              (K)           [cal/(g·mol)]     [cal/(g·mol)]
36   ------------------------------------------------------------------------
37          1.5             470.4316          -352.1580          211.2948
38          2               527.8032          -639.0158          383.4095
39          2.5             577.0800          -885.3998          531.2399
40          3               620.7382         -1103.6911          662.2147
41          3.5             660.2178         -1301.0888          780.6533
42          4               696.4405         -1482.2023          889.3214
43          4.5             730.0372         -1650.1861          990.1116
44          5               761.4616         -1807.3079         1084.3847
45
46
A37: (G) [W19] 1.5
29-May-90   10:03 AM                                              EDIT
```

(a)

FIGURE 5.13 Naming a block or range of cells. (a) Specifying the range name: VOL2. (b) Painting the range.

Hiding Rows and Columns Yet another feature of spreadsheets is the ability to hide desired rows and columns. For example, if a template contains a number of columns for storing intermediate results during a computation, they can be hidden from view using the Hide Column feature. The display screen or the printout will not be cluttered by data that may not be required for analysis. This feature can also be used to hide some sensitive information in the spreadsheet, e.g., salaries in a financial template.

5.4.2 Creating and Displaying Graphs

The major types of graphs that can be produced using spreadsheet programs are: bar, stacked-bar, *X-Y*, line and, pie. The selection of a particular type of graph will depend on the problem domain, the number of data points, and the intended purpose.

Types of Graphs We now discuss the major types of graphs that can be produced using spreadsheet programs.

```
Enter Name to Create/Modify : VOL2
Specify Coordinate for the block : A37..A44
            A                      B                    C                  D
27
28
29
30    Cp [cal/(g·mol·K)]                        3 V1 (liter)                    1
31    Cv [cal/(g·mol·K)]                        5 T1 (K)                      400
32    Gamma                       0.6
33    ------------------------------------------------------------------------
34         V2                     T2                    W              DELTA H
35       (liter)                  (K)             [cal/(g·mol)]      [cal/(g·mol)]
36    ------------------------------------------------------------------------
37         1.5                  470.4316            -352.1580           211.2948
38         2                    527.8032            -639.0158           383.4095
39         2.5                  577.0800            -885.3998           531.2399
40         3                    620.7382           -1103.6911           662.2147
41         3.5                  660.2178           -1301.0888           780.6533
42         4                    696.4405           -1482.2023           889.3214
43         4.5                  730.0372           -1650.1861           990.1116
44         5                    761.4616           -1807.3079          1084.3847
45
46
A44: (G) [W19] +A43+0.5
29-May-90   10:03 AM                                                      POINT
```

(b)

FIGURE 5.13 (continued).

TABLE 5.6 QUICK REFERENCE TABLE FOR RANGE OR BLOCK NAME COMMAND

Task	1-2-3	Quattro	Excel

Creating a range name	/Range
	  Name
	    Create
	      \<name\>
	        \<enter\>
	          <..>
		/Block
		  Advanced
		    Create name
		      \<name\>
		        \<enter\>
		          <..>
			Select range of cells
			/Format
			  Define name
			    \<name\>
			      Refers to
			        \<cell address\>
			          OK

Erasing a range name	/Range
	  Name
	    Delete
	      \<name\>
	        \<enter\>
		/Block
		  Advanced
		    Delete name
		      \<name\>
		        \<enter\>
			Select range of cells
			/Format
			  Define name
			    \<name\>
			      Delete

Note: <..> denotes a range of cells.

138 SPREADSHEET COMMANDS AND FEATURES

C_p [cal/(g·mol·K)]		3	V1 (liter)	1
C_v [cal/(g·mol·K)]		5	T1 (K)	400
Gamma		0.6		

V2 (liter)	T2 (K)	W [cal/(g·mol)]	Delta H [cal/(g·mol)]
1.5	470.4316	−352.1580	211.2948
2	527.8032	−639.0158	383.4095
2.5	577.0800	−885.3998	531.2399
3	620.7382	−1103.6911	662.2147
3.5	660.2178	−1301.0888	780.6533
4	696.4405	−1482.2023	889.3214
4.5	730.0372	−1650.1861	990.1116
5	761.4616	−1807.3079	1084.3847

FIGURE 5.14 Thermodynamics template: Tabular display of results.

```
Source block of cells : B37..B37
Destination for cells : B37
```

	A	B	C	D
27				
28				
29				
30	Cp [cal/(g·mol·K)]		3 V1 (liter)	1
31	Cv [cal/(g·mol·K)]		5 T1 (K)	400
32	Gamma		0.6	
33	----------	----------	----------	----------
34	V2	T2	W	DELTA H
35	(liter)	(K)	[cal/(g·mol)]	[cal/(g·mol)]
36	----------	----------	----------	----------
37	1.5	470.4316	−352.1580	211.2948
38	2	527.8032	−639.0158	383.4095
39	2.5	577.0800	−885.3998	531.2399
40	3	620.7382	−1103.6911	662.2147
41	3.5	660.2178	−1301.0888	780.6533
42	4	696.4405	−1482.2023	889.3214
43	4.5	730.0372	−1650.1861	990.1116
44	5	761.4616	−1807.3079	1084.3847
45				
46				

```
B37: (F4) [W23] +$D$31*($D$30/A37)^($B$32-1)
29-May-90  10:04 AM                                              POINT
```

(a)

FIGURE 5.15 The Move command. (a) Specifying the source range to be moved. (b) Contents of B37 have been moved to cell B45. Cell references in B45 remain the same (see bottom of screen).
(c) Formula in C37 has changed; so has the formula in D37.

	A	B	C	D
27				
28				
29				
30	Cp [cal/(g·mol·K)]		3 V1 (liter)	1
31	Cv [cal/(g·mol·K)]		5 T1 (K)	400
32	Gamma		0.6	
33	---			
34	V2	T2	W	DELTA H
35	(liter)	(K)	[cal/(g·mol)]	[cal/(g·mol)]
36	---			
37	1.5		-352.1580	211.2948
38	2	527.8032	-639.0158	383.4095
39	2.5	577.0800	-885.3998	531.2399
40	3	620.7382	-1103.6911	662.2147
41	3.5	660.2178	-1301.0888	780.6533
42	4	696.4405	-1482.2023	889.3214
43	4.5	730.0372	-1650.1861	990.1116
44	5	761.4616	-1807.3079	1084.3847
45		470.4316		
46				

B45: (F4) [W23] +$D$31*($D$30/A37)^($B$32-1)
29-May-90 10:06 AM   READY

(b)

FIGURE 5.15 (continued).

TABLE 5.7 QUICK REFERENCE TABLE FOR MOVE COMMAND

Task	1-2-3	Quattro	Excel
Move a cell	/Move <Enter> Destination <enter>	/Block Move <enter> Destination <enter>	Select cell /Edit Cut Select target Paste
Move a range of cells	/Move <..> <enter> Destination <enter>	/Block Move <..> <enter> Destination <enter>	Select group of cells Edit Cut Select target Paste

Note: <..> denotes a range of cells.

140 SPREADSHEET COMMANDS AND FEATURES

	A	B	C	D
27				
28				
29				
30	Cp [cal/(g·mol·K)]	3	V1 (liter)	1
31	Cv [cal/(g·mol·K)]	5	T1 (K)	400
32	Gamma	0.6		
33	-------------------	----------------	----------------	----------------
34	V2	T2	W	DELTA H
35	(liter)	(K)	[cal/(g·mol)]	[cal/(g·mol)]
36	-------------------	----------------	----------------	----------------
37	1.5		-352.1580	211.2948
38	2	527.8032	-639.0158	383.4095
39	2.5	577.0800	-885.3998	531.2399
40	3	620.7382	-1103.6911	662.2147
41	3.5	660.2178	-1301.0888	780.6533
42	4	696.4405	-1482.2023	889.3214
43	4.5	730.0372	-1650.1861	990.1116
44	5	761.4616	-1807.3079	1084.3847
45		470.4316		
46				

C37: (F4) [W15] -$B$31*(B45-$D$31)
29-May-90 10:06 AM   READY

(c)

FIGURE 5.15 (continued).

Bar Graph In a **bar graph,** the data points are displayed as rectangular bars. The length of the bar represents the magnitude of the data point. This type of graph is used for discrete data points and is effective if the number of points is less than 12. A bar graph has two axes: horizontal, or X, and vertical, or Y. The X axis has labels that identify the bars.

Table 5.8 shows data pertaining to the evolution of Intel's microprocessors.[1] Figure 5.16 shows the bar graph relating the microprocessor to the number of transistors on the chip. Note that the horizontal or X-axis points are discrete. Herein lies the strength of the bar graph. It is a simple yet very effective means of representing data graphically.

An extension of the bar graph is the **stacked-bar graph;** in this type of graph, the bars corresponding to two or more Y-axis values for a given X-axis value are stacked on top of each other. For example, the cost of bringing a product to market can be

---

[1]From Ann Lewnes, "The Intel 386 Architecture: Here to Stay," *Microcomputer Solutions,* July/August 1989.

TABLE 5.8 THE INTEL MICROPROCESSOR CHIP: THE GROWTH DATA

Year	Chip	Number of transistors per chip
1971	4004	2.3K
1974	8080	6K
1978	8086	29K
1982	80286	134K
1985	80386	275K
1990	80860	1M
1990	80486	1.2M
1993	80586	4M
2000	Micro 2000	100 M

From: Ann Lewnes, "The Intel 386 Architecture: Here to Stay," *Microcomputer Solutions,* July/August 1989, pp. 3–5. Published by Intel Corp., Calif.

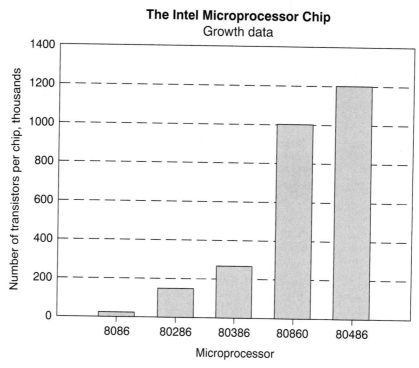

FIGURE 5.16 Bar graph of Intel microprocessor evolution.

split into five major components, viz., research and development, raw material, manufacturing, marketing and overhead. The data for five different products (A, B, C, D, and E) from a company are shown in Table 5.9. The corresponding stacked-bar graph is shown in Figure 5.17. Thus the stacked-bar graph not only gives a comparative picture of the total cost for each of the products but also the relative distributions of the component or subcosts.

TABLE 5.9 DISTRIBUTION OF PRODUCT COSTS

Product	R&D, $	Raw Material, $	Manufacturing, $	Marketing, $	Overhead, $
A	1.25	1.15	4.75	1.45	3.50
B	1.45	5.25	2.34	1.25	5.25
C	0.95	2.45	0.95	0.75	3.20
D	1.15	2.00	1.00	0.50	2.00
E	2.45	5.45	3.45	1.65	7.25

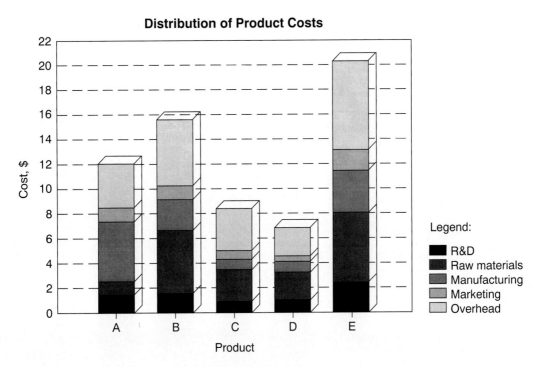

FIGURE 5.17 Stacked-bar graph of distribution of product costs.

X-Y and Line Graphs In an *X-Y* **graph**, the data points are displayed in an *X-Y* plane. The *X*-axis, or abscissa, values are equally spaced. This type of graph is suitable for plotting continuous data typically obtained from experiments. For example, the variation of viscosity of a liquid with temperature or the relationship between stress and strain of a polymeric material can be depicted in the form of *X-Y* graphs. Figure 5.18*a* shows an *X-Y* graph relating the angle of departure of a projectile to its range.

In a **line graph,** on the other hand, the abscissa values are not equally spaced. A line graph relating Intel's microprocessor chip density to the year of introduction is shown in Figure 5.18*b*. Note that the years on the *X*-axis do not have equal intervals.

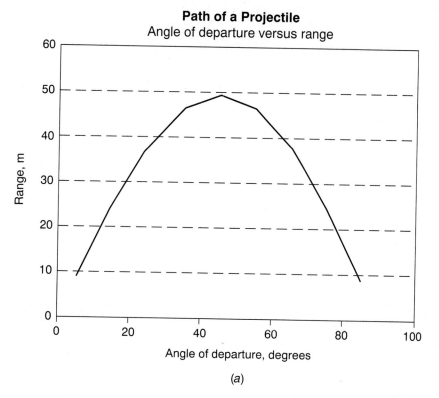

FIGURE 5.18 *X-Y* and line graphs. (*a*) *X-Y* graph of angle of departure versus range. (*b*) Line graph of year versus growth in Intel's chip density.

However, the growth trend—increasing chip density with years—is clearly depicted in the figure.

Pie Chart As the name implies, a **pie chart** is a circle divided into slices. Each slice of the pie corresponds to a data point. It is commonly used for representing the relative contribution of several parts to the whole.

Figure 5.19*a* shows the time spent by a typical engineer in various tasks during a 40-h workweek. From the corresponding pie chart in Figure 5.19*b*, it is clear that the engineer is spending a sizable chunk of the time (19 percent) in meetings. Using this chart, you can investigate if these meetings are indeed productive and contribute to the overall task objectives.

Spreadsheet as a Data Entry and Graphing Tool For creating the examples of the bar, stacked-bar, and pie charts discussed here, we just entered the data in a spreadsheet in different columns and used the graphing feature. Thus the spreadsheet can also serve as a tool for quick data entry and graphical representation of the entered data.

Generating a Graphical Representation Figure 5.20 shows the major sequence of steps in creating graphical representations of data. The initial or planning phase

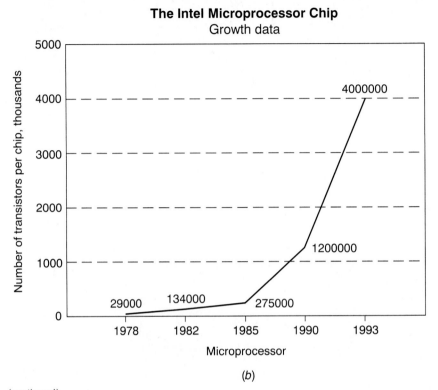

FIGURE 5.18 *(continued)*.

consists of the identification of independent and dependent variables and the selection of type of graph (e.g., bar, line, or *X-Y*). The execution phase depends on the particular spreadsheet program. The specific sequence of commands for creating a graph in the three major spreadsheet programs is shown in Quick Reference Table 5.10.

We will follow the procedure in Figure 5.20 for generating a graphical representation of the data in the thermodynamics template in Figure 5.12.

Identify Independent Variables From our problem statement in Section 5.1, it is clear that we want to find the effect of V_2 on the other variables. Therefore, V_2 is the independent variable.

Identify Dependent Variables The dependent variables are T_2, W, and ΔH.

Deciding on a Graph Type Of the different types of graphs available, the *X-Y* graph will be the appropriate one to depict the relationships. Recall that the bar graph is most suited for discrete values, the line graph is for unequally spaced *X*-axis values, and the pie chart is for expressing relative proportions of the variables.

Selecting the Graph We invoke the graph menu of the spreadsheet (/G in 1-2-3 or Quattro) and the set of options is shown in Figure 5.21*a*. We select the Type option and choose X-Y from the menu.

Task	Hours spent
Design	15.5
Analysis	13
Meetings	7.5
Reports	4

(a)

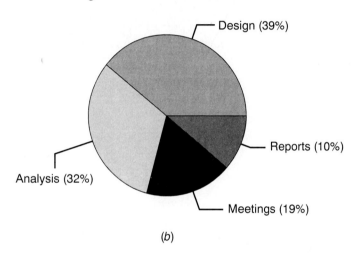

(b)

FIGURE 5.19 An engineer's time-task distribution (40 h/week). (a) Data points. (b) Pie chart: shows the relative proportions of the tasks.

Mark Independent (or X-Axis) Variable Data Range We then select the X-Axis option on the graph menu to specify the values for the X axis. In response to the prompt for the range (A37..A44), we type the range name VOL2 and press the Enter key (see Figure 5.21b). Recall that we created this range name in Section 5.3.1 for the values of V_2 in cells A37..A44.

Mark Dependent (or Y-Axis) Variable Data Range A spreadsheet typically allows you to specify up to six sets of values for the Y axis. We select the range B37..B44, the values of T_2, as the first set of values (Figure 5.21c).

View the Graph When we select the View option, the relationship between final volume (V_2) and the final temperature (T_2) is displayed as shown in Figure 5.22.

Customizing the Graph Spreadsheet programs have a whole set of features and functions for customizing graphs. These include adding labels to the values, titles for the axes and graph, displaying legends, selecting appropriate colors and fonts, and modifying the axes scales. Using these features you can produce presentation-quality graphs. Figure 5.23 shows the customized version of the graph in Figure 5.22.

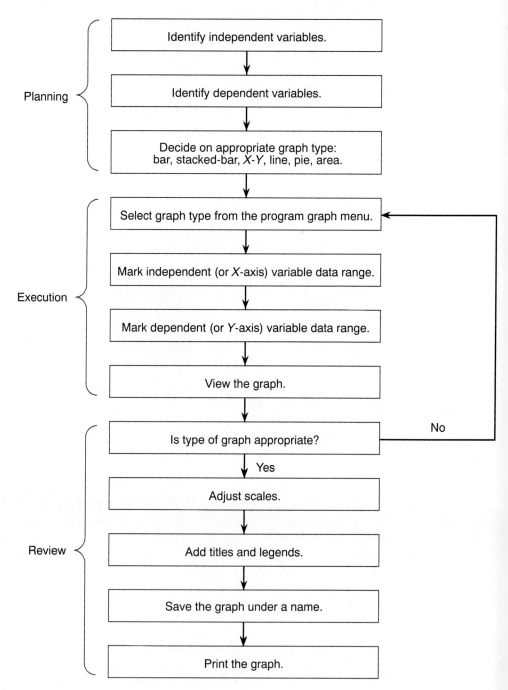

FIGURE 5.20 Creating a graph: sequence of steps.

TABLE 5.10　QUICK REFERENCE TABLE FOR GRAPH COMMAND

Task	1-2-3	Quattro	Excel
Selecting a graph type	/Graph 　Type 　　Line 　　Bar 　　Stacked bar 　　Pie 　　X-Y	/Graph 　Graph 　　Line 　　Bar 　　Stacked bar 　　Pie 　　X-Y 　　Area 　　Rotated bar 　　3-D bar	Select range of cells F11 　Gallery 　　Area 　　Bar 　　Column 　　Line 　　Pie 　　Scatter
Selecting X-axis range	/Graph 　X 　　<..>	/Graph 　X 　　<..>	Select range of cells
Selecting Y-axis range(s)	/Graph 　ABCDEF 　　<..>	/Graph 　Series 　　123456 　　<..>	Select range of cells
Viewing a graph	/Graph 　View	/Graph 　View	Select range of cells F11

Note: <..> denotes a range of cells.

Naming the Graph　During the course of problem solving, we are typically assessing the effect of one or more variables on the others. In doing so, we examine several combinations of variables, and it is common practice to create several graphs, each with a different set of variables. Spreadsheet programs provide the facility to save each of these graphs under a separate name. Graphs, once named, can be redisplayed, modified, or erased.

For example, we will use the Graph Name feature and save the graph in Figure 5.23 under the name "temp." The commands for naming a graph for the three major spreadsheet programs are shown in Quick Reference Table 5.11.

Plotting Another Graph　Let's now plot the effect of V_2 on the work done (W) and enthalpy change (ΔH). Again the X-Y graph would be appropriate for this

TABLE 5.11　QUICK REFERENCE TABLE FOR GRAPH NAME COMMAND

Task	1-2-3	Quattro	Excel
Naming a graph	/Graph 　Name 　　Create 　　Edit 　　Delete 　　Use	/Graph 　Name 　　Display 　　Store 　　Erase 　　Reset	From Graph window /File 　Save 　　<name> 　　OK

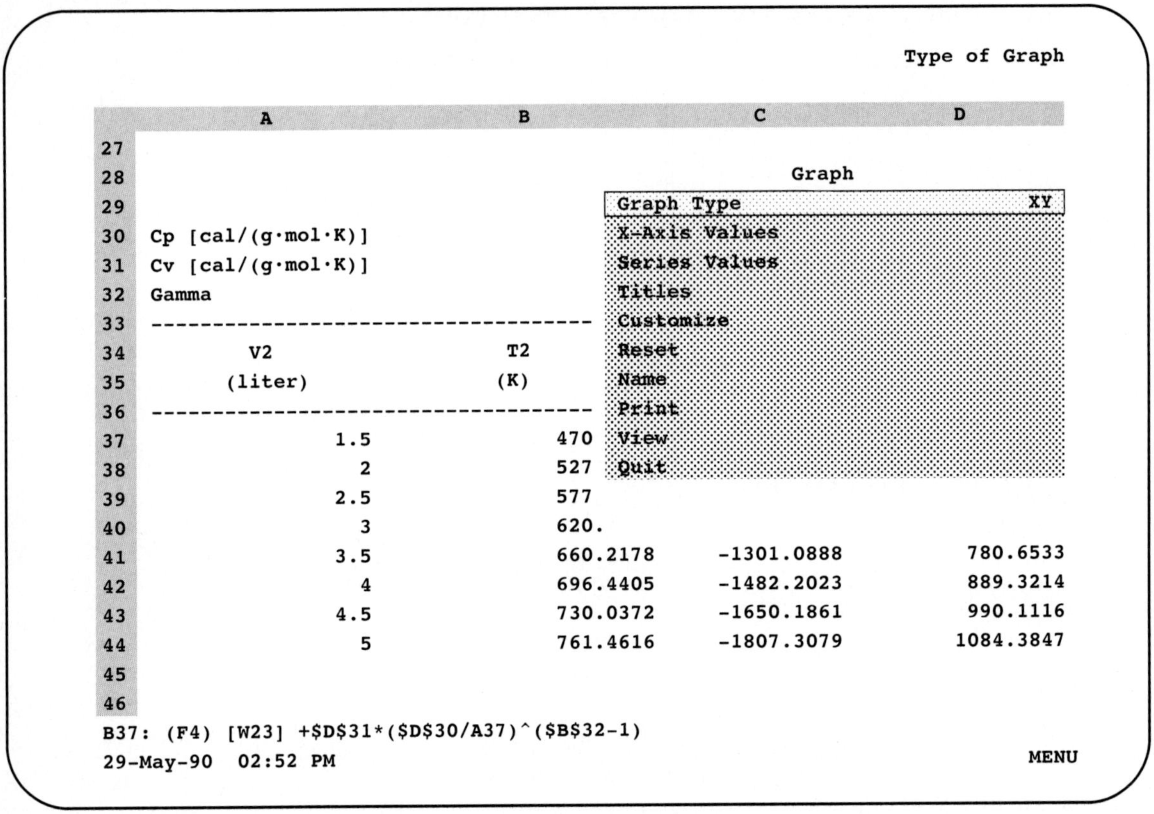

```
                                                            Type of Graph

             A                    B                 C                 D
    27
    28                                           Graph
    29                              Graph Type                         XY
    30  Cp [cal/(g·mol·K)]          X-Axis Values
    31  Cv [cal/(g·mol·K)]          Series Values
    32  Gamma                       Titles
    33  ----------------------------Customize
    34        V2               T2   Reset
    35      (liter)           (K)   Name
    36  ----------------------------Print
    37          1.5           470   View
    38          2             527   Quit
    39          2.5           577
    40          3             620.
    41          3.5           660.2178      -1301.0888          780.6533
    42          4             696.4405      -1482.2023          889.3214
    43          4.5           730.0372      -1650.1861          990.1116
    44          5             761.4616      -1807.3079         1084.3847
    45
    46
    B37: (F4) [W23] +$D$31*($D$30/A37)^($B$32-1)
    29-May-90   02:52 PM                                            MENU
```

(a)

FIGURE 5.21 Creating an *X-Y* graph. *(a)* Options in a graph menu (Quattro). *(b)* Painting the *X*-axis range (V2). *(c)* Painting the *Y*-axis range (T2).

purpose. Since the *X*-axis variable remains the same as before, the task is very simple. First, we disassociate the values of T_2 from the first *Y*-axis variable (using the Reset option in the Graph command). Then we assign the two ranges (C37..C44 and D37..D44) to the *Y*-axis variables. The resulting graph is shown in Figure 5.24. We name this graph WH. As you can see, it is extremely easy to create graphical representations of relationships using spreadsheets.

Printing Graphs Depending on the spreadsheet program, it may or may not be possible to print hard copies of the graphs from the spreadsheet program itself. For example, in 1-2-3, a graph has to be saved using the Graph Save command and printed using the PrintGraph program. In contrast, graphs can be printed directly from Quattro and Excel.

5.5 OTHER SPREADSHEET COMMANDS

We will now discuss some other spreadsheet commands that are commonly used during the template development process.

```
     Values to plot on or label the x-axis (y for rotated bar) : A37..A44

              A                B                C                D
    27
    28
    29
    30   Cp [cal/(g·mol·K)]                 3    V1 (liter)                    1
    31   Cv [cal/(g·mol·K)]                 5    T1 (K)                      400
    32   Gamma                            0.6
    33   ------------------------------------------------------------------------
    34              V2                    T2                W            DELTA H
    35           (liter)                  (K)         [cal/(g·mol)]   [cal/(g·mol)]
    36   ------------------------------------------------------------------------
    37             1.5                470.4316         -352.1580         211.2948
    38               2                527.8032         -639.0158         383.4095
    39             2.5                577.0800         -885.3998         531.2399
    40               3                620.7382        -1103.6911         662.2147
    41             3.5                660.2178        -1301.0888         780.6533
    42               4                696.4405        -1482.2023         889.3214
    43             4.5                730.0372        -1650.1861         990.1116
    44               5                761.4616        -1807.3079        1084.3847
    45
    46
    A44: (G) [W19] +A43+0.5
    29-May-90  03:01 PM                                                      POINT
```

(b)

FIGURE 5.21 (continued).

5.5.1 The Erase and Delete Commands

During the course of problem solving, it often becomes necessary to blank or erase the contents of one or more cells or even the entire spreadsheet. Spreadsheet programs provide the **Erase command** for erasing the cell contents. This command prompts you for the range of cells to be erased. The commands for erasing cell contents for the three major spreadsheet programs are shown in Quick Reference Table 5.12.

The cell formats and attributes (e.g., protection) are typically retained, only the contents are erased. The Erase command does not "delete" the affected rows or columns. In Figure 5.25a, the contents of the range of cells B37 through B44 have been marked for being erased. When the command is executed, the other cells using the values of B37..B44 display erroneous results (see Figure 5.25b). This is because an empty or blank cell referenced in a formula is assumed to have a value of zero. Since B37 = 0, the result of the formula in cell C37, −$B$31 * (B37 − $D$31), is −(−5)(0 − 400) = −2000 cal/(g·mol). Likewise, in cell D31, the formula +$B$30 * (B37 − $D$31) results in 3(0 − 400) = −1200 cal/(g·mol).

```
Values for this series : B37..B44

             A                    B                   C                   D
27
28
29
30   Cp [cal/(g·mol·K)]                         3 V1 (liter)                       1
31   Cv [cal/(g·mol·K)]                         5 T1 (K)                         400
32   Gamma                          0.6
33   ----------------------------------------------------------------------------
34        V2                    T2                   W                  DELTA H
35     (liter)                   (K)           [cal/(g·mol·K)]     [cal/(g·mol·K)]
36   ----------------------------------------------------------------------------
37         1.5             470.4316             -352.1580            211.2948
38         2               527.8032             -639.0158            383.4095
39         2.5             577.0800             -885.3998            531.2399
40         3               620.7382            -1103.6911            662.2147
41         3.5             660.2178            -1301.0888            780.6533
42         4               696.4405            -1482.2023            889.3214
43         4.5             730.0372            -1650.1861            990.1116
44         5               761.4616            -1807.3079           1084.3847
45
46
B44: (F4) [W23] +$D$31*($D$30/A44)^($B$32-1)
29-May-90   03:01 PM                                                     POINT
```

(c)

FIGURE 5.21 (continued).

TABLE 5.12 QUICK REFERENCE TABLE FOR ERASE COMMAND

Task	1-2-3	Quattro	Excel
Erasing the entire worksheet	/Worksheet Erase Yes	/Erase	Select entire worksheet /Edit Clear All Formats Formulas Notes OK
Erasing a range of cells	/Range Erase <..> <enter>	/Block Erase <..> <enter>	Select range of cells /Edit Clear All Formats Formulas Notes OK

Note: <..> denotes a range of cells.

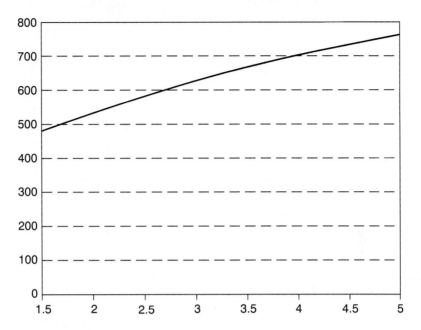

FIGURE 5.22 Graph of V_2-T_2 relationship.

FIGURE 5.23 Customized version of graph in Figure 5.22. Note the title and axes labels.

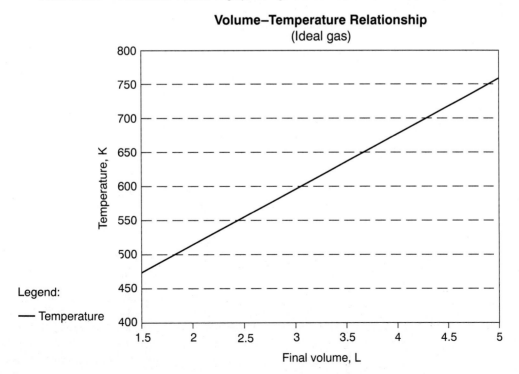

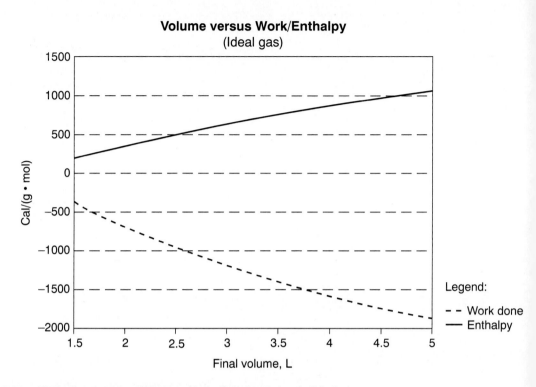

FIGURE 5.24 Final volume versus work done (dashed) and enthalpy (solid) change.

The Delete Command In contrast, the **Delete command** is used for deleting rows or columns. Whereas, the Erase command can act on a single cell, the Delete command typically acts on entire rows or columns. The formulas containing references to the deleted cells will be affected. For example, in Figure 5.26a, when row 32, containing the computed value of γ in B32, is deleted, the other formulas are affected (Figure 5.26b). The commands for deleting rows or columns for the three major spreadsheet programs are shown in Quick Reference Table 5.13.

Once the cell contents are lost, either due to the Erase command or the Delete command, they cannot be retrieved. So considerable care should be exercised when invoking these commands.

5.5.2 The Window or Layout Command

The Window or Layout command enables you to view different segments of the spreadsheet simultaneously in separate windows on the screen. The number of windows that can be created depends on the particular spreadsheet program. The screen can be split horizontally (Horizontal window) or vertically (Vertical window) and you can set the size of the windows. When the cursor is moved in one window, the other

```
Block to be modified : B37..B44

         A                    B                    C                 D
27
28
29
30   Cp [cal/(g·mol·K)]                      3 V1 (liter)                    1
31   Cv [cal/(g·mol·K)]                      5 T1 (K)                      400
32   Gamma                                 0.6
33   ------------------------------------------------------------------------
34        V2                   T2                   W              DELTA H
35      (liter)                (K)             [cal/(g·mol)]     [cal/(g·mol)]
36   ------------------------------------------------------------------------
37        1.5               470.4316            -352.1580           211.2948
38        2                 527.8032            -639.0158           383.4095
39        2.5               577.0800            -885.3998           531.2399
40        3                 620.7382           -1103.6911           662.2147
41        3.5               660.2178           -1301.0888           780.6533
42        4                 696.4405           -1482.2023           889.3214
43        4.5               730.0372           -1650.1861           990.1116
44        5                 761.4616           -1807.3079          1084.3847
45
46
B44: (F4) [W23] +$D$31*($D$30/A44)^($B$32-1)
01-Jun-90   10:18 PM                                                    POINT
```

(a)

FIGURE 5.25 The Range or Block Erase command. (a) Cells B37..B44 are marked for being erased. (b) When the contents are erased, the cells are treated as having 0s; values of W and ΔH are all the same.

TABLE 5.13 QUICK REFERENCE TABLE FOR DELETE COMMAND

Task	1-2-3	Quattro	Excel
Deleting row(s)	/Worksheet Delete Row <..> <enter>	/Row Delete <..> <enter>	Select row(s) /Edit Delete
Deleting column(s)	/Worksheet Delete Column <..> <enter>	/Column Delete <..> <enter>	Select column(s) /Edit Delete

Note: <..> denotes a range of cells.

```
                A                    B                 C                    D
27
28
29
30    Cp [cal/(g·mol·K)]                        3  V1 (liter)                        1
31    Cv [cal/(g·mol·K)]                        5  T1 (K)                          400
32    Gamma                                   0.6
33    ------------------------------------------------------------------------
34            V2                     T2                W              DELTA H
35         (liter)                   (K)         [cal/(g·mol)]     [cal/(g·mol)]
36    ------------------------------------------------------------------------
37            1.5                                   2000.0000        -1200.0000
38            2                                     2000.0000        -1200.0000
39            2.5                                   2000.0000        -1200.0000
40            3                                     2000.0000        -1200.0000
41            3.5                                   2000.0000        -1200.0000
42            4                                     2000.0000        -1200.0000
43            4.5                                   2000.0000        -1200.0000
44            5                                     2000.0000        -1200.0000
45
46
B37: (F4) [W23]
01-Jun-90  10:19 PM                                                       READY
```

(b)

FIGURE 5.25 (continued).

window can be left stationary or it can be set to scroll synchronously with the first window. The F6 Function key (see keyboard overlay in Figure 4.8) can be used to move the cursor from one window to the other.

The window feature is extremely useful when analyzing large templates. For example, in Figure 5.27, the screen has been split vertically, and it is possible to view the values in columns H and beyond, alongside columns A through C. This helps us in getting a complete picture of the problem being solved. In a similar manner, horizontal windows can be created. The commands for creating windows in the three major spreadsheet programs are shown in Quick Reference Table 5.14.

Freezing Variable Names and Titles When you are trying to display certain segments of a large template in a window, other parts of the template that contain titles, variable descriptions, and other identifying information may scroll out of the window causing a great deal of confusion. Without the variable descriptions and column headers, the values will be meaningless and you cannot do anything with the template (see Figure 5.28a). This is analogous to losing your bearings while navigating

5.5 OTHER SPREADSHEET COMMANDS   155

```
Delete all rows in the selected block : A32..A32
```

	A	B	C	D
27				
28				
29				
30	Cp [cal/(g·mol·K)]	3	V1 (liter)	1
31	Cv [cal/(g·mol·K)]	5	T1 (K)	400
32	Gamma		0.6	
33	--			
34	V2	T2	W	DELTA H
35	(liter)	(K)	[cal/(g·mol)]	[cal/(g·mol)]
36	--			
37	1.5	470.4316	-352.1580	211.2948
38	2	527.8032	-639.0158	383.4095
39	2.5	577.0800	-885.3998	531.2399
40	3	620.7382	-1103.6911	662.2147
41	3.5	660.2178	-1301.0888	780.6533
42	4	696.4405	-1482.2023	889.3214
43	4.5	730.0372	-1650.1861	990.1116
44	5	761.4616	-1807.3079	1084.3847
45				
46				

```
A32: (G) [W19] 'Gamma
01-Jun-90  10:20 PM                                    POINT
```

(a)

FIGURE 5.26 The Delete Row command. (a) Row 32 is marked for deletion. (b) When row 32 is deleted, the formulas for T_2, W, and ΔH are affected.

TABLE 5.14 QUICK REFERENCE TABLE FOR THE WINDOW COMMAND

Task	1-2-3	Quattro	Excel
Creating a window	/Worksheet Windows Horizontal Vertical	/Layout Windows Horizontal Vertical	ALT – Split Use position bar to create panes
Controlling Window Display	/Worksheet Windows Synchronous Unsynchronize	/Layout Windows Synchronous Unsynchronize	From multiple panes /Options Freeze panes OR /Options Unfreeze panes
Closing a window	/Worksheet Windows Clear	/Layout Windows Clear	ALT – Split Use position bar to remove panes

```
           A                    B                 C                 D
27
28
29
30    Cp [cal/(g·mol·K)]                    3 V1 (liter)                1
31    Cv [cal/(g·mol·K)]                    5 T1 (K)                  400
32    ---------------------------------------------------------------------
33           V2                T2                 W              DELTA H
34         (liter)             (K)         [cal/(g·mol)]      [cal/(g·mol)]
35    ---------------------------------------------------------------------
36           1.5                                 ERR               ERR              ERR
37           2                                   ERR               ERR              ERR
38           2.5                                 ERR               ERR              ERR
39           3                                   ERR               ERR              ERR
40           3.5                                 ERR               ERR              ERR
41           4                                   ERR               ERR              ERR
42           4.5                                 ERR               ERR              ERR
43           5                                   ERR               ERR              ERR
44
45
46
A32: (G) [W19] \-
01-Jun-90   10:20 PM                                                          READY
```

(b)

FIGURE 5.26 *(continued)*.

in an ocean. However, you can use one of the features of spreadsheet programs to freeze this identification information on the screen and retain your bearings.

The Window or Layout Title command in the spreadsheet program can be used to fix titles in both the horizontal and vertical directions. Once the titles are fixed, they remain on the screen regardless of the portion of the template being displayed in the window (Figure 5.28b). When the Title Horizontal option is chosen, rows above the cursor location are frozen; similarly when the Title Vertical option is selected, columns to the left of the cursor are frozen. The commands for freezing titles for the three major spreadsheet programs are shown in Quick Reference Table 5.15.

5.5.3 The File Command

The File command in the spreadsheet program is used for carrying out operations on template files, e.g., saving and retrieving templates, combining templates, and extracting portions of a template.

```
    J3: 75                                                              READY

              A              B         C           H         I         J
    1   Path of a Projectile             1
    2   -------------------------------- 2         -----------------------
    3   Angle of departure  [deg]      5 3         55        65        75
    4   Initial velocity    [m/sec]   22 4         22        22        22
    5   Gravity             [m/sec^2] 9.8 5        9.8       9.8       9.8
    6   Maximum height      [m]   0.187577 6       16.570    20.283    23.040
    7   Range               [m]   8.576093 7       46.409    37.833    24.694
    8   Time                [sec]  0.391311 8       3.678    4.069     4.337
    9                                    9
    10                                   10
    11                                   11
    12                                   12
    13                                   13
    14                                   14
    15                                   15
    16                                   16
    17                                   17
    18                                   18
    19                                   19
    20                                   20
    19-Jul-90   10:04 PM
```

FIGURE 5.27 The Window command: displaying a vertical window.

TABLE 5.15 QUICK REFERENCE TABLE FOR THE TITLES COMMAND

Task	1-2-3	Quattro	Excel
Freezing titles	/Worksheet Titles Horizontal Vertical	/Layout Titles Horizontal Vertical	From multiple panes /Options Freeze panes
Clearing titles	/Worksheet Titles Clear	/Layout Titles Clear	From multiple panes /Options Unfreeze panes

As we saw in the previous chapter, the Save option of the File command is used for saving files. Likewise the Retrieve option of the File command is used for retrieving a previously saved template. When this command is used, the existing contents of the spreadsheet are erased and the new template is loaded.

```
K8: (F3) 2*K4*@SIN(K3*@PI/180)/K5                                    READY


           D         E         F         G         H         I         J         K
 3        15        25        35        45        55        65        75        85
 4        22        22        22        22        22        22        22        22
 5       9.8       9.8       9.8       9.8       9.8       9.8       9.8       9.8
 6     1.654     4.410     8.124    12.347    16.570    20.283    23.040    24.506
 7    24.694    37.833    46.409    49.388    46.409    37.833    24.694     8.576
 8     1.162     1.897     2.575     3.175     3.678     4.069     4.337     4.473
 9
10
11
12
13
14
15
16
17
18
19
20
21
22
19-Jul-90  10:04 PM
```

(a)

FIGURE 5.28 The Window or Layout Title command. (a) Variable descriptions and titles are scrolled out of view. (b) "Frozen" descriptions and titles are displayed regardless of the segment of the template in view.

Combining Templates There are situations, however, where it is necessary to combine one or more templates into a single one. For example, the manufacturing manager may be responsible for consolidating the production figures from the various plants and reporting it to the upper management. Alternately, the manager may want to assess the relative performances of the various plants to take appropriate actions. The File Combine command of the spreadsheet program can be used to combine several templates into one.

There are several suboptions to the File Combine command. The values in the cells of the current template can be replaced by the values, or they can be added to or subtracted from the values in the templates being combined. Therefore, it is important that sufficient care be exercised in the initial design of the individual templates to ensure compatibility when they are consolidated into a single template.

```
K4: (G) 22                                                              READY


              A                B           H         I         J         K
   1   Path of a Projectile
   2   ----------------------------------------------------------------------
   3   Angle of departure   [deg]         55        65        75        85
   4   Initial velocity     [m/sec]       22        22        22        22
   5   Gravity              [m/sec^2]    9.8       9.8       9.8       9.8
   6   Maximum height       [m]       16.570    20.283    23.040    24.506
   7   Range                [m]       46.409    37.833    24.694     8.576
   8   Time                 [sec]      3.678     4.069     4.337     4.473
   9
  10
  11
  12
  13
  14
  15
  16
  17
  18
  19
  20
19-Jul-90  10:05 PM
```

(b)

FIGURE 5.28 (continued).

TABLE 5.16 QUICK REFERENCE TABLE FOR THE FILE COMMAND

Task	1-2-3	Quattro	Excel

Combining files	/File
	  Combine
	    Add
	    Subtract
	    Copy
	      \<f_name\>
		/File
		  Combine
		    Copy
		    Add
		    Subtract
		      \<f_name\>
			Select range of cells
			/Edit
			  Copy
			    \<Select destination\>
			    paste

Extracting from a file	/File
	  eXtract
	    \<f_name\>
		/File
		  eXtract
		    \<f_name\>
			Select range of cells
			/Edit
			  Copy
			    \<Select destination\>
			    paste

Note: \<f_name\> denotes a file name.

Extracting Portions of a Template Another common feature of spreadsheets is the ability to extract and save portions of a template. These partial templates can be used "as is" or integrated into other templates using the File Combine command. For example, if a template contains the values of the elastic modulus for various materials and this information is needed in another template where the stresses in the beam are being analyzed, then these values can be extracted using the File Extract command and brought into the analysis template with the File Combine command.

The commands for combining templates and extracting portions of templates for the three major spreadsheet programs are shown in Quick Reference Table 5.16.

SUMMARY

We solved a problem in thermodynamics with the help of a spreadsheet. In the process, we looked at commands for copying cells and discussed the concepts of absolute and relative cell addresses. We also used some spreadsheet features to name ranges, create graphs, and print tables of data. We explored some other useful commands in a spreadsheet: moving cells, deleting rows and columns, combining and extracting files, and viewing select portions of the spreadsheet.

You can see that the spreadsheet is indeed a very powerful and easy-to-use tool for quantitative problem solving.

CHAPTER 6

SPREADSHEET APPLICATION EXAMPLES (CASE STUDIES)

In this chapter, we present three application examples to illustrate the power of spreadsheets in quantitative problem solving. We reinforce the methodology introduced in Chapter 4 for using spreadsheets. We explore the macro features in spreadsheets while developing a template for the reliability analysis of electronic circuits.

For the case study on mass balance computations, we create a template and discuss ways to resolve circular references in cells with the help of the iteration feature. We also look at means of avoiding circular references. In the case study on engineering economy, we develop a single template for evaluating investment alternatives using three different methods. We also discuss the @IF and financial functions in spreadsheets when developing the template.

Finally, we briefly discuss some limitations of spreadsheets in scientific and engineering problem solving.

6.1 RELIABILITY ANALYSIS OF ELECTRONIC COMPONENTS

Quality control or quality assurance in manufacturing deals with the conformance of new products to specifications and standards. Reliability, on the other hand, deals with the service life of products. A high-quality product does not always guarantee high reliability. For example, a car that has been built to specifications and has passed all the quality-control standards can break down every 3 months after the first 6 months. In this case, the reliability of the car is poor (over the long run) though it met all the quality requirements and was trouble-free during the initial period. In contrast, if the car has a breakdown in the first week and runs trouble-free thereafter, its reliability is said to be high over the long run.

The three major parameters associated with reliability of products are the mean time between failures (MTBF), the anticipated service (or evaluation) time, and the environment in which the product is operating. For example, it is likely that the reliability of an electronic circuit mounted on a lunar module will be more greatly affected by the severe conditions on the moon than those on the earth. Moreover, the reliability of a circuit is far more critical on the space shuttle in orbit than it is in a musical clock at home.

We will develop a template for assessing the reliability of a transistor control circuit. It is based on the example in Hicks's *Standard Handbook of Engineering Calculations* (McGraw-Hill, New York, 1972, page 5-19). For details on reliability theory see Grosh.[1]

The Transistor Control Circuit Figure 6.1 shows the transistor control circuit used in an aircraft electronic device. The typical electronic component failure rates are given in Table 6.1 for three failure rates: low, mean, and high. The reliability of the component is affected by the environment, and Table 6.2 shows the typical environmental weighting factors. If the mission time is 200 h, calculate the total failure rate, MTBF, and the circuit reliability.

6.1.1 Problem Analysis and Template Layout

Following our spreadsheet problem-solving methodology discussed in Section 4.3, the first step is to develop the mathematical model. Figure 6.2 shows the mathematical relationships and the steps involved in solving the problem. The paper-and-pencil solution assuming mean failure rates is shown in Figure 6.3.

Need for a Spreadsheet Template When we examine the solution in Figure 6.3, we see that several of the known values can change. First, if the failure rate is high for the same product, the calculations have to be repeated with the new set of values.

FIGURE 6.1 The transistor control circuit. (After Hicks.)

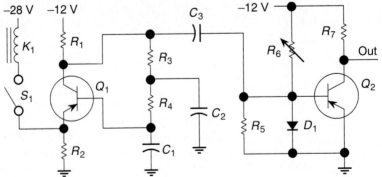

[1]Dorish Lloyd Grosh, *A Primer of Reliability Theory,* Wiley, New York, 1989.

TABLE 6.1 TYPICAL ELECTRONIC COMPONENT FAILURE RATES
Per 10^6 h

Component	Low	Mean	High
Capacitor	0.006	0.010	0.014
Diode	0.150	0.200	0.240
Potentiometer	0.100	0.250	0.750
Resistor	0.110	0.250	0.570
Solenoid	0.036	0.050	0.910
Switch*	0.015	0.060	0.123
Transistor	0.310	0.500	0.840
Prn-ckt sol jnt	0.004	0.008	0.060

*Per 10^6 cycles.

TABLE 6.2 TYPICAL ENVIRONMENTAL WEIGHTING FACTORS

Environment	Weighting factor
Laboratory	1
Ground	10
Shipboard	20
Trailer	30
Rail	40
Bench	60
Aircraft	150
Missile	1000

If the circuit is operating in a different environment (say, in a missile instead of an aircraft), the overall failure rate and other parameters have to be recomputed using the new weighting factor. If the number of components in the circuit changes, the calculations have to be repeated. However, the set of steps involved in the computation is essentially the same as shown in Figure 6.2.

Because of the nature of the computations and the potential benefits that can be derived including what-if analysis, it is worthwhile to develop a template for solving this problem.

Designing the Layout Figures 6.4 through 6.7 show the layout of the template. We have divided the template into the following areas: input, data table storage, and results. The objective behind this layout is that we enter the known values in the input screen and trigger the defined sequence of computations (or procedure): Depending on the specified failure rate (low, mean, or high) and the environment (aircraft, missile, etc.), select the appropriate failure rates and weighting factor from the data tables and use them in the computations. Once the calculations are complete, display the results in a table and let the template prompt us if we want to continue the calculations.

In effect, we are automating or programming the solution process with the help of one of the powerful features of spreadsheets: **macro programming language.** The

Inputs:

1. The number of components of each type (nct) in the circuit.
2. Failure rate (low, mean, or high).
3. Environment in which the circuit is operating.
4. Mission time (t).

Algorithm:

1. Based on failure rate, obtain the corresponding component failure rates (ifr) from Table 6.1.
2. Compute component failure rates: $nct_i \times ifr_i$
3. Compute unadjusted failure rate:

$$\sum_{i=1}^{n} nct_i \, ifr_i$$

4. Obtain environmental weighting factor from Table 6.2.
5. Compute adjusted total failure rate (ATFR):

$$\text{ATFR} = \sum_{i=1}^{n} nct_i \, ifr_i \times \text{weighting factor}$$

6. Compute MTBF:

$$\text{MTBF} = \frac{10^6}{\text{ATFR}} \quad (h)$$

7. Compute circuit reliability (R):

$$R = e^{-t/\text{MTBF}} \times 100 \quad (\%)$$

8. Repeat computations, if necessary; else end.

Outputs:

ATFR (per million h)
MTBF (h)
Circuit reliability (%)

FIGURE 6.2 The solution procedure.

sequence of instructions or procedure is known as a **macro**. Moreover, the use of macros in problem solving with spreadsheets lets an expert in an area create a template that can be used on a routine basis by others in the organization. The formulas and program sequences can be protected so that they are not accidentally overwritten by an unsuspecting user. Capsule 6.1 provides a comprehensive summary of macros in spreadsheets.

6.1 RELIABILITY ANALYSIS OF ELECTRONIC COMPONENTS

Name	Number (Input)	Indl F. R. (from Table 6.1, mean failure rate)	Total F. R.
Capacitor	3	0.010	3 × 0.010 = 0.030
Diode	1	0.200	1 × 0.200 = 0.200
Potentiometer	1	0.250	1 × 0.250 = 0.250
Resistor	6	0.250	6 × 0.250 = 1.500
Solenoid	1	0.050	1 × 0.050 = 0.050
Switch	1	0.060	1 × 0.060 = 0.060
Transistor	2	0.500	2 × 0.500 = 1.000
Prn-ckt sol jnt	22	0.008	22 × 0.008 = 0.176

Unadjusted total failure rate = sum of F. R. = 3.266

Weighting factor (K) = 150 (choose from Table 6.2, for aircraft)

Total failure rate = 3.266 × 150 = 489.90 per million hours

Mean time between failures = 10^6/489.90 = 2041.23 h

Circuit reliability = $e^{-(200/2041.23)}$ × 100 = 90.67 percent

Note: For simplicity we assume that the failure rate for the switch is also per 10^6 h (instead of per 10^6 cycles).

FIGURE 6.3 Reliability analysis: the paper-pencil solution.

CAPSULE 6.1

MACROS IN SPREADSHEETS

A **macro** is a storehouse of a set of instructions consisting of keystrokes and/or commands designed to carry out specific tasks. Since a set of instructions for the computer is a program (see Section 3.1), a macro is a program. A spreadsheet macro can range from a simple one that shortens a sequence of commands (/Block, Display Format, Fixed 3) into a single keystroke (ALT-F) to one that fully automates the solution process and displays customized menus and prompts—in other words, serving as a stand-alone program!

Just as a programming language has its own syntax, macro programming languages in spreadsheets have their own syntax. These include the major constructs of conditionals, looping, branching, and subroutine calls. It is important to bear in mind that these tools are provided to enhance the usefulness of the spreadsheet framework.

They are not intended to be substitutes for programming languages. It is not uncommon to find that a macro enthusiast has created a template so replete with macros that you begin to wonder why you need the spreadsheet framework, for any reason other than its easy-to-use data entry format. While there is nothing wrong with such a template, you should try to match the proper tool with the task at hand and not get carried away by the tool itself. With this caveat in mind, let's look at some features of macros that will help you design more powerful templates.

Creating a macro

The sequence of steps involved in creating and using macros is shown in Figure 6.8. The first step is to define the need and purpose of the macro. For example, if a set of commands such as formatting the display of numbers is used frequently, it may be worthwhile to have a macro because it will reduce the number of keystrokes, speed up the process of invoking the command, and reduce the likelihood of errors. Alternatively, you may want to execute a series of commands and perform computations on a large template including displaying a customized menu. A macro is ideal for this situation as well.

Placing a macro

A macro is stored as a label entry in a cell in the spreadsheet. Some programs, such as Excel, provide separate macro sheets on which you can store the macros. The macro should be placed in such a way that it does not detract from the information in the spreadsheet. Remember, the macro is only an assistant and is better positioned in the background. Moreover, care should be taken to avoid placing macros in rows or columns that are likely to be deleted or modified during the course of template development. Placing them in the *extreme* right corner of the spreadsheet will result in a waste of memory space. Also, placing them in remote parts of the template might cause you to forget them and unintentionally delete rows or columns. Although there is no one single best place to store a macro, it is common practice to store the macro in the lower right-hand corner of the spreadsheet.

Naming a macro

Once the placement of the macro is established, an appropriate name for the macro should be chosen. A macro can be invoked (or executed) in one of two ways: either by a keystroke combination (ALT or CTRL key with an alphabet character) or from the macro menu of the program. Depending on how the macro will be invoked, you should select an appropriate name, preferably one that is mnemonic. For example, ALT-C is appropriate to initiate *c*alculations, while ALT-M is appropriate for *m*oving cells and so on. Naming macros is similar to naming blocks of cells in spreadsheets (see Section 5.3.1).

Entering a macro

A macro can be entered as a label by typing in the appropriate keystrokes along with the keywords from the macro programming language. Alternatively, and more appropriately for macros that shorten commands, you can use the **macro record mode** and create the macro. It is important to understand how macros are executed by the spreadsheet. The execution begins with the first cell in the named macro block and moves on to the cell below and so on till an empty cell or the keyword {QUIT} or {RETURN} is encountered. In other words, macro execution is top-down in the column beginning with the first cell in the macro. Therefore, while entering the macro, care should be taken to end the macro with a blank cell or one of the keywords. Failure to do this can lead to some unexpected surprises.

You can invoke one or more macros within a macro; this is known as **chaining**, or sequencing, macros. You can do this with the help of a branching statement {BRANCH macro name} which is similar to the GO TO statement in programming languages, or call it as a subroutine {macro name}. You can test the value of a variable (i.e., the cell contents) using the {IF} statement and invoke an appropriate macro to carry out the task — analogous to the IF-THEN-ELSE clause of programming languages. You should enter the various macros in an order that resembles the anticipated flow of execution; this will make it easier for you to understand and debug the macro. Since creating macros is similar to programming in a conventional language, it is advisable to adopt good practices associated with programming (see Section 3.6).

Documenting the macro

This step is crucial, yet often ignored, in creating and using macros. A macro is more cryptic than a program in a conventional language. Consequently, it is important to document the macro, explaining its purpose and what it does. Otherwise, a macro that is very obvious to you when you create it will not make any sense after a week. You can use the {;} feature provided in some spreadsheets to document the macro. Alternatively, you can enter the text as labels, just as you would document the template.

Testing the macro

You can invoke the macro either by using a combination keystroke (e.g., ALT or CTRL and alphabet letter) or using the /Macro Execute command in spreadsheets. Either way, it is good practice to save the current template before you test a macro, especially if the macro includes statements for some major steps such as deleting or inserting rows or columns. This way, you will minimize any loss due to errors in the macro. Just as you would test select portions of a program in a conventional language, it is advisable to test the macros individually before executing the main macro that uses the individual macros. This will greatly help in the debugging process: It is easier to identify and correct errors if you are dealing with a small segment of the code.

Debugging the macro

Spreadsheet programs are typically provided with a macro debugger that can be used to trace the execution of the macro step by step to see what happens on the screen; this is a valuable tool for debugging macros. Figure 6.9 shows the execution of a macro and the debugger window. Depending on the spreadsheet program, you can set the debugger to execute a specific number of instructions in the macro and wait for your input before resuming execution; these stop points in the macro are known as **breakpoints**. This feature speeds up the debugging process because the step-by-step action can be confined only to certain portions of the macro that are suspect.

Once the macro is tested and debugged, it is a safe practice to use the Range or Block Protect feature of spreadsheets to protect the area containing the macros. This will guard against accidental erasures or overwriting. It will also prevent unauthorized changes to the macro, especially from inexperienced users.

Using the macro

A well-designed and tested macro has several advantages; it will simplify your task while improving your productivity. You can create a library of macros for frequently encountered tasks that you can use in all your spreadsheets. You can set the spreadsheet to load such utility macros automatically along with the spreadsheet. Depending on the complexity of the macro, you can create customized templates for use in the organization. For example, you can create a set of templates for use in the design department where the engineers will just enter product data and analyze the results, all with just a few keystrokes. Indeed, a new breed of spreadsheet macro programmers is steadily growing; these programmers are developing customized templates especially for business applications such as loan and real estate analysis, cost accounting, and inventory management.

FIGURE 6.4 The input area. Note the code for environments and prompts for data entry.

```
         A              B              C         D         E            F              G
 9
10              Enter Circuit Components Below & Press <ALT-C> to Calculate.
11
12              Name              Number                 Environment         Code
13              --------------------------                 -----------------------------
14              |      Capacitor              |           Laboratory           0
15              |         Diode               |           Ground               1
16              |     Potentiometer           |           Shipboard            2
17              |        Resistor             |           Trailer              3
18              |        Solenoid             |           Rail                 4
19              |         Switch              |           Bench                5
20              |       Transistor            |           Aircraft             6
21              |    Prn-ckt sol jnt          |           Missile              7
22              --------------------------                 -----------------------------
23      Enter:
24              | Failure rate ==>            |           (0=Low, 1=Mean, 2=High)
25              | Mission time ==>            |           (in hours)
26              | Environ code ==>            |           (choose from table above)
27              --------------------------
28
A28:
19-Aug-90   08:06 AM                                                              READY
```

```
         E           F            G            H          I          J
  9
 10              Typical Environmental Weighting Factors
 11
 12              Environment        Code      Weighting Factor
 13              ------------------------------------------------
 14              Laboratory          0             1
 15              Ground              1            10
 16              Shipboard           2            20
 17              Trailer             3            30
 18              Rail                4            40
 19              Bench               5            60
 20              Aircraft            6           150
 21              Missile             7          1000
 22              ------------------------------------------------
 23
 24
 25
 26
 27
 28
E28:
19-Aug-90  08:06 AM                                              READY
```

FIGURE 6.5 Part of the data table storage area: weighting factors.

6.1.2 Creating the Template

We create the template using the layout shown in Figures 6.4 through 6.7. We enter the variable descriptions, the values of the individual failure rates and the environmental weighting factors. For brevity, we do not show the template header information (identification, scope, and restrictions in Figure 4.11).

Associating Names with Cells We use the Range or Block Name feature (see Section 5.3.1) to name several cells in the spreadsheet (see Table 6.3). These block or range names are a lot easier and meaningful to handle than cell addresses in a macro. This will also help us to perform the various operations on blocks of cells. For example, in Figure 6.4, when the number of individual components is entered in the input screen in cells C14..C21 (*inum*), these values can be copied to cells C37..C44 (*onum*) in the results area in Figure 6.7a and used in the computations. Likewise, depending on the value of the failure rate code *fr*, the required individual failure rates (in *frlow*, *frmean*, and *frhigh*) in Figure 6.6 can be copied to *ifr* in the results area. The input and results screen areas are suitably named *begin* and *display*, respectively.

```
           J            K           L          M          N          O          P
  8
  9
 10            Typical Electronic Component Failure Rates
 11                        (per 10^6 h)
 12                  Component      Low       Mean        High
 13            ------------------------------------------------
 14                  Capacitor      0.006     0.010       0.014
 15                      Diode      0.150     0.200       0.240
 16               Potentiometer     0.100     0.250       0.750
 17                   Resistor      0.110     0.250       0.570
 18                   Solenoid      0.036     0.050       0.910
 19                  Switch [*]     0.015     0.060       0.123
 20                 Transistor      0.310     0.500       0.840
 21              Prn-ckt sol jnt    0.004     0.008       0.060
 22            ------------------------------------------------
 23            [*] per 10^6 cycles
 24
 25
 26
 27
L27: (F3)
19-Aug-90   10:50 AM                                CALC                       READY
```

FIGURE 6.6   Part of the data table storage area: component failure rates.

TABLE 6.3   RANGE OR BLOCK NAMES AND CELLS IN THE SPREADSHEET

Name	Range of cells	Explanation
begin	A9..G28	Input screen area
display	A32..F53	Results screen area
fr	C24	Failure rate code
frlow	L14..L21	Failure rate: low values
frmean	M14..M21	Failure rate: mean values
frhigh	N14..N21	Failure rate: high values
ifr	E37..E44	Individual failure rate
inum	C14..C21	Input number of components
messarea	B28	Error message area
onum	C37..C44	Number of components: output area
tfr	F37..F44	Total component failure rates

We can use cell names in the formulas (see Figure 6.7a). Notice the similarity between the flow of calculations in Figure 6.3 and that in Figure 6.7. We use the built-in functions @SUM, @EXP, and @CHOOSE in the formulas. The Sum function is used for summing the values in the cells in the argument, *tfr* in this case, while Exp is the Exponential function.

170 SPREADSHEET APPLICATION EXAMPLES (CASE STUDIES)

```
        A         B              C           D         E              F            G
32
33                             * * *     Results    * * *
34
35                 Name         Number              Indl F.R.    Total F.R.
36              ------------------------------------------------------------
37              Capacitor                                        +C37*E37
38              Diode                                            +C38*E38
39              Potentiometer                                    +C39*E39
40              Resistor                                         +C40*E40
41              Solenoid                                         +C41*E41
42              Switch                                           +C42*E42
43              Transistor                                       +C43*E43
44              Prn-ckt sol jnt                                  +C44*E44
45              ------------------------------------------------------------
46              Unadjusted total failure rate                    @SUM(TFR)
47              ------------------------------------------------------------
48              Weighting factor (K)      ==>        ERR
49              Total failure rate        ==>        ERR per million h
50              Mean time btw failures    ==>        ERR h
51              Circuit reliability       ==>        ERR %

D33: [W5]
19-Aug-90   10:36 AM                                                   READY
```

(a)

Formulas used in column E of the template:

48	Weighting factor (K)	→	@CHOOSE(C26,H14,H15,H16,H17,H18,H19,H20,H21)
49	Total failure rate	→	+F46*E48
50	Mean time between failures	→	10^6/E49
51	Circuit reliability	→	@EXP(−C25/E50)*100

(b)

FIGURE 6.7 The output area. (a) The results screen. The error message is due to blank cells referenced in the formulas (see Section 4.4.1). (b) Formulas in calculations.

The Input and Results Screen Notice the design of the input screen (Figure 6.4): The various choices for the environment are listed along with the code and a prompt to select the code from the table (and enter in C26).

The built-in @CHOOSE function is used in Figure 6.7 for selecting the appropriate environmental weighting factor (from Figure 6.5) based on the code entered in cell C26 in Figure 6.4. The entry in cell E48 of the results screen:

@CHOOSE(C26,H14,H15,H16,H17,H18,H19,H20,H21)

6.1 RELIABILITY ANALYSIS OF ELECTRONIC COMPONENTS

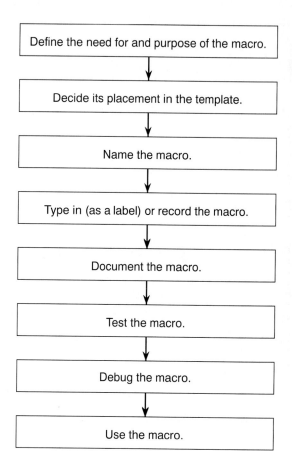

FIGURE 6.8 The sequence of steps in creating and using a macro.

works as follows: Depending upon the value in cell C26 (the first entry in the function arguments), one of the cells in the list of cells H14..H21 will be selected and its contents entered in E48. In other words, the first argument in the @CHOOSE function, cell C26, can be thought of as an **index cell** because it serves as an index into a list of cells in the remaining arguments. A value of 0 in the index cell selects the first cell in the list, a value of 1 selects the second cell, and so on. Thus, if C26 contains 0, the value in H14 will be selected; note that this corresponds to Laboratory in Figure 6.5. If C26 contains 1, H15 will be selected, and so on. For instance, if C26 contains 7, the value in H21 will be chosen (corresponds to Missile in Figure 6.5).

This also explains the reasons for using codes in specifying the environment in the template. It simplifies the data entry process while the entry maps directly into the table. Likewise, a prompt is displayed for entering the code appropriate to the failure rate (low, mean, or high) in cell C24. The value assigned (entered) in the cell (0, 1, or 2) will determine which of the values from the failure rate tables (Figure 6.6) will be used in the computations. This type of user interface is friendly and helpful, especially if the user is a novice. Note that rendering the model easy to use corresponds to one of the steps listed in Figure 4.10 on using spreadsheets in problem solving.

```
Source block of cells : frmean

           A         B              C          D        E         F            G
    9      ████
   10                Enter Circuit Components Below & Press <ALT-C> to Calculate.
   11
   12                Name                Number              Environment      Code
   13                ------------------------------          ------------------------------
   14               |    Capacitor         3 |               Laboratory         0
   15               |    Diode             1 |               Ground             1
   16               |    Potentiometer     1 |               Shipboard          2
   17               |    Resistor          6 |               Trailer            3
   18               |    Solenoid          1 |               Rail               4
   19               |    Switch            1 |               Bench              5
   20               |    Transistor        2 |               Aircraft           6
                   -------------------------DEBUG Window-------------------------------
 B73: [W18]  '{if fr=1}{copM}{branch calc1}
 B85: [W18]  '/bcfrmean{cr}ifr{cr}{return}
 B86: [W18]  '/bcfrhigh{cr}ifr{cr}{return}
                   -------------------------TRACE Window-------------------------------
            Space = Step by Step, Enter = Stop at next bkpt
  A9:
 20-Aug-90   12:02 PM            DEBUG MACRO        CALC                       EDIT
```

FIGURE 6.9 The macro debugger.

Mode of Recalculation Recall from Capsule 5.1 that the default mode of recalculation in the spreadsheet is automatic. This means each time an entry is made in the input screen, the spreadsheet will recalculate all the cells — a time-consuming process. Since we want to initiate the computations only after all the data have been entered, we set the recalculation mode to manual.

Creating Macros We follow the methodology discussed in Capsule 6.1 and create macros for carrying out the various actions listed in the algorithm in Figure 6.2. Since the macro represents a sequence of spreadsheet instructions, the advantage of creating and using a macro is that the entire sequence of computations can be invoked with a single keystroke. This will greatly simplify the use of the template and speed up the solution process. Since the locations of the input and results screen have been identified, we will use the area below the results screen for storing the macros to minimize the risk of the macros being lost if a row or column is deleted. We will begin creating the macros a few rows below the last row in the results screen.

The first step is to identify the purpose of each macro. Table 6.4 shows the results of the analysis, the names of the individual macros, and their locations in the macro

TABLE 6.4   LIST OF MACROS IN THE SPREADSHEET

Name	Location	Algorithm steps	Purpose
\c	B64	[0]	Initiate solution (the main macro).
select	B72	[1]	Select computation branch.
copL	B84	[1]	Copy low failure rate values.
copM	B85	[1]	Copy mean failure rate values.
copH	B86	[1]	Copy high failure rate values.
calc1	B94	[2–7]	Perform calculations and display results.
errloc	B77		Handle errors due to incorrect failure rate.
errmes	B79		Error message to be displayed.
\k	B100	[8]	Handle the menu choices.
menu	B102	[8]	Display menu after computations.
sure	C106	[8]	Make sure before quitting.
suresub	C107	[8]	Handle exit from menu.
end	B111	[8]	End of interaction: Ask to save template.

Note: The Cell Name command is used to assign the name of the macro to the cell containing the specific macro instructions. As a means of documenting the macro location in the template, the macro name is entered in the neighboring cell. Thus, the name \c is assigned to cell B64 since it contains the macro instructions; for documentation, the macro name \c is itself entered in A64.

area of the spreadsheet. The table also shows the steps in Figure 6.2 handled by each of the macros. Figures 6.10 through 6.12 show the specific instructions in each of the macros. Figure 6.13 shows a flowchart of the actions in the template when the main macro named \C is invoked.

The \C Macro and {Branch} Statement   The \C macro in cell B64 (Figure 6.10) is the main macro that triggers the solution process; this is indicated at the top of the input screen in Figure 6.4. We enter the macro as a label entry (by prefixing it with an apostrophe). Since we want the macro to be invoked by a keystroke combination (ALT-C), we name it \C.[1] As a good practice of documenting the macro, we enter the name of the macro (\C) in the neighboring cell, A64. The statement **{Branch** *select*} in the macro transfers control to location *select* (cell B72 in the template). Note the syntax of the macro statement; the keyword Branch is enclosed in braces {}, and **{Branch location}** corresponds to GO TO <statement number> in a conventional programming language.

The Select Macro for Component Failure Rates   The Select macro in cell B72 tests the contents of cell C24 (named *fr*) and calls the appropriate subroutine (*copL*, *copM* or *copH*) to copy component failure rates corresponding to low, mean, or high failure rates, respectively. The **{IF condition=true}** statement is used for the test. A subroutine call is made by enclosing the name of the subroutine in braces: **{***copL***}** in the example. If the first condition is satisfied, *copL* is called, and when it returns, the macro branches to location *calc*1, the **{Branch calc1}** instruction.

If the first condition is not satisfied, the condition on the next line is tested; if this also fails, control passes on to the next one. If this also fails (for example, if a value

---

[1]All entries that we type at the computer are underlined in the text.

```
       A         B          C        D         E          F         G
57                        * * *  M a c r o  A r e a  * * *
58
59
60
61     * * *  Trigger Solution   * * *
62     Initiate solution.
63
64     \c        {branch select}
65
66
67     * * *  Control Routines   * * *
68     Depending on the type of failure entered, call the subroutine for copying
69     individual failure rates from the table; then branch to routine for
70     calculations (calc).
71
72     select    {if fr=0}{copL}{branch calc1}
73               {if fr=1}{copM}{branch calc1}
74               {if fr=2}{copH}{branch calc1}
75               {branch errloc}
76
B76: [W18]
19-Aug-90  10:38 AM                                                    READY
```

FIGURE 6.10 The main and select macros. Note the comments for the macros and the name of the macro to its left which helps identification.

of 4 is entered in C24), the macro branches to location *errloc,* an error handling routine. It is always a good practice to take such defensive measures to trap and handle unforeseen errors.

The Errloc Macro and Error Messages The Errloc macro in cell B77 (Figure 6.11) causes the input screen to be displayed followed by an error message (stored in location *errmes*) in cell B28 (*messarea*). The message "Invalid failure rate; press <enter> to continue" is accompanied by two beeps to signal an error. At this point, the macro waits for the Enter key to be pressed; this is indicated by the {?} in the macro. This ensures that the process will continue only after we acknowledge the error from the keyboard. Once we press the Enter key, the macro resumes; the error message is erased (/be in Quattro), and the screen is ready for input.

The CopL Macro The CopL macro in cell B84 (Figure 6.11) is used for copying the component failure rates in *frlow* to *ifr*, the block of cells in the results screen (see Table 6.3). This is a simple Block Copy command (/bc in Quattro) where the source range is *frlow* and the target range is *ifr*. Note the use of **{cr}** to indicate a carriage return or Enter after the source and target ranges are specified. The **{Return}** keyword is used to signal the end of the subroutine—similar to the Return statement in

```
         A         B              C          D          E           F            G
77   errloc    {goto}$begin{cr}{down 19}{right}/bcerrmes{cr}{cr}{beep}{beep}{?}/be
78
79   errmes    Invalid failure rate; press <enter> to continue.
80
81   * * *  Copy Subroutines   * * *
82   Copy appropriate failure rates from failure rates table to results area.
83
84   copL      /bcfrlow{cr}ifr{cr}{return}
85   copM      /bcfrmean{cr}ifr{cr}{return}
86   copH      /bcfrhigh{cr}ifr{cr}{return}
87
88
89
90   * * *  Calculation Routine   * * *
91   Copy number of components from input to results area; trigger calculations;
92   display the results table and invoke the menu for continuation.
93
94   calc1     /bcinum{cr}onum{cr}{calc}{home}{goto}$display{cr}{branch \k}
95
96
B96: [W18]
19-Aug-90  10:38 AM                                                       READY
```

FIGURE 6.11 The error handling, copying, and calculation macros.

FORTRAN. Likewise, the *CopM* and *CopH* macros are used for copying the values for mean and high failure rates, respectively.

The Calc1 Macro   The Calc1 macro in cell B94 is responsible for the actual calculations. At first, the number of individual components entered in *inum* are copied into the results screen, *onum*. The calculations are initiated by the {**Calc**} keyword, which is equivalent to pressing the Calc function key (F9). Using the {**Goto**} keyword, the cursor is moved to the *display* region, the results screen. Finally, the macro branches to \k, the macro for displaying the menu.

The Continuation Menu Macro   Since our objective is to automate the entire solution process, we want the template to provide a menu of options after the calculations are performed. The two basic options are: (1) solving another problem and (2) terminating the solution process, respectively. We want to display a customized menu that will be similar to the regular command menu of the spreadsheet in its operation. The options will be displayed with an explanation in the control panel; when an option is selected, appropriate actions will be initiated by the template.

Customized Menu   Customized menus can be built using the {**Menubranch location**} statement in a macro. For example, in the \k macro in cell B100 (Figure 6.12),

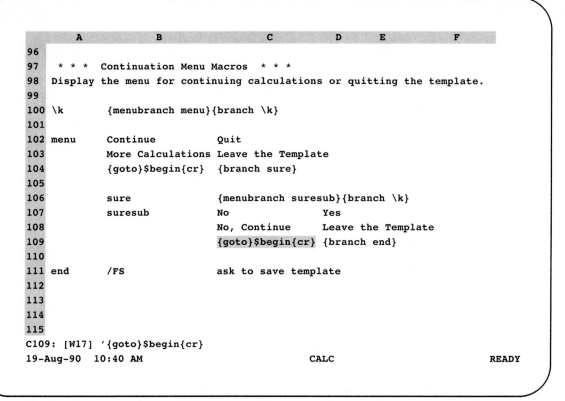

FIGURE 6.12 The continuation menu macros. *Note:* These macros can be used in any template with designated begin and display regions.

control is transferred to the macro *menu*, which displays the options shown in B102 and C102 with the corresponding explanations below these cells, respectively. The options can be selected either by typing the first letter of the option (C or Q) or by highlighting the option and pressing the Enter key. If any other key is pressed (e.g., t), the statement **{Branch \k}** in cell B100 will redisplay the menu. You cannot exit the menu without selecting one of the options. When the Continue option is selected, the input screen is displayed using the **{Goto $begin}** statement. Note the use of the absolute cell address ($begin) in the block or range name. This ensures that the complete input screen will be displayed. Otherwise, only a part of the screen may be displayed, and this will not be helpful for entering data to solve the next problem.

If the Quit option is chosen, the macro branches to another macro *Sure* which displays another menu seeking another confirmation with the macro *Suresub*. Note the similarity between this and the behavior of the spreadsheet program when you attempt to leave the program. This extra confirmation process will help avoid exiting the template by mistake. If the No option is selected, the input screen is displayed; on the other hand, if the Yes option is selected, the macro branches to *End*, which invokes

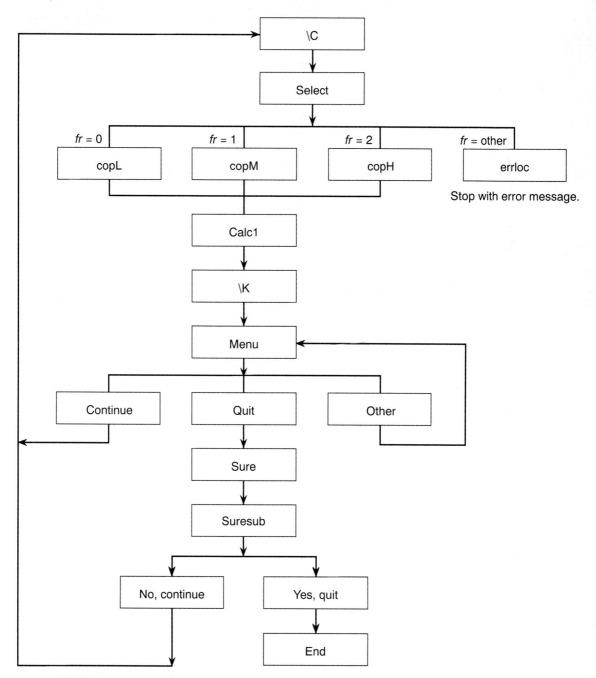

FIGURE 6.13 Flowchart of actions in the macro.

the File Save command of the spreadsheet. At this point, the macro terminates its execution.

This set of continuation menu macros has been designed in such a way that it can be used in any spreadsheet. In fact, we chose the name \k (for *continue*) since the letter c had already been used in the macro \c. Thus, a macro created for one template can be used in others if it is generic in nature.

Table 6.5 lists some major keywords or constructs frequently used in creating macros in spreadsheet applications.

Protecting Cells After testing and debugging the individual macros, we use the cell protect feature in the spreadsheet program to protect the areas containing macros. An important caveat: When the contents of a protected cell are copied to an unprotected cell, the latter acquires the attributes of the former cell and becomes protected. For example, let's say cell B79 containing the error message is protected. When an error is encountered, the contents of this cell are copied to cell *messarea* (cell B28); cell B28 becomes protected in the process. When we press the Enter key to acknowl-

TABLE 6.5   THE MAJOR KEYWORDS IN MACRO PROGRAMMING LANGUAGES

Keyword	Description
{CR} or ~	Enter.
{BS} or {Backspace}	Backspace.
{Esc}	Escape.
{Del}	Delete.
{Ins}	Insert.
{Edit}	Edit.
{Up}	Up arrow.
{Down}	Down arrow.
{Left}	Left arrow.
{Right}	Right arrow.
{GoTo}	GoTo (F5).
{Calc}	Calc (F9).
{?}	Wait for Enter key to be pressed.
{Beep}	Produce a beep sound.
{Menubranch <location>}	Branch to a custom menu at <location>.
{Menucall <location>}	Subroutine call to a custom menu at <location>.
{Branch <location>}	Transfer control to macro at <location>.
{FOR countLOC, init, end, step, startLOC}	Execute subroutine specified number of times.
{IF <condition>}	Test condition for TRUE or FALSE.
{ONERROR brloc, messloc}	On error, continue execution at brloc; display message stored in messloc.
{Quit}	Terminate macro execution.
{Return}	Return from subroutine.
{Close}	Close an open file.
{;}	Prefix for a remark statement.

edge the error, the macro tries to blank the cell and fails to do so with a "protected cell encountered" error message. At this point the macro is terminated, unless there is an **{ONERROR branch}** statement to handle the error. Therefore, care should be exercised when protecting cells in the spreadsheet, especially when the contents of protected cells are expected to be copied to other cells.

6.1.3 Solving the Problem

We find the number of components of each type in the circuit in Figure 6.1 and enter it in the input screen as shown in Figure 6.14. We also specify the failure rate (1 for mean), the environment code (6 for aircraft), and the mission time (200 h). When we press ALT-C as suggested by the prompt at the top of the screen, the various macros take over and the results screen is displayed as shown in Figure 6.15. The weighting factor used in the calculations is 150, and the total failure rate is 489.90 per million h. The circuit reliability is 90.67 percent while MTBF is 2041.23 h. Note the display of the customized menu on the screen with the options to continue or quit.

When we select the Continue option from the menu, the input screen is displayed and the template is ready for solving another problem. This time, we enter 4 in cell

FIGURE 6.14 The input screen with values.

```
           A              B              C         D      E          F              G
  9
 10              Enter Circuit Components Below & Press <ALT-C> to Calculate.
 11
 12              Name              Number                Environment       Code
 13              -------------------------             -----------------------------
 14         |    Capacitor           3 |                Laboratory          0
 15         |    Diode               1 |                Ground              1
 16         |    Potentiometer       1 |                Shipboard           2
 17         |    Resistor            6 |                Trailer             3
 18         |    Solenoid            1 |                Rail                4
 19         |    Switch              1 |                Bench               5
 20         |    Transistor          2 |                Aircraft            6
 21         |    Prn-ckt sol jnt    22 |                Missile             7
 22              -------------------------             -----------------------------
 23    Enter:
 24         |    Failure rate ==>    1 |                (0=Low, 1=Mean, 2=High)
 25         |    Mission time ==>  200 |                (in hours)
 26         |    Environ code ==>    6 |                (choose from table above)
 27              -------------------------
 28
C26: U [W9] 6
19-Aug-90   10:45 AM                        CALC                          READY
```

```
                                                              More Calculations

         A           B              C         D        E            F              G
  32
  33                         * * *    Results    * * *
  34
  35                  Name         Number          Indl F R    Total F R  Continue
  36              ------------------------------------------------------- Quit
  37                Capacitor         3            0.010         0
  38                   Diode          1            0.200         0
  39             Potentiometer       1            0.250         0.250
  40                 Resistor         6            0.250         1.500
  41                 Solenoid         1            0.050         0.050
  42                   Switch         1            0.060         0.060
  43                Transistor        2            0.500         1.000
  44           Prn-ckt sol jnt       22            0.008         0.176
  45              ----------------------------------------------------------
  46           Unadjusted total failure rate                     3.266
  47              ----------------------------------------------------------
  48           Weighting factor (K)   ==>         150
  49           Total failure rate     ==>         489.90 per million h
  50           Mean time btw failures ==>         2041.23 h
  51           Circuit reliability    ==>          90.67 %
 A32:
 19-Aug-90   10:46 AM              MACRO                                    READY
```

FIGURE 6.15 The results screen. Note the display of the customized menu with its explanation at the top of the screen.

C24 for the failure rate and initiate the solution. Since 4 is an invalid entry, the *Errloc* macro is triggered; the spreadsheet beeps and displays the error message as shown in Figure 6.16. When we respond with the Enter key, the message disappears and the input screen is ready for input.

After solving a few more problems involving different circuits and environments, we select the Quit option from the menu, save the template under the name RELIAB1 and then leave the spreadsheet. This template can even be used by reliability analysis experts who are not necessarily well versed in the use of spreadsheets.

Thus, using the macro programming language in spreadsheets, we have created an easy-to-use and powerful template for problem solving.

6.2 THE ITERATION FEATURE IN MASS BALANCE CALCULATIONS

Engineers and scientists are involved in the design and operation of manufacturing facilities for the production of chemicals and materials such as sulfuric acid, paper, and polyester. The production process consists of several steps carried out in sequence. At each stage of the process, ingredients (e.g., chemicals, solvents) flow into and out of

```
        A              B                C        D         E           F              G
 9
10              Enter Circuit Components Below & Press <ALT-C> to Calculate.
11
12                    Name              Number             Environment        Code
13                 ------------------------------          ---------------------------
14             |     Capacitor            3    |          Laboratory            0
15             |       Diode              1    |          Ground                1
16             |   Potentiometer          1    |          Shipboard             2
17             |      Resistor            6    |          Trailer               3
18             |      Solenoid            1    |          Rail                  4
19             |       Switch             1    |          Bench                 5
20             |     Transistor           2    |          Aircraft              6
21             |   Prn-ckt sol jnt       22    |          Missile               7
22                 ------------------------------          ---------------------------
23      Enter:
24             |  Failure rate ==>        4    |          (0=Low, 1=Mean, 2=High)
25             |  Mission time ==>      200    |          (in hours)
26             |  Environ code ==>        6    |          (choose from table above)
27                 ------------------------------
28                 Invalid failure rate; press <enter> to continue.
B28: U [W18] 'Invalid failure rate; press <enter> to co
21-Aug-90   12:01 AM                     MACRO        CALC                        READY
```

FIGURE 6.16 The error message in the input screen. The error is due to the invalid failure rate.

a reaction vessel or reactor. The plant has to be designed in such a way that under steady-state conditions, the mass inflow equals the mass outflow for each of the reactors. Failure to maintain this mass balance will cause the process to go out of control.

The related computations are commonly referred to as *mass*, *material*, and *energy balance calculations*. They involve the solution of sets of equations and, depending on the process, either linear or nonlinear algebraic equations. For example, given the flow rates, we may want to estimate the concentration of chemicals in each of the reactors or calculate the flow rates for attaining a certain level of concentration of the chemical in the reactor.

In this application example, we develop a template for carrying out mass balance calculations in a chemical plant with three reactors. We will discuss the iteration feature in spreadsheets and use it for solving the set of simultaneous linear mass balance equations.

The Chemical Reactors Problem Figure 6.17a shows three of the reactors in a chemical plant connected to each other with pipes. In steady state, the mass balance has to be maintained and the mass transfer into one reactor should equal the transfer out of the reactor. The rate of transfer of chemicals through the pipes is given by the

182 SPREADSHEET APPLICATION EXAMPLES (CASE STUDIES)

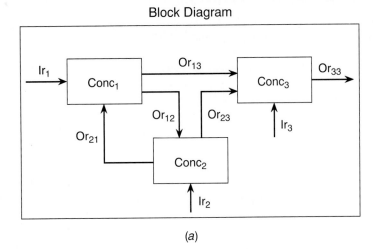

Flow rate	Value, m³/h	Concentration	Value, kg/m³
Ir_1	40	Inc_1	10
Ir_2	105	Inc_2	3
Ir_3	100	Inc_3	2.5
Or_{12}	60	$Conc_1$	?
Or_{13}	30	$Conc_1$	?
Or_{21}	20	$Conc_2$	?
Or_{23}	80	$Conc_2$	?
Or_{33}	90	$Conc_3$	?

Note: Rate of transfer of chemical = flow rate × concentration.

(b)

FIGURE 6.17 Mass balance in a chemical plant. (*a*) The block diagram. (*b*) Flow rates and concentrations.

product of the flow rate and the concentration of the chemicals. If the flow rates and concentrations are as shown in Figure 6.17*b*, estimate the concentrations in the three reactors.

6.2.1 Problem Analysis and Template Layout

According to the methodology for problem solving with spreadsheets, discussed in Section 4.3, the first step is to develop the mathematical model. To derive the mass balance equations, we write the expression for the mass entering the reactor and equate it to the mass leaving the reactor. The resulting relationships are shown in Figure 6.18.

These relationships represent a set of simultaneous linear equations in three unknowns, $Conc_1$, $Conc_2$, and $Conc_3$. This is because the value of $Conc_1$ depends on $Conc_2$, while that of $Conc_2$ depends on $Conc_1$, and $Conc_3$ depends on $Conc_1$ and $Conc_2$. To solve the equations using a spreadsheet program, we have to first rearrange the

For each reactor: Mass in = mass out.

Reactor 1:

$$Ir_1\ Inc_1 + Or_{21}\ Conc_2 = Or_{12}\ Conc_1 + Or_{13}\ Conc_1 \tag{1}$$

Reactor 2:

$$Ir_2\ Inc_2 + Or_{12}\ Conc_1 = Or_{21}\ Conc_2 + Or_{23}\ Conc_2 \tag{2}$$

Reactor 3:

$$Ir_3\ Inc_3 + Or_{13}\ Conc_1 + Or_{23}\ Conc_2 = Or_{33}\ Conc_3 \tag{3}$$

Rearranging Equations (1 through 3) to isolate the unknowns to the left we have:

$$Conc_1 = \frac{Ir_1\ Inc_1 + Or_{21}\ Conc_2}{Or_{12} + Or_{13}} \tag{4}$$

$$Conc_2 = \frac{Ir_2\ Inc_2 + Or_{12}\ Conc_1}{Or_{21} + Or_{23}} \tag{5}$$

$$Conc_3 = \frac{Ir_3\ Inc_3 + Or_{13}\ Conc_1 + Or_{23}\ Conc_2}{Or_{33}} \tag{6}$$

FIGURE 6.18 The mass balance relationships.

three equations so that each of the unknowns is explicitly isolated on the left-hand side of the equation. This is shown in the second set of equations (4 through 6) in the figure.[1] Since one of the objectives behind this example is to explain the concept of circular reference and ways to resolve it, we don't simplify the equations further at this stage.

Designing the Template Layout Figure 6.19 shows the layout of the template. We have assigned a separate column to indicate the units for the variables. All the known variables are followed by the unknown variables, $Conc_1$, $Conc_2$, and $Conc_3$.

6.2.2 Circular References and Iteration

When we enter the formulas in the spreadsheet, the interdependencies of the three unknowns will lead to indirect circular references (see Section 5.1.2): the value of $Conc_1$ will change if that of $Conc_2$ changes, even as the value of $Conc_2$ depends on the value of $Conc_1$, while that of $Conc_3$ depends on $Conc_1$ and $Conc_2$. This simultaneity or circular

---

[1]Additional algebraic manipulation of the equations can be carried out to avoid an iterative solution. See Section 6.2.4.

```
            A                          B                              C
  1                        Mass Balance Example
  2                    Solving a System of Linear Equations
  3    ----------------------------------------------------------------
  4    Parameter              Value                         Unit
  5    ----------------------------------------------------------------
  6    Ir1                                             40   m^3/h
  7    Ir2                                            105   m^3/h
  8    Ir3                                            100   m^3/h
  9    Or12                                            60   m^3/h
 10    Or13                                            30   m^3/h
 11    Or21                                            20   m^3/h
 12    Or23                                            80   m^3/h
 13    Or33                                            90   m^3/h
 14
 15    Inc1                                            10   kg/m^3
 16    Inc2                                             3   kg/m^3
 17    Inc3                                           2.5   kg/m^3
 18    Conc1                                                kg/m^3
 19    Conc2                                                kg/m^3
 20    Conc3                                                kg/m^3
A14:
31-Aug-90   04:02 PM                                               READY
```

FIGURE 6.19 Layout of the mass balance template.

reference can be resolved using the **Iteration feature** in the spreadsheet. When we select this option, the spreadsheet will recalculate the formulas involved in the circularity a specified number of times. Each time the formulas are recalculated, the results might get closer and closer to the correct value — a process known as **convergence**. There are instances when the iteration may not converge, resulting in no solution; this is known as **divergence**.

Therefore, iteration is a trial-and-error process in which the parameters (i.e., some variable values) are changed repeatedly until certain predefined solution criteria are met. The process is triggered by an initial guess or estimate for one or more of the parameters. The number of steps taken to arrive at the solution (also known as the number of iterations) depends on the guess values. The iterative process is also discussed in detail in Section 7.5.

We can set the number of iterations in the spreadsheet. The default value is 1, and a value between 10 and 20 is reasonable for most computations. The upper limit on the number of iterations that can be specified depends on the particular program.

6.2.3 Creating the Template

We create the template using the layout shown in Figure 6.19. We enter the variable names in the first column and the corresponding known values in the second column. Until now, we have been using the cell address while entering formulas in cells. For example, in Figure 6.7, the formula in cell E49 reads +F46 * E48. Instead of using cell addresses, we can use variable or label names in the formulas. This will make it easy to interpret the formula. For instance, if the values for sales and costs in a spreadsheet are entered in cells B1 and B2 and profit is computed in B3, we could enter "sales-costs" in B3 instead of +B1 − B2. To do this, however, we need to associate the variable names or labels with cells containing the corresponding variable values.

Associating Label Names with Cells We will use the **Label Name** feature in the spreadsheet for associating label names with cells. Note that this is similar to the Range or Block Name feature discussed in Section 5.3.1, with the exception that the association is only with respect to the neighboring cell. This means that a label name can be associated with a cell which is to the right, left, above, or below the cell containing the name.

For example, in Figure 6.19, we need to associate the label in A6, viz., Ir_1 with the cell to the right, viz., B6; the label Ir_2 with cell B7, Ir_3 with B8, and so on till $Conc_3$ with cell B20. With the cursor highlighting the label in A6, we select the Label Name option (under the Range or Block command) and select the Right option from the menu (see Figure 6.20a). We then highlight cells A6..A20 (Figure 6.20b) and press the Enter key. The command is executed and the labels in A6..A20 are associated with the corresponding cells in B6..B20. The commands for the Label Name feature for the three major spreadsheets are shown in Quick Reference Table 6.6.[1]

We use these names and enter the formulas for $Conc_1$, $Conc_2$, and $Conc_3$ in B18, B19 and B20, respectively. The CIRC indicator is displayed on the screen signaling the circular reference discussed earlier. Figure 6.21 shows the screen with all the formulas.

TABLE 6.6 QUICK REFERENCE TABLE FOR ASSOCIATING LABEL NAMES WITH CELLS

Associate	1-2-3	Quattro	Excel
Label name with cell	/Range Name Labels	/Block Advanced Labels Right Down Left Up	/Format Define name <name> <cell> OK

[1] While the specific command sequences and menus will vary from program to program, the underlying logic for executing the commands remains essentially the same across all programs. To enhance the usefulness of the text, the specific command sequences for Lotus 1-2-3, Quattro, and Excel are given in quick reference tables. As of this writing, Quattro Pro, an enhanced version of Quattro, has been released. Though Quattro Pro's command structure differs from that of Quattro, we have chosen to include the commands for Quattro since Quattro Pro's commands are similar to that of Excel.

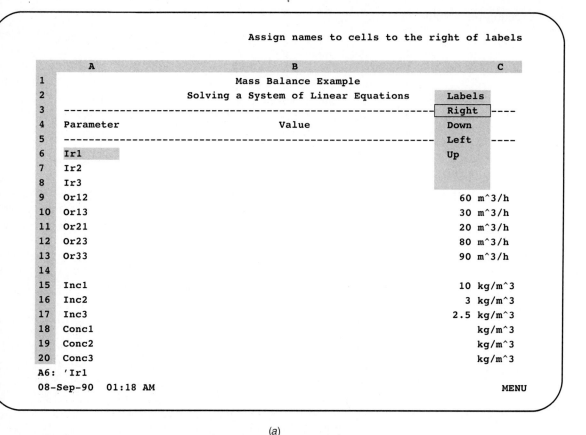

(a)

FIGURE 6.20 Associating label names with cells. (a) The Block Label Name menu. (b) The highlighted block of labels; note the prompt at the top of the screen.

Since we need to use the Iteration feature, we change the recalculation mode from automatic to manual. Otherwise, each time a value is changed, the spreadsheet will be recalculated, and because of the iterations, the spreadsheet will be recalculated several times; this can be a very time-consuming process for large spreadsheets.

Specifying Number of Iterations We invoke the command for setting the number of iterations and specify 20 in response to the prompt. This means that the program will recalculate the template several times until either a solution is obtained (for a given tolerance) or the specified number of iterations (20, in the present example) is carried out. The commands for specifying the number of iterations for the three major spreadsheet programs are shown in Quick Reference Table 6.7.

Solving the Problem We trigger the solution process by pressing the CALC key (Figure 4.8). The program immediately goes into the iteration mode, and the values for the three unknowns change rapidly until the solution shown in Figure 6.22 is reached. The concentrations in the three reactors are 5.94, 6.71, and 10.72 kg/m^3,

```
Assign names to cells to the right of labels : A6..A20

              A                        B                              C
 1                           Mass Balance Example
 2                      Solving a System of Linear Equations
 3      ------------------------------------------------------------
 4      Parameter                    Value                        Unit
 5      ------------------------------------------------------------
 6      Ir1                                                     40  m^3/h
 7      Ir2                                                    105  m^3/h
 8      Ir3                                                    100  m^3/h
 9      Or12                                                    60  m^3/h
10      Or13                                                    30  m^3/h
11      Or21                                                    20  m^3/h
12      Or23                                                    80  m^3/h
13      Or33                                                    90  m^3/h
14
15      Inc1                                                    10  kg/m^3
16      Inc2                                                     3  kg/m^3
17      Inc3                                                   2.5  kg/m^3
18      Conc1                                                       kg/m^3
19      Conc2                                                       kg/m^3
20      Conc3                                                       kg/m^3
A20:  'Conc3
08-Sep-90   01:18 AM                                              POINT
```

(b)

FIGURE 6.20 (continued).

respectively. These values are reasonable; an examination of the individual flow rates (Figure 6.17) shows that the concentration increases progressively from reactor 1 to reactor 3, confirming the results obtained.

The figure also shows a simple block diagram for the three reactors. We created a single block at first and used the Range or Block Copy feature to make up the other blocks. We save this template under the name *Masball*.

Caveat When Using Iteration   A major drawback of the Iteration feature in a spreadsheet, however, is that the spreadsheet does not signal the reason for terminating the recalculations, i.e., whether a solution has been reached or the specified number of iterations has been carried out *without* reaching a solution. So it is up to us to evaluate the results displayed and determine their validity. One approach to reconfirming the results is to increase the number of iterations to, say, 100, and check to see if the same results are obtained. A different approach to ensuring the correctness of the results is to try and simplify the equations to avoid circular references and the iterative process altogether.

```
                A                       B                           C
  1                          Mass Balance Example
  2                     Solving a System of Linear Equations
  3     ----------------------------------------------------------------
  4     Parameter                 Value                          Unit
  5     ----------------------------------------------------------------
  6     Ir1                                                      40  m^3/h
  7     Ir2                                                      105 m^3/h
  8     Ir3                                                      100 m^3/h
  9     Or12                                                     60  m^3/h
 10     Or13                                                     30  m^3/h
 11     Or21                                                     20  m^3/h
 12     Or23                                                     80  m^3/h
 13     Or33                                                     90  m^3/h
 14
 15     Inc1                                                     10  kg/m^3
 16     Inc2                                                     3   kg/m^3
 17     Inc3                                                     2.5 kg/m^3
 18     Conc1      (IR1*INC1+OR21*CONC2)/(OR12+OR13)                 kg/m^3
 19     Conc2      (IR2*INC2+OR12*CONC1)/(OR21+OR23)                 kg/m^3
 20     Conc3      (IR3*INC3+OR13*CONC1+OR23*CONC2)/OR33             kg/m^3
A14:
31-Aug-90  04:02 PM                    CIRC                                READY
```

FIGURE 6.21 The mass balance template. Cell names are used in the formulas.

TABLE 6.7 QUICK REFERENCE TABLE FOR SETTING THE NUMBER OF ITERATIONS

Action	1-2-3	Quattro	Excel
Set number of iterations	/Worksheet Global Recalculation Iteration <#> <enter>	/Block Default Iteration <#> <enter>	/Options Calculations Iteration <#> OK

6.2.4 Avoiding Circular References and Iterative Solutions

Let's take a second look at Equations (4 through 6) in Figure 6.18 and attempt to simplify them to resolve the circular reference. Substituting for the unknown $Conc_2$ from Equation (5) in (4), we have:

$$Conc_1 = \frac{Ir_1\ Inc_1\ (Or_{21} + Or_{23}) + Ir_2\ Inc_2\ Or_{21}}{(Or_{21} + Or_{23})(Or_{12} + Or_{13}) - Or_{21}\ Or_{12}} \quad (1)$$

6.2 THE ITERATION FEATURE IN MASS BALANCE CALCULATIONS

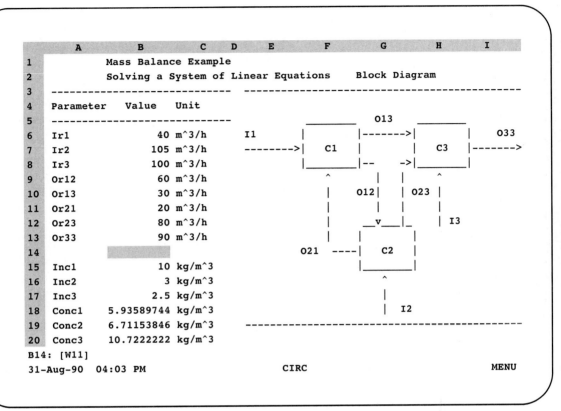

FIGURE 6.22 The solution and the flow diagram. (The flow diagram is also in the template itself.)

It is clear that $Conc_1$ can now be evaluated without any iteration since all the variables on the right-hand side of Equation (1) are known. Once $Conc_1$ is evaluated, its value can be used in Equation (5) in Figure 6.18 to evaluate $Conc_2$, and both these values can be used to evaluate $Conc_3$ from Equation (6).

We modify the formula for $Conc_1$ in Figure 6.22 (cell B18) using the Edit feature to read:

$$(IR1 * INC1 * (OR21 + OR23) + IR2 * INC2 * OR21)/$$
$$((OR21 + OR23) * (OR12 + OR13) - OR21 * OR12)$$

Since the circular reference has been resolved, the CIRC indicator disappears; when we press the CALC key, the unknowns are evaluated. The solution shown in Figure 6.23 is the same as the one in Figure 6.22.

By solving the same problem with and without iteration, we have shown that spending time analyzing the problem *prior* to using the spreadsheet cannot only simplify and speed up the solution process (for the spreadsheet) but can also assure us of the confidence in the results obtained.

```
              A                        B                              C
   1                          Mass Balance Example
   2                     Solving a System of Linear Equations
   3      ----------------------------------------------------------------
   4      Parameter                  Value                          Unit
   5      ----------------------------------------------------------------
   6      Ir1                                                40     m^3/h
   7      Ir2                                               105     m^3/h
   8      Ir3                                               100     m^3/h
   9      Or12                                               60     m^3/h
  10      Or13                                               30     m^3/h
  11      Or21                                               20     m^3/h
  12      Or23                                               80     m^3/h
  13      Or33                                               90     m^3/h
  14      
  15      Inc1                                               10     kg/m^3
  16      Inc2                                                3     kg/m^3
  17      Inc3                                              2.5     kg/m^3
  18      Conc1                                  5.935897435897     kg/m^3
  19      Conc2                                  6.711538461538     kg/m^3
  20      Conc3                                 10.72222222222      kg/m^3
  B14: [W58]
  06-Sep-90   07:21 AM                                              READY
```

FIGURE 6.23 The solution obtained without iteration. Note the absence of the CIRC indicator found in Figure 6.22.

6.2.5 Generalizing the Template

Let's expand the scope of the original template shown in Figure 6.22 (using iteration) to solve a wider range of problems. For example, in Figure 6.17, what if there is mass transfer from reactor 3 to the other two reactors, or there are outputs at reactors 1 and 2 (similar to Or_{33}). In other words, each of the reactors has outputs to the other and also to the outside. It is very easy to modify the mass balance equations to account for these changes. The new set of relationships is shown in Figure 6.24. Since the three reactors are "identical" in terms of flows, it shows up in the similarity of the expressions.

Modifying the Template We insert rows for the additional variables (Or_{11}, Or_{22}, Or_{31}, and Or_{32}) using the Insert Row command. Then we associate the new labels with their corresponding cells using the Label Name command (see Section 6.2.3). We enter the edit mode (using the Edit key) and modify the formulas for $Conc_1$, $Conc_2$, and $Conc_3$ as shown in Figure 6.24.

	A	B	C
5	-----	---	------
22	Conc1	(IR1*INC1+OR21*CONC2+OR31*CONC3)/(OR12+OR13+OR11)	kg/m^3
23	Conc2	(IR2*INC2+OR12*CONC1+OR32*CONC3)/(OR21+OR23+OR22)	kg/m^3
24	Conc3	(IR3*INC3+OR13*CONC1+OR23*CONC2)/(OR31+OR32+OR33)	kg/m^3

FIGURE 6.24 Expressions for generalizing the mass balance template in Figure 6.21. Note that there are outputs from each of the reactors to the others and to the outside.

	A	B	C
5	-----	--------------	------
6	Ir1		40 m^3/h
7	Ir2		105 m^3/h
8	Ir3		100 m^3/h
9	Or11		0 m^3/h
10	Or12		60 m^3/h
11	Or13		30 m^3/h
12	Or21		20 m^3/h
13	Or22		0 m^3/h
14	Or23		80 m^3/h
15	Or31		0 m^3/h
16	Or32		0 m^3/h
17	Or33		90 m^3/h
18			
19	Inc1		10 kg/m^3
20	Inc2		3 kg/m^3
21	Inc3		2.5 kg/m^3
22	Conc1		5.935897435897 kg/m^3
23	Conc2		6.711538461538 kg/m^3
24	Conc3		10.72222222222 kg/m^3

FIGURE 6.25 Solution of problem in Figure 6.17a. Zeroes are entered for Or_{11}, Or_{22}, Or_{31}, and Or_{32}.

Validating the Generalized Template It is extremely important to ensure the integrity of the template after any modifications and before solving new problems. Testing the template with a known problem is not only easy but also helps in isolating problems in the template, viz., debugging.

We will again solve the problem in Figure 6.17; we assign a value of 0 to Or_{11}, Or_{22}, Or_{31}, and Or_{32} (since these outputs are not present in the block diagram) and press the CALC key. The program starts iterating and arrives at the solution shown in

Figure 6.25. This solution is identical to the one in Figure 6.22; this validates the new template.

Another Example   Figure 6.26 shows the block diagram and data for another problem. The corresponding solution obtained with the new template is shown in Figure 6.27. We save the template under the name *Masbal2*.

Thus, we have utilized the Iteration feature in spreadsheets to resolve circular references in solving a set of linear simultaneous equations dealing with mass balance in a chemical plant.

6.3 BUILT-IN FUNCTIONS AND ENGINEERING ECONOMICS

In the real world of engineering design and analysis, economics plays as big a role as technical considerations in the evaluation and ultimate selection of a specific design. Likewise, decisions on investments in a particular technology, product, or process involve economic considerations. For example, if an organization wants to invest in a computer-aided design (CAD) system, there are several systems in the market, each with its own advantages and disadvantages. The differences include the capital and maintenance costs, training time, and potential productivity gains. The engineer's task is to compare each alternative and select the one that best meets the selection criteria.

During the evaluation process, costs are estimated or predicted for each of the alternatives, and they are judged against anticipated benefits. The structured approach or methodology for handling this class of problems is known as **engineering economy.** The basic concepts associated with engineering economy are integral to the practice of engineering.

We will briefly discuss the fundamental concepts of engineering economy and develop a template for analyzing investment alternatives. We will also explore some built-in functions in spreadsheets for financial and logical computations.

6.3.1 Time Value of Money and Evaluation Criteria

A major consideration underlying investment decisions is that money has a time value. This means that paying $10,000 for a machine today is more expensive than paying $10,000 a year from now. This is because $10,000 invested today at 6 percent interest is worth more than $10,000 a year later. Conversely, if money is borrowed for the capital investment, the organization is paying interest on the loan; the potential benefits accruing from the investment should be weighed against the investment cost and the cost of capital itself. The mathematics behind these calculations is simple and straightforward.

Cash-Flow Diagram   The various transactions involving flow of cash into and out of an organization are represented graphically on a **cash-flow diagram.** As shown

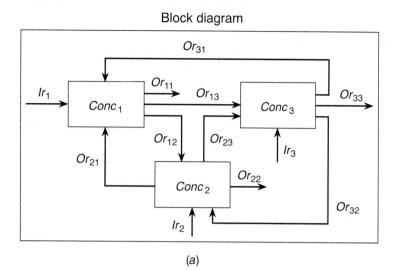

FIGURE 6.26 Another mass balance problem. (*a*) The block diagram. (*b*) Flow rates and concentrations. *Note:* Rate of transfer of chemical = flow rate × concentration.

in Figure 6.28, the horizontal line represents the time scale, typically in years. The arrows below the time scale indicate cash outflows from the organization; these are also referred to as *disbursements* or *costs* to the organization. The arrows above the horizontal line indicate cash inflows, also known as *receipts* or *income*. Thus, the cash outflows in Figure 6.28 are $105,000 initially, followed by $4000/yr, while inflows are $700/yr and $20,000 at the end of 5 yr.

Basic Definitions and Mathematical Relationships The **initial** or **first cost** refers to the capital cost incurred at the time an asset is acquired. Typically, this includes the sales tax, the freight, and installation charges. Since money is spent by

	A	B	C
5	----	----	----
6	Ir1		40 m^3/h
7	Ir2		105 m^3/h
8	Ir3		100 m^3/h
9	Or11		10 m^3/h
10	Or12		60 m^3/h
11	Or13		30 m^3/h
12	Or21		20 m^3/h
13	Or22		15 m^3/h
14	Or23		80 m^3/h
15	Or31		20 m^3/h
16	Or32		15 m^3/h
17	Or33		90 m^3/h
18			
19	Inc1		10 kg/m^3
20	Inc2		3 kg/m^3
21	Inc3		2.5 kg/m^3
22	Conc1	7.258681429291	kg/m^3
23	Conc2	7.653246099649	kg/m^3
24	Conc3	8.640161046806	kg/m^3

B5: [W53] \-
31-Aug-90 03:47 PM   CIRC   READY

FIGURE 6.27   Solution of problem in Figure 6.26a.

the organization, the initial cost represents a cash outflow. The **useful life** of an asset or product refers to the expected length of service from the asset, generally specified in years. The **salvage value** of an asset is the net sum realized by the organization

FIGURE 6.28   A cash-flow diagram. Cash outflows (or disbursements) are below the time scale; cash inflows (or receipts) are above the time scale.

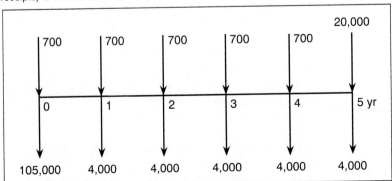

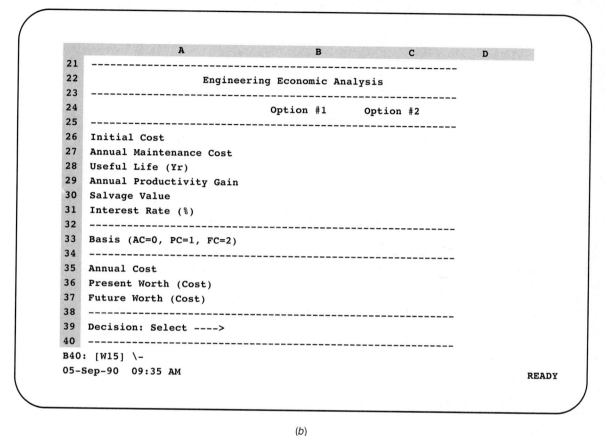

(b)

FIGURE 6.29 (continued).

for a principal and is based on Equation (1) in Capsule 6.2. We make use of these functions in creating the template.

6.3.3. Creating and Using the Template

We create the template using the layout shown in Figure 6.29. Figure 6.30 shows the formulas entered in cells B35, B36, B37, and B39, respectively. The interpretation of the formula in B35 is also given in the figure.

Formula Interpretation Based on the selection code, the formula in cell B39 compares the values in B35 and C35 (or B36 and C36 or B37 and C37) and returns the contents of cell B24 or C24; these cells contain the names of the alternatives being evaluated. In our example, they are option #1 and option #2, respectively. We use @@("B24") and @@("C24") for this purpose. If the quotation marks are omitted, e.g., @@(B24), the contents of B24 are used as the cell address for returning a value.

CAPSULE 6.3

THE @IF STATEMENT

As we mentioned in Section 4.4.1, the spreadsheet program has a whole range of built-in functions that can be used in formulas. These functions are typically classified as **mathematical, logical, financial, counting, string, date and time,** and **database** functions. As the names imply, each class of functions is appropriate for carrying out a specific set of tasks. For instance, we used the @SIN function from the mathematical category in the projectile template in Section 4.4, the @SUM function in Figure 6.7 for summing up a range of cells, and so on. We will now look at the @IF function from the set of logical functions.

The @IF function

The **@IF** function is similar to the IF-THEN-ELSE clause in conventional programming languages. The complete format of the @IF function is as follows:

@IF(Test_cond, True_expr, False_expr)

where Test_cond is the condition being tested.
True_expr is the value returned if the test condition is true.
False_expr is the value returned if the test condition is false.

Any of the relational operators (greater than, less than, equal to, etc.) can be used in the Test_cond expression (see Table 3.1 for a complete list). In addition, the boolean operators — AND, OR, and NOT — can be used. Typically, these boolean operators are specified as #AND#, #OR#, and #NOT#, respectively.

For example, in a spreadsheet, if cell B2 = 21, B3 = 44, B4 = 65, B5 = Yes, B6 = No, and B7 = Can't, then,

@IF(B3 < B4, B5, B6) returns Yes.
@IF(B3 = B4, B5, B6) returns No.
@IF(B3 + B4 > B4, B5, B6) returns Yes.
@IF(B3 < B4, @IF(B2 > B3, B6, B5), B7) returns Yes.

In each instance, if the tested condition is true, the result of the True_expr is returned, else that of False_expr. In the first example, since B3 is less than B4, the contents of B5, Yes, are returned.

It is possible to nest the @IF functions as shown in the last example. Here, since the first condition (B3 < B4) is true, it tries to evaluate @IF(B2 > B3, B6, B5); since B2 is not greater than B3, the contents of B5, Yes, are returned.

Copying the Formulas We use the Copy command in the spreadsheet to copy the formulas in cells B35..B37 and B39 to cells C35..C37 and C39, respectively.

Solving the Problem We assign the known values from Table 6.8 and choose the annual cost method by entering 0 in cells B33 and C33. The results are shown in Figure 6.31. The values are formatted to two decimal places, and where appropriate, the currency format has been used. The annual cost of option #1 is $29,279.83 while that of option #2 is lower at $27,518.68. Hence the resulting decision in cell B39: select option #2. The other cells display the "Not applicable" message.

Figure 6.32 shows the results when the present value method is selected by entering 1 for the selection code. Here again, the present cost is lower for option #2, and the decision is the same as before. In Figure 6.33, the results of the future worth method (selection code = 2) are shown with the same decision as before. That is, whichever method is used for evaluating the two workstation alternatives, option #2 is the suggested choice. We save the template under the name *Engecon1*.

Thus, using some built-in functions, we have created a template for solving problems in engineering economy, and in particular for the evaluation of investment alternatives.

FORMULAS IN:

Cell B35:

$$@IF(B33 = 0, @PMT(B26, B31/100, B28) + B27 - B29$$
$$- (B30 * B31/100)/((1 + B31/100)^{\wedge} - B28 - 1), \text{"Not Applicable"})$$

Cell B36:

$$@IF(B33 = 1, +B26 + @PV((B27 - B29), B31/100, B28)$$
$$- (B30 * (1 + B31/100)^{\wedge} - B28), \text{"Not Applicable"})$$

Cell B37:

$$@IF(B33 = 2, B26 * (1 + B31/100)^{\wedge}B28$$
$$+ @FV((B27 - B29), B31/100, B28) - B30, \text{"Not Applicable"})$$

Cell B39:

$$@IF(B33 = 0, @IF(C35 < B35, @@(\text{"C24"}), @@(\text{"B24"})),$$
$$@IF(B33 = 1, @IF(C36 < B36, @@(\text{"C24"}), @@(\text{"B24"})),$$
$$@IF(C37 < B37, @@(\text{"C24"}), @@(\text{"B24"}))))$$

(a)

INTERPRETATION OF FORMULA IN CELL B35

The first argument in the @IF statement checks to see if the selection code in cell B33 is 0.

IF B33 = 0,
    THEN compute the following:

Expression	Corresponds to
@PMT(B26, B31/100, B28)	Equation (1) in Capsule 6.2
+B27	MC_{pw}
-B29	APG_{pw}
$-(B30 * B31/100)/((1 + B31/100)^{\wedge} -B28 - 1)$	Equation (2) in Capsule 6.2

ELSE
    Return "Not Applicable."

(b)

FIGURE 6.30 The formulas and their interpretation. (a) The formulas in the cells. (b) Interpretation of the formula in cell B35.

```
                    A                        B                    C              D
21  ----------------------------------------------------------------
22                       Engineering Economic Analysis
23  ----------------------------------------------------------------
24                                      Option #1           Option #2
25  ----------------------------------------------------------------
26  Initial Cost                        $105,000.00         $75,000.00
27  Annual Maintenance Cost             $4,000.00           $8,000.00
28  Useful Life (Yr)                    5.00                5.00
29  Annual Productivity Gain            $700.00             $500.00
30  Salvage Value                       $20,000.00          $5,000.00
31  Interest Rate (%)                   12                  12
32  ----------------------------------------------------------------
33  Basis (AC=0, PC=1, FC=2)            0                   0
34  ----------------------------------------------------------------
35  Annual Cost                         $29,279.83          $27,518.68
36  Present Worth (Cost)                Not Applicable      Not Applicable
37  Future Worth (Cost)                 Not Applicable      Not Applicable
38  ----------------------------------------------------------------
39  Decision: Select ---->              Option #2
40  ----------------------------------------------------------------
B40: [W15] \-
05-Sep-90   09:35 AM                                                     READY
```

FIGURE 6.31 The annual cost method. The decision is to select option 2.

6.4 LIMITATIONS OF SPREADSHEETS

Inasmuch as the spreadsheet is a very powerful and flexible tool for problem solving, it does have a few limitations. First and foremost, we need to rewrite the equations in the mathematical model to explicitly isolate the unknown variable to the left-hand side of the equation and enter the right-hand-side expression in the cell for calculating the unknown. This is very similar to writing assignment statements in a programming language. This might require extensive algebraic manipulation, and in some cases, it may not be possible to isolate the unknown on the left-hand side of the equation. From this viewpoint, using a spreadsheet is not very different from using a conventional programming language.

Predetermined Combination of Known and Unknown Variables Second, every template is designed and created for estimating a certain set or combination of unknowns from a given set of known variables. Once this is done, it is easy to carry out what-if analysis by changing the values of the known variables. However, if we want to make one of the known variables an unknown and vice versa, the solution cannot be obtained without extensive modifications, or even complete redesign of the

```
                    A                    B                    C               D
21  ----------------------------------------------------------------
22                         Engineering Economic Analysis
23  ----------------------------------------------------------------
24                                   Option #1            Option #2
25  ----------------------------------------------------------------
26  Initial Cost                     $105,000.00          $75,000.00
27  Annual Maintenance Cost            $4,000.00           $8,000.00
28  Useful Life (Yr)                        5.00                5.00
29  Annual Productivity Gain             $700.00             $500.00
30  Salvage Value                     $20,000.00           $5,000.00
31  Interest Rate (%)                         12                  12
32  ----------------------------------------------------------------
33  Basis (AC=0, PC=1, FC=2)                   1                   1
34  ----------------------------------------------------------------
35  Annual Cost                   Not Applicable      Not Applicable
36  Present Worth (Cost)             $105,547.22          $99,198.69
37  Future Worth (Cost)           Not Applicable      Not Applicable
38  ----------------------------------------------------------------
39  Decision: Select ---->        Option #2
40  ----------------------------------------------------------------
B40: [W15] \-
05-Sep-90   09:35 AM                                             READY
```

FIGURE 6.32 The present worth method. The decision is to select option 2.

template in some instances. The inability to interchange the status of variables from known(s) to unknown(s), and vice versa, can be a major limitation especially in engineering design problems.

For example, the mass balance template (Figure 6.27) has been designed to estimate the concentrations in the three reactors for any combination of flow rates or input concentrations. However, if we want to determine the flow rates for maintaining specific concentrations in the reactors, we need to modify the relationships in the template. The engineers in the real world encounter such problems on a regular basis; these are the so-called **how-can problems** because they are trying to resolve how a certain objective can be achieved. This is different from the what-if class of problems where we try to determine the effect of variables on the objective.

Limitations of the Iteration Feature Third, spreadsheets provide the iteration feature, which is extremely useful and essential for solving certain types of engineering problems (see Section 6.2). However, the lack of explicit control over the iterative process and the absence of feedback from the program on the status of the iterative calculations are limitations of the iteration feature.

```
            A                      B              C              D
21  ----------------------------------------------------------
22                   Engineering Economic Analysis
23  ----------------------------------------------------------
24                          Option #1       Option #2
25  ----------------------------------------------------------
26  Initial Cost            $105,000.00     $75,000.00
27  Annual Maintenance Cost   $4,000.00      $8,000.00
28  Useful Life (Yr)              5.00           5.00
29  Annual Productivity Gain    $700.00        $500.00
30  Salvage Value            $20,000.00      $5,000.00
31  Interest Rate (%)              12             12
32  ----------------------------------------------------------
33  Basis (AC=0, PC=1, FC=2)        2              2
34  ----------------------------------------------------------
35  Annual Cost             Not Applicable Not Applicable
36  Present Worth (Cost)    Not Applicable Not Applicable
37  Future Worth (Cost)     $186,010.27    $174,821.98
38  ----------------------------------------------------------
39  Decision: Select ---->  Option #2
40  ----------------------------------------------------------
B40: [W15] \-
05-Sep-90   09:35 AM                                      READY
```

FIGURE 6.33 The future worth method. The decision is to select option 2.

This sets the stage for the next major category of problem-solving tools, viz., equation solvers, which we discuss in the following chapters.

SUMMARY

In this chapter, we discussed some specific applications of spreadsheets in quantitative problem solving. In each example, we explored some additional features of spreadsheets. While developing a template for the reliability analysis of electronic circuits, we discussed the basic concepts behind macros and methods for creating and using them in a spreadsheet. We also discussed the development of customized menus similar to the spreadsheet program menu to enhance the human-computer interface.

We then presented the basic concepts of mass balance computations and solved a set of simultaneous linear equations using the spreadsheet. We looked at circular references in templates and ways of resolving them: by using the iteration feature and by algebraic manipulation. We also discussed some important points pertaining to the use of the iteration feature in spreadsheets.

Turning to the area of engineering economy, we introduced the basic concepts and discussed three methods for evaluating investment alternatives. We discussed the @IF function along with some financial built-in functions. We developed a template for evaluating investment alternatives and arriving at a decision based on the results. Finally, we briefly discussed the limitations of spreadsheets for engineering problem solving.

CHAPTER 7

EQUATION SOLVING

In this chapter, we introduce the concept of problem solving without having to program in a conventional programming language. This concept of problem solving without programming forms the basis for a category of tools known as *equation solvers*. We use an example from mechanical engineering to illustrate the basics of an equation solver and discuss its role in quantitative problem solving. We develop a structured approach to the use of equation solvers based on the problem-solving methodology discussed in Chapter 2. A brief history of the evolution of equation solvers concludes the chapter.

7.1 HELICAL SPRING DESIGN

Mechanical springs are used in machines to cushion impact and shock loading, to exert force, to provide flexibility, and to maintain contact between machine members. The design of springs primarily involves the relationships between force, stress, torque, and deflection. Helical springs are usually made of circular cross-section wire as shown in Figure 7.1; they are made to resist tensile, compressive, or torsional loads.[1] Figure 7.2 shows the equations relating material properties and design parameters for helical springs.

Example Consider the following problem: A helical spring made from a wire of circular cross-section, $\frac{1}{16}$ in in diameter, is subjected to an axial load of 10 lb. If the modulus of rigidity of the material is 11.5×10^6 lb/in^2, the spring index is 6, and the spring constant is 10 lb/in, calculate the number of coils, the shear stress and the deflection of the spring.

---

[1] For details, see J. E. Shigley, *Mechanical Engineering Design,* McGraw-Hill, New York, 1977.

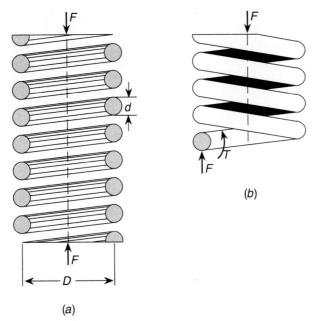

FIGURE 7.1 Helical spring design. (*a*) An axially loaded helical spring. (*b*) Free-body diagram showing the wire being subjected to a direct shear and a torsional shear. (After J. E. Shigley, *Mechanical Engineering Design*, McGraw-Hill, New York, 1977.)

The solution obtained with the help of a calculator and paper and pencil is shown in Figure 7.3. The spring deflection is 1 in, the shear stress is 48,990.183 lb/in^2, and the number of coils is 41.59. If we want to use a different material for the spring or if the spring constant changes, we have to repeat all the computations. In engineering design, we often need to estimate the values of different sets or combinations of variables to meet certain performance criteria. For example, we may need to find the values of the other variables if the maximum allowable shear stress were 45,000 lb/in^2. Here's where an equation solving tool or **equation solver** comes in handy.

Figure 7.4 shows the mathematical relationships for the design of helical springs entered in TK Solver Plus. TK Solver Plus exemplifies the class of equation solving tools; its predecessor, TK!Solver, pioneered this software category (more on history in Section 7.7). Once you create a model by defining the mathematical relationships, the equation solver lets you change the values of the desired variables, and the unknowns will be automatically computed. The advantages of this approach to problem solving include speed and creative approaches to problem solutions; the results can also be displayed in the form of tables and plots. The set of relationships and associated information, known as a **model**, can be saved, retrieved, modified, and used at a later date. This is where the apparent similarity with spreadsheets ends.

Beyond Spreadsheets In a spreadsheet template, once the formulas are defined, the values of variables referenced in the formulas can be changed during what-if

$$S_s = K\frac{8FC}{\pi d^2} \tag{1}$$

$$\frac{4C-1}{4C-4} + \frac{0.615}{C} = K \tag{2}$$

$$C = \frac{D}{d} \tag{3}$$

$$Defl = \frac{8FC^3n}{dG} \tag{4}$$

$$k\,Defl = F \tag{5}$$

$$Energy = \frac{S_s^2}{4G} \tag{6}$$

where S_s = shear stress, lb/in^2
K = Wahl factor
F = axial load, lb
C = spring index
d = wire diameter, in
D = coil diameter, in
$Defl$ = spring deflection, in
n = number of coils
G = modulus of rigidity, lb/in^2
k = spring constant, lb/in
$Energy$ = energy stored, in·lb/in^3

FIGURE 7.2 Helical spring design: mathematical relationships.

Given: $F = 10$ lb
$C = 6$
$d = \frac{1}{16}$ in
$G = 1.15 \times 10^7$ lb/in^2
$k = 10$ lb/in
Equations in Figure 7.2.

Solution: From Equation (2): $K = \dfrac{(4*6-1)}{(4*6-4)} + \dfrac{0.615}{4} = 1.2525$

From Equation (3): $D = 6 * 0.0625 = 0.375$ in

From Equation (5): $Defl = 10/10 = 1$ in

From Equation (1): $S_s = \dfrac{1.2525 * 8 * 10 * 6}{\pi * 0.0625^2} = 48{,}990.183$ lb/in^2

From Equation (4): $n = \dfrac{1 * 0.0625 * 1.15 * 10^7}{8 * 10 * 6^3} = 41.5943$

From Equation (6): $Energy = \dfrac{48{,}990.183^2}{4 * 1.15 * 10^7} = 52.17474$ in·lb/in^3

FIGURE 7.3 Paper-and-pencil solution of spring design problem.

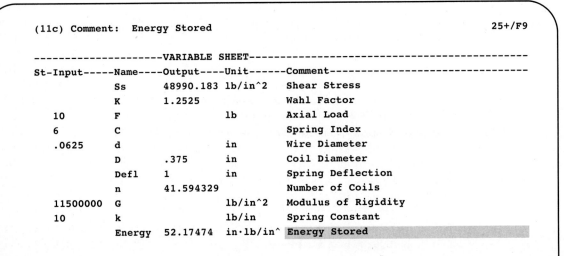

FIGURE 7.4 A TK Solver Plus model for helical spring design. (*a*) Rule Sheet. (*b*) Variable Sheet.

analysis and the new values computed (see Section 4.3.2). However, as discussed in Section 6.4, the inability to interchange the status of variables from known(s) to unknown(s) and vice versa is a major limitation of spreadsheets. Equation solvers go beyond spreadsheets and enable you to swap inputs for outputs and vice versa without modifying the equations.

For example, using the same model in Figure 7.4 and without rearranging the equations, we can find answers to the following question: "How should the spring parameters change so that the spring can withstand a maximum allowable stress of 45,000 lb/in^2?" In other words, equation solvers enable you to perform **how-can** analysis, in addition to the traditional what-if analysis. This means that you can find out how one or more of the parameters should change to achieve a desired performance. This flexibility of equation solvers opens up powerful and challenging avenues for scientists and engineers engaged in quantitative problem solving.

7.2 WHAT IS AN EQUATION SOLVER?

Equation-solving tools or equation solvers represent a class of software programs that enable you to solve sets of algebraic equations without programming in a conventional language such as BASIC, FORTRAN, Pascal, or C. An equation solver typically lets you enter the algebraic equations or formulas "as is" and uses a unique method of propagating the solution through the set of equations. With an assortment of built-in functions, the ability to define your own functions, instantaneous error diagnostics and help facility, the equation solver provides an extremely flexible and easy to learn and use interactive problem-solving environment for the scientist and engineer.

Since TK Solver Plus, or TK for short (from Universal Technical Systems, Inc.), is the most advanced and popular of the various tools in this category, we will use the TK syntax in our discussions. Other equation-solving programs include Eureka from Borland and SolverQ from the University of Wisconsin. Moreover, TK provides the additional capability of defining your own procedures in a BASIC-like syntax; the result is a truly integrated problem solving and knowledge management environment in which sets of equations can be combined seamlessly with sets of procedures. Since quantitative problem solving generally involves both declarative and procedural computations, TK fills the need.

Major programs related to the equation-solving category include MathCAD, an electronic scratchpad from MathSoft, and Mathematica, a symbolic manipulation tool from Wolfram Research.

7.2.1 Basics of an Equation Solver

As scientists and engineers, we typically organize our information in terms of formulas, equations, variables, units of measurement and conversions, mathematical functions, graphs, and tables. All this together constitutes the mathematical **model**. TK Solver Plus and other equation-solving tools provide several **sheets** for storing and organizing information in much the same way as we do in real life. The number, size, and layout of the sheets depends on the specific program; so does the number of equations and variables that can be handled by the program.

Different aspects of the information in the model are entered and viewed in sheets. For instance, TK has a Rule Sheet for entering and viewing equations; a Variable Sheet for entering and viewing variable names, values, and associated units; a Unit Sheet for defining unit conversions, a Plot Sheet for defining the graphs to be plotted; and so on (see Figure 7.5). The information in the various sheets is linked to each other, resulting in a well-organized collection of knowledge about a specific domain, e.g., design of helical springs or ac circuit analysis. This information can be manipulated through two major problem-solving mechanisms: the **Direct** and **Iterative Solvers.** And, TK can handle up to 32,000 equations in a single model.

7.2.2 The TK Sheets

Let's look a little closer at the TK start-up screen shown in Figure 7.6a. The screen is split into two windows. The Variable Sheet is displayed in the top window, while the Rule Sheet is displayed in the bottom window. The highlighted bar, known as the **cursor**, appears in the first row of the Rule Sheet. Each row in the Rule Sheet is made up of two fields: **Status field,** where the status of the rule is displayed and **Rule field,** where the rule is entered. The role of the Status field is subsequently explained in Section 7.4.1. The location or address of the cursor is displayed in the **message area** at the top left-hand corner of the screen as shown in Figure 7.6b. The address 1r on the **status line** indicates that the cursor is in the first row of the Rule field. This address helps us navigate through TK. The contents of the field are also displayed along with the address.

The solution status is displayed at the top right-hand corner of the status line. An OK indicates that the equations are solved or satisfied, while an ! or F9 denotes the need to solve the model.[1] The second line of the message area, known as the **prompt/ error line,** is used for responding to some commands, particularly the file storage and retrieval commands.

The Variable Sheet Figure 7.6a also shows the various fields in the Variable Sheet. The field names point to the information being stored and displayed in them. Variable names are entered in the **Name field,** the input values in the **Input field,** the results of the calculations are displayed in the **Output field,** the units of measurement are entered in the **Unit field,** and so on.

Moving around the TK Sheets We can move the cursor around the TK sheets in the windows with the help of the **Arrow keys, Function keys, Switch key, GoTo command,** or the **Scroll bars.**

The Arrow Keys The cursor can be moved from the current location to the adjacent field using the Arrow keys. For example, if we press the Down Arrow key thrice from the first row of the Rule Sheet, it is moved to the fourth row (4r); then,

---

[1] The ! is the legacy from the predecessor of TK Solver Plus; you had to press ! to solve the model in TK!Solver. This in turn was inherited from Software Arts' VisiCalc, the first electronic spreadsheet.

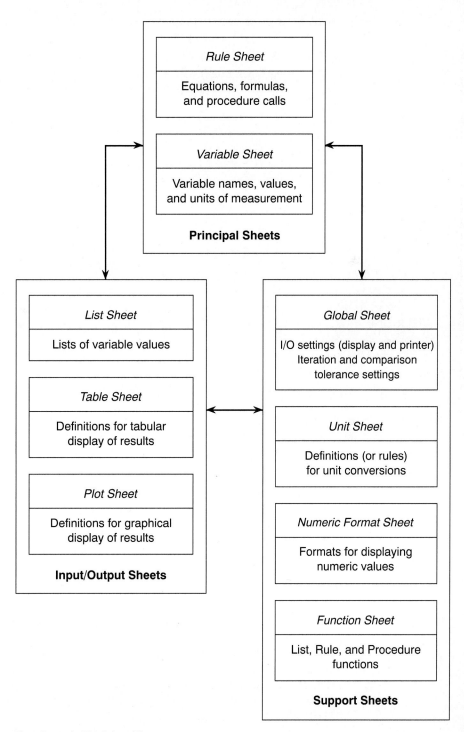

FIGURE 7.5 The sheets in TK Solver Plus.

```
(1r) Rule:                                                          25+/OK
For Help, type ? or press F1
--------------------VARIABLE SHEET------------------------------------------
St-Input-----Name----Output----Unit------Comment---------------------------










--------------------RULE SHEET---------------------------------------------
S-Rule---------------------------------------------------------------------







 F1 Help  F2 Cancel  F5 Edit  F9 Solve  / Commands  = Sheets  ; Window switch
```

(a)

FIGURE 7.6 The TK screen layout. (a) Start-up screen. (b) The message area.

pressing the Left Arrow key positions the cursor in the Status field of the fourth row (4s). Note that the status line displays the current location of the cursor and the contents of the cell (Figure 7.7a). When we move the cursor down past the ninth row, the rows at the top disappear from the window and the new rows (and row numbers) appear in the window (Figure 7.7b).

Using the Arrow keys to move from one part of the sheet to another (e.g., from the first row to the 44th) can be rather tedious. This is where the function keys on the numeric keypad come in handy. The PgUp (page-up) and PgDn (page-down) keys enable us to scroll the TK sheets vertically page by page, i.e., seven rows at a time.

The Switch Key When two sheets are displayed on the screen, we can press the ; key, known as the **Switch key** in TK, to move the cursor from one sheet to the other. For example, when we press the Switch key, with the cursor in the 15th row of the Rule Sheet, it moves to the Variable Sheet as shown in Figure 7.7c. If we press the Switch key again, the cursor returns to the previous position (the 15th row) in the Rule Sheet.

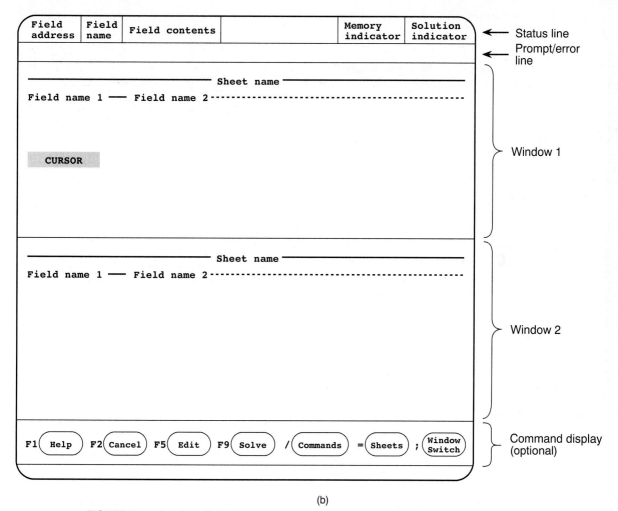

FIGURE 7.6 *(continued)*.

The Function Keys The function keys (F1, F2, ..., F10) on the personal computer keyboard are utilized by TK for executing some frequently used commands. The specific key bindings are shown in Figure 7.8. Thus, F1 is used for invoking the TK Help facility, while F9 is used for solving the model, and so on. A few commands are displayed at the bottom of the screen as shown in Figure 7.6. This display can be turned off, if desired.

The GoTo Command We can use the **GoTo command** to move the cursor to a desired location on the sheet. When we press :, the GoTo command, TK displays a prompt for the destination as shown in Figure 7.9*a*. When we respond with the address, 9c in the figure, the cursor is moved to the Comment field on the ninth row. Thus, the GoTo command enables us to rapidly display a specific part of the sheet.

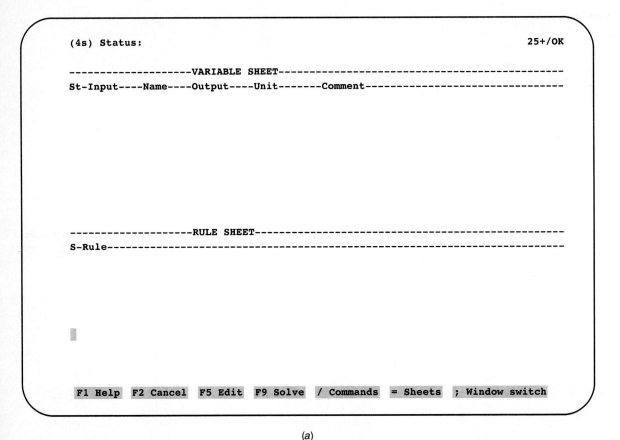

(a)

FIGURE 7.7 Moving around TK. (a) Cursor in the Status field of fourth row. (b) Scrolling: The top rows of the Rule Sheet disappear. (c) Switch key: Cursor moves from sheet to sheet.

The Select Command The **Select command** can be used for selecting and displaying a specific sheet in the current window. For example, if the Variable Sheet is displayed on the screen and you want to display the Unit Sheet instead, you invoke the Select command by typing = from any sheet (see command display at the bottom of the TK screen). From the resulting menu displaying the various sheet names, you select Unit Sheet.

Thus, using the various commands, we can bring any TK sheet or any part of it into the display windows for working.

7.3 A STRUCTURED APPROACH TO PROBLEM SOLVING WITH EQUATION SOLVERS

The use of equation solvers in problem solving follows the methodology discussed in Chapter 2 and shown in Figure 2.1. The first step in problem solving is the analysis of the problem and the development of the mathematical model (see Figure 2.3). Then the model is incorporated in the framework of the equation-solving tool. Unlike using

```
    (15s) Status:                                                          25+/OK

    ---------------------VARIABLE SHEET---------------------------------------
    St-Input-----Name----Output----Unit------Comment-------------------------










    ---------------------RULE SHEET-------------------------------------------
    S-Rule--------------------------------------------------------------------









    F1 Help  F2 Cancel  F5 Edit  F9 Solve  / Commands  = Sheets  ; Window switch
```

FIGURE 7.7 (continued).

(b)

a programming language or a spreadsheet, however, this is a fairly simple step because there is no need to develop the solution strategy, viz., the algorithm per se. Consequently, it is also easier and faster than the other two methods. Once equations and associated information are entered, the model is tested and debugged. It is then documented and saved for later use.

7.3.1 Developing the TK Model

Figure 7.10 shows the major steps in the development of a TK model from a mathematical model. As in programming, a structured approach to the development of a TK model will result in one that is easy to understand, use, and modify.

Design the Model Layout The structure and composition of the nine sheets in TK lend themselves to an elegant and well-defined layout for the model. In contrast to using a spreadsheet, TK's structure greatly simplifies the layout design process.

```
   (1i) Input:                                                      25+/OK

   ---------------------VARIABLE SHEET------------------------------------------
   St-Input------Name----Output----Unit------Comment---------------------------




   ---------------------RULE SHEET----------------------------------------------
   S-Rule-----------------------------------------------------------------------






   F1 Help  F2 Cancel  F5 Edit  F9 Solve  / Commands  = Sheets  ; Window switch
```

(c)

FIGURE 7.7   (continued).

Thus the equations in the model are entered in the Rule Sheet, the units in the Unit Sheet, and so on.

Enter the Mathematical Relationships   One of the hallmarks of TK is that you can enter the equations "as is," i.e., you can have expressions on both sides of the equal sign—there is no need to isolate the unknown variable on the left-hand side of the equation! This, of course, is a major departure from programming languages and spreadsheets. In TK,

$$\text{<expression 1>} = \text{<expression 2>}$$

is a valid construct and it does not matter on which side of the equal sign the unknown variable appears. TK is capable of evaluating the unknown variable, provided the values of the other variables in the equation are either given as inputs or can be computed from other equations. The equal sign *is* a true equal sign and *not* an assignment operator.

F1	Display help
F2	Cancel entry
F3	Load file
F4	Save file
F5	Edit contents
F6	Single window
F7	Display plot
F8	Display table
F9	Solve model
F10	List solve

FIGURE 7.8 Function key bindings in TK.

Moreover, there is no need to enter the equations in any particular order. TK tries to solve the equations in a model one by one and cycles through the set of equations until one of the following conditions is met: (1) All the unknowns are evaluated, (2) no more unknowns can be evaluated, and/or (3) an error is encountered.

Enter Associated Information When you enter the equations in the Rule Sheet, TK automatically displays the variable names in the Name field of the Variable Sheet. You enter the units of measurement associated with the variables in the Unit field of the Variable Sheet. You also define any required unit conversions on the Unit Sheet. You can use the Function Definition feature to define relationships between sets of values, e.g., material and elastic modulus or salary and tax rate. At this stage, the model is ready for testing.

Assign Values to Known Variables and Solve Testing the model with sample data is the next step in the development process (Figure 7.10). The results obtained must be confirmed, or validated, independently. The equations in the Rule Sheet should be checked for correctness along with the unit conversions defined on the Unit Sheet. The solution path must be traced to ensure the integrity of the computational flow (see Section 7.4.1 for tracing methods). Recall that this is similar to testing a program by "playing the computer."

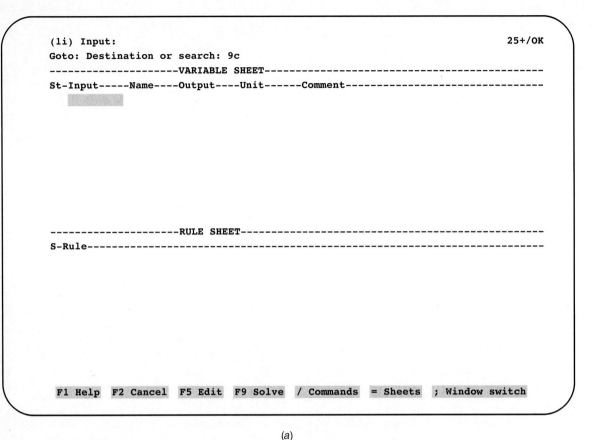

(*a*)

FIGURE 7.9 The GoTo command. (*a*) TK prompt for the destination. (*b*) Cursor positioned in the destination.

Correct Errors There are four main sources of common errors in TK models. They are: (1) different spellings of the same variable name in the equations; (2) errors in entering equations; (3) different spellings for the same unit name and/or absence of unit conversions; and (4) semantic or logic errors caused by illegal function arguments (e.g., the sine function) or overdetermined equations, i.e., the left-hand side of the equation is not equal to the right-hand side due to too many or inconsistent values for variables. Consequently, the results you see on the screen may be different from what you expect. Errors in the model should be traced to the source(s) and corrected, and then the test-modify-test cycle should be repeated till all the bugs are eliminated.

In addition to these errors that TK can detect, there are some errors that TK cannot detect. Examples include use of incorrect or inappropriate formulas or equations, invalid range of variable values, and incorrect unit conversion definitions.

Document the Model Once the model is validated, it should be documented with appropriate comments. Several of the TK sheets provide **Comment fields** that can be

```
(9c) Comment:                                                         25+/OK

     ---------------------VARIABLE SHEET-------------------------------------
     St-Input-----Name----Output----Unit------Comment-----------------------










     ---------------------RULE SHEET-----------------------------------------
     S-Rule-----------------------------------------------------------------








     F1 Help  F2 Cancel  F5 Edit  F9 Solve  / Commands  = Sheets  ; Window switch
```

(b)

FIGURE 7.9 (continued).

used for this purpose. The documentation should cover the purpose and scope of the model, how it works, and other pertinent information associated with its identification and limitations (e.g., range of variable values). This part of the model documentation is typically created in the Rule Sheet.

Improve the User Interface Here again, the structure of TK obviates the need for extensive effort in designing the user interface to make the model easy to use. It is a good idea to add helpful hints on the number of variables that should be specified and/or when a specific solving mechanism should be used. The Numeric Formatting feature should be used to format the values for improving their appearance. At the end of this stage, the model is ready for future use.

7.3.2 Using the TK Model: What-If and How-Can Analyses

TK is a very powerful and easy to use analytical tool. Consequently, the time spent in initially developing a model is not much more than when using a calculator, even

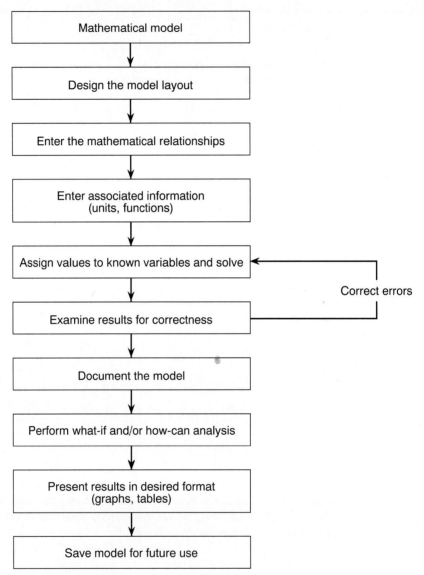

FIGURE 7.10 Steps in the development and use of a TK model in problem solving.

for a simple problem. But the ability to evaluate various alternatives with very little additional time and effort often justifies the incremental initial time. Indeed, the ease with which TK can be used makes it a "supercalculator." Of course, for complex problems involving several equations, units of measurements, and functions, the comparison with calculators is not fair to the latter.

The ability to perform how-can analysis, in addition to what-if analysis, is one of the distinct advantages of TK over other problem-solving tools. Some individual sheets in the model such as the Unit Sheet and Function Sheet can be stored separately and used in other models. Thus, you can create a library of unit conversions, procedures, etc., and load them in any model. On a larger scale, you can select a specific domain, e.g., electric circuits, and create (and store) a library of models covering the field. These models can be subsequently used in design, analysis, and problem solving.

Equation Solvers, Spreadsheets, Programming, and Calculators The time taken for solving a problem depends on the chosen technique or tool. Moreover, certain types of problems lend themselves well to a specific tool. Consequently, it would be incorrect to categorically state that one technique is *always* better than the others. In general, however, equation solvers take less time than others in quantitative problem solving. This is because, between spreadsheets and equation solvers, the latter are specifically designed for science and/or engineering applications. Calculators, except for the simplest of problems, take longer than other techniques, and the ability to solve problems repeatedly, with very little incremental effort, is limited. By its very nature, programming in a conventional language takes longer than software tools.

Consequently, an equation solver (e.g., TK) is a safe choice among the various alternatives. It also affords avenues for creative approaches to a problem, avenues not readily found in other tools. The cost-benefit ratio, i.e., the ensuing advantages when compared to the effort required for learning and using a tool or technique, also favors TK. Programming requires a great deal of study, effort, and time and the development of an algorithm prior to coding, whereas in TK you can go directly from the model to the solution. And finally, you don't have to rearrange or algebraically manipulate the equations each time you want to investigate a different aspect of the problem, a definite advantage over programming and spreadsheets.

Should You Use an Equation-Solving Program? The time spent in the development of a TK model should be regarded as an investment with a long-term payoff. Therefore, you should carefully weigh the pros and cons before developing a model for the problem at hand. Some issues to bear in mind are: Will it be useful at a later date and/or for others? Is there a need to perform what-if and how-can analyses? Should the results be presented in the form of tables and/or graphs? Are there any features in the equation solver, such as built-in functions, that will simplify the solution process? Does the solution require extensive procedural computation? Can procedures, created earlier for other models, be readily used for the present problem? Once you are convinced that the equation-solving framework is the appropriate one, it helps to follow the structured methodology discussed here.

7.4 THE HELICAL SPRING DESIGN EXAMPLE

We will illustrate the problem-solving methodology by creating a TK model for the design of helical springs discussed earlier in Section 7.1.

7.4.1 Creating a TK Model

The first step in creating a TK model from the mathematical relationships shown in Figure 7.2 is to design the layout of the model. However, as discussed earlier, the structure of TK obviates the need for this step.

Enter the Mathematical Relationships In the TK model, we will use the same variable names as in Figure 7.2. We load the TK Solver Plus program; with the cursor in the first Rule field of the Rule Sheet, we type the first equation Ss = K*8*F*C / (pi()*d^2).[1] We use the Backspace key to correct errors while typing. When we finish typing the equation, we press the Enter key. The equation is entered into the Rule field, and the variable names are displayed in the Name field of the Variable Sheet as shown in Figure 7.11a. Since the cursor is in the Rule field, the contents of the field are also displayed at the top of the screen. An asterisk (*) appears in the Status field of the rule indicating that the rule has not been solved. This is also indicated by the solution status indicator in the top right-hand corner of the Status line as it changes from OK to F9. Note that F9 corresponds to the Function key for solving the model.

The Rule Status When we move the cursor to the Status field of the rule (using the Left Arrow key), the display on the status line reads "Unsatisfied" as shown in Figure 7.11b. This indicates that the constraint has not been resolved, i.e., the equation has not been solved. This message is very helpful when there are several equations in the model and we want to identify the equations that have not been solved. When the equation or constraint is resolved, the asterisk disappears from the Status field, and if the cursor is moved to the Status field, a message "Satisfied" will appear on the status line. On the other hand, if an error is encountered in the equation during the solution process, e.g., when the value of the left-hand side is found to be not equal to the right-hand side (due to overspecification of variable values), a > appears in the Status field, and an appropriate error message is displayed on the status line. Thus, the Status field is a helpful aid in debugging the model.

Built-in Functions As shown in Figure 7.11, we entered pi() to denote the π in the first equation in Figure 7.2. This is one of a wide range of trigonometric, hyperbolic, financial, and TK-specific functions built into TK Solver Plus. The arguments of the functions (zero, one or more arguments) are specified in parentheses. Where no arguments are required in a function as in the π function, a pair of empty parentheses () follows the function name.[2] The function names are mnemonic; thus, Sin stands for the trigonometric *sine* function, Max for the maximum function, etc. The function names are case-insensitive, i.e., they can be specified in either upper or lowercase or in any combination. Thus, SIN, sin, Sin, SIn and other variations are all valid.

Arithmetic Operators and Precedence The following operators are used to specify the major arithmetic operations:

Operator	Operation	Precedence
^	Exponentiation (raising to a power)	Highest
* /	Multiplication and division	↓
+ -	Addition and subtraction	Lowest

---

[1] All entries that we type at the computer are shown underlined in text.

[2] The origins can be traced to the early development work on TK in which software engineers schooled in LISP where involved. Also, the first version of TK was written in IL (Implementation Language), an in-house Software Arts language derived from LISP and PL/I.

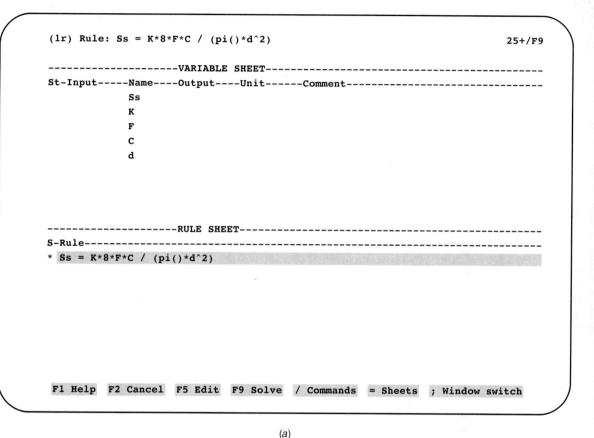

(a)

FIGURE 7.11 Entering an equation. (a) The Variable and Rule Sheets. (b) The rule status: unsatisfied.

In conventional programming the precise order of computations is predetermined by the characteristics of the language and the program. Moreover, the computations are performed only on the right-hand side of the assignment statement and the result is assigned to a variable on the left-hand side. On the other hand, in TK, there is no predetermined sequence, and the computational flow depends on which of the variables is known. It follows the logic that we humans adopt in solving equations. This means the equations are scanned and operations performed (on both sides of the equal sign) whenever operand values are available.

However, when the equation is entered, TK **parses** it and determines the precedence of the various operations. **Parsing** refers to the process of scanning the equations and checking for syntax errors such as missing operators or variables, illegal function names, etc. TK's parser works similar to the ones in conventional languages such as BASIC or FORTRAN. It looks for the subexpressions inside the innermost pairs of parentheses first. Once a group of operands and operators is isolated, the parser interprets them in the following order of precedence: unary plus and minus have the highest priority and are performed first; this is followed by the exponentiation

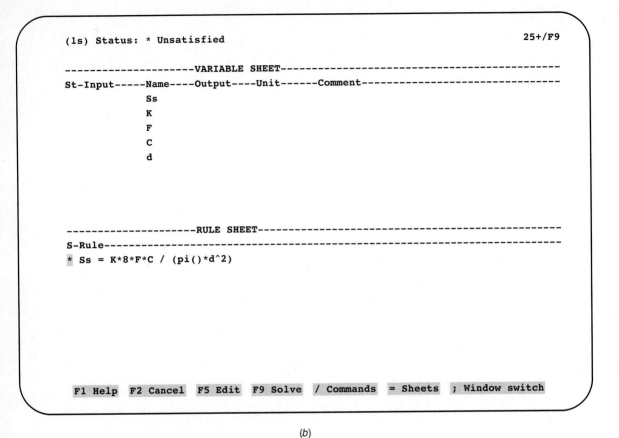

(b)

FIGURE 7.11 (continued).

operation, the multiplication and division operations, and finally the addition and subtraction operations. Operations of equal level of precedence (+ and − or * and /) are interpreted from left to right; here too, parentheses may affect the order of parsing and the results of the computations. Thus, it is important to keep in mind the precedence of operations when expressing an equation as a TK rule.

Detecting and Correcting Syntax Errors We use the Down Arrow key and move the cursor to the second row of the Rule Sheet and type the second equation from Figure 7.2. When we press the Enter key, TK beeps and displays the message "Expected operator not found" and places us in the edit mode as shown in Figure 7.12a. The TK parser detects the missing operator between 0.615 and C, and it positions the cue appropriately as shown in the figure. When we type the missing / and press the Enter key, the equation is parsed correctly and the asterisk appears in the Status field (Figure 7.12b). Since the variables in this equation (C and K) are already present in the Variable Sheet, they are not added to the Variable Sheet. TK ensures that variables appear only once in the Variable Sheet. Note the similarity between this rule and Equation (2) in Figure 7.2. TK lets you enter equations as you know them (from

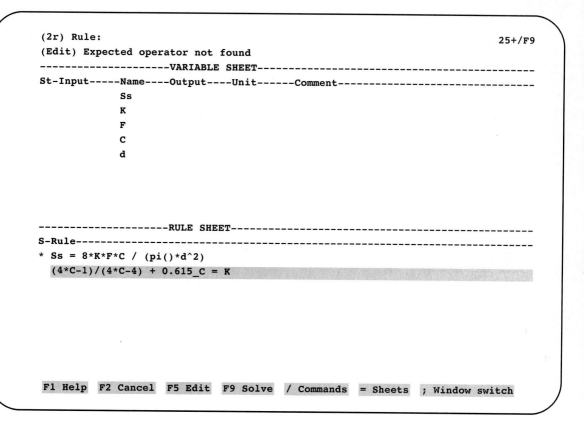

(a)

FIGURE 7.12 The TK parser and syntax errors. (a) TK detects a missing operator in the second equation (the cue is positioned between 0.615 and C). (b) The error is corrected. *Note:* A TK purist would enter the expression in the denominator as 4 * (C − 1), not as 4 * C − 4.

textbooks or self-derived) including expressions on the left-hand side of the equal sign—a departure from programming language statement.

If TK spots any syntax errors when you enter an equation, it will display an appropriate message and place you in the edit mode. When you correct the error and press the Enter key, the equation is parsed and if no other errors are found, the equation will be accepted in the Rule field and an asterisk displayed in its Status field.

When we enter the remaining equations from Figure 7.2, the Variable and Rule Sheets look as shown in Figure 7.13a. The new variables in the rules are automatically added to the Variable Sheet by TK. Only the first nine variable names are seen in the Variable Sheet in the figure. Using the Switch command (;), we move the cursor from the Rule Sheet to the Variable Sheet. When we scroll the sheet through the window with the Down Arrow key, the other variables are brought into view (Figure 7.13b). However, the variables at the top disappear as shown in the figure. By pressing the Function key F6, we can display a single window so that all the variables are seen (Figure 7.13c).

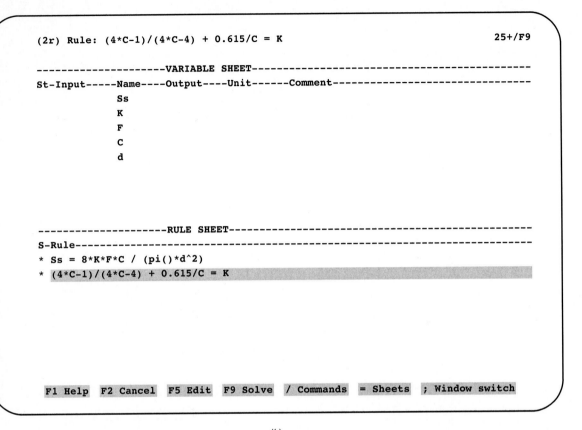

(b)

FIGURE 7.12 (continued).

Enter Associated Information The next step in the problem-solving process is to associate units of measurement with the variables in the model. We position the cursor in the Unit field of the first variable, S. From Figure 7.2 we know that its associated unit is lb/in^2; we enter lb/in^2 in the Unit field and press the Enter key.

Calculation and Display Units of a Variable The first time a specific unit is associated with a variable, it is treated as the **Calculation unit.** The Calculation unit is the unit in which the value is substituted for the variable. This means the calculations will be performed with the value of the variable in this unit. The **Display unit** is the unit in which a variable value is entered and displayed in the Variable Sheet. During the solution process, TK converts the value in the Display unit to the Calculation unit (more on this in Section 8.2.1).

In the present example, lb/in^2 is both the Calculation and Display unit for Ss. If the two units are different, the relationships or conversion between the two units should be defined on the Unit Sheet. For example, if the Calculation unit for a variable is meter and the Display unit is centimeter, the conversion between the two units should be defined on the Unit Sheet. If a conversion is not defined and a value is

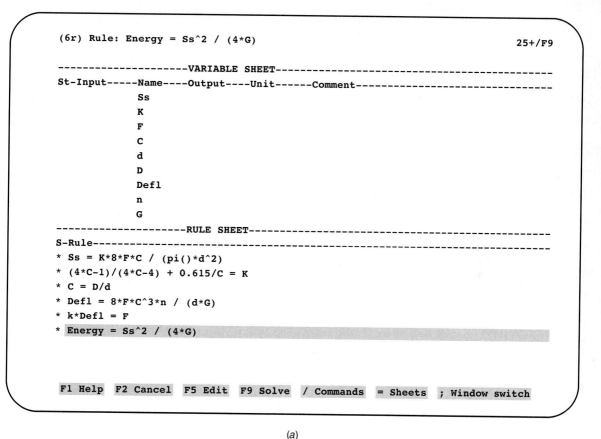

(a)

FIGURE 7.13 The variable and rules in the model. (*a*) Variable and Rule Sheets after rule entry. (*b*) Variables at the top scroll out of view. (*c*) Single variable window with F6.

assigned to the variable, TK displays a ? next to the variable, indicating its inability to convert the value to the Calculation unit. This error message helps in debugging the model.

TK, however, does not perform any dimensional analysis. It is up to us to ensure consistency in the set of units associated with the variables in the model. Otherwise, the solutions obtained will be incorrect.

For now, we will use the set of units in Figure 7.2 and associate them with the different variables; the resulting Variable Sheet with the unit names is shown in Figure 7.14.

Assign Values to Known Variables and Solve The model is now ready for testing and use. We will solve the problem shown in Figure 7.3. We move the cursor to the Input field of the known variables and assign the values. For example, the wire diameter is given as 1/16 in. We can enter the fraction 1/16 in the Input field of d, and TK will compute the result and display it as .0625 (see Figure 7.15*a*). Thus, the

```
(12n) Name:                                                          25+/F9

---------------------VARIABLE SHEET----------------------------------------
St-Input-----Name----Output----Unit------Comment--------------------------
             C
             d
             D
             Defl
             n
             G
             k
             Energy

---------------------RULE SHEET--------------------------------------------
S-Rule---------------------------------------------------------------------
* Ss = K*8*F*C / (pi()*d^2)
* (4*C-1)/(4*C-4) + 0.615/C = K
* C = D/d
* Defl = 8*F*C^3*n / (d*G)
* k*Defl = F
* Energy = Ss^2 / (4*G)
```

(b)

```
(1n) Name: Ss                                                        25+/F9

---------------------VARIABLE SHEET----------------------------------------
St-Input-----Name----Output----Unit------Comment--------------------------
             Ss
             K
             F
             C
             d
             D
             Defl
             n
             G
             k
             Energy
```

(c)

FIGURE 7.13   *(continued)*.

```
(11u) Unit: in·lb/in^3                                              25+/F9

----------------------VARIABLE SHEET----------------------------------------
St-Input-----Name----Output----Unit------Comment---------------------------
             Ss                lb/in^2
             K
             F                 lb
             c
             d                 in
             D                 in
             Defl              in
             n
             G                 lb/in^2
             k                 lb/in
             Energy            in·lb/in^
```

FIGURE 7.14 The Variable Sheet with unit information. *Note:* Since the Unit field is only 9 characters wide, the complete unit name for *Energy* is displayed at the top on the status line.

Input field of the Variable Sheet acts as a **calculator**. In fact, fields in other sheets that accept numeric values as input act as calculators. This feature is extremely useful for simple arithmetic operations.

Solving the Model After entering all the values, we press the Solve key (F9). TK displays the message "Direct Solver" on the prompt/error line and instantaneously displays the solution shown in Figure 7.15b. The number of coils is 41.59, the shear stress is 48,990.183 lb/in², and the spring deflection is 1 in. The Direct Solver is one of the two problem-solving mechanisms built into TK, and it works on the principle of consecutive substitution.

Note that the solution status indicator at the top right corner of the screen has changed from F9 to OK indicating that the equations have been solved. The asterisks have disappeared from the Status field of the rules; when we move the cursor to the Status field of the first rule, the message "Satisfied" is displayed on the status line as shown in Figure 7.15c. Thus, we have created our first TK model and solved the spring design problem! It is worth noting that the Solve key (F9) can be pressed from any sheet to trigger the solution process.

How Does TK Tick? Each constraint or equation is understood by TK as designating multidirectional relationships between the variables in the equation. This is in contrast to a statement in a conventional programming language where the relationship implication is unidirectional—the computed value on the right-hand side of the equals sign is assigned to the variable on the left-hand side. TK treats the equations as a network of constraints and attempts to resolve each of the constraints for a given set of variable values. For this reason, TK can be thought of as a **constraint resolution system**.

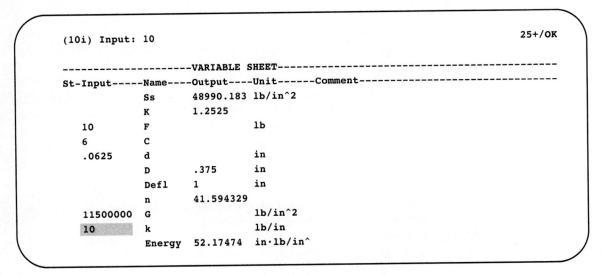

FIGURE 7.15 The helical spring problem. (a) Problem formulation: Input variables assigned. (b) Solution: Unknown variables computed. (c) The rule status: satisfied.

The way TK ticks, or solves the equations, is very similar to how we solve sets of equations: substituting values consecutively in the equations and evaluating the unknowns. As each equation is solved, TK adds the newly computed values to the set of other known values and moves on to the next equation; this consecutive substitution and solution process, or **propagation of solution,** continues, until any one of the

```
(1s) Status:    Satisfied                                              25+/OK

-----------------------RULE SHEET----------------------------------------------
S-Rule------------------------------------------------------------------------
 Ss = K*8*F*C / (pi()*d^2)
  (4*C-1)/(4*C-4) + 0.615/C = K
  C = D/d
  Defl = 8*F*C^3*n / (d*G)
  k*Defl = F
  Energy = Ss^2 / (4*G)
```

(c)

FIGURE 7.15 (continued).

following conditions is reached: all the equations are solved, no more equations can be solved, or an error is encountered.

Tracing the Solution Path   Let's trace the workings of the Direct Solver on the set of equations and known variable values in this model. We can do this by simply examining the variable values and equations in Figure 7.15 using a variable-equation coincidence (VEC) matrix. The VEC matrix is an $m \times n$ array or matrix with the m rows representing equations and the n columns representing variables.[1] The occurrence of the ith variable in the jth equation is indicated by a check (x) in the cell at the intersection of the jth row and the ith column. Figure 7.16a shows the VEC matrix for the rules in Figure 7.15. The 11 columns correspond to the variables and the 6 rows to the equations in the Rule Sheet. For example, five variables occur in the first equation, and the occurrences are indicated by an x in the corresponding cells. If a variable occurs more than once in an equation, it is denoted by a pound (#) in the appropriate cell. This symbol indicates that the marked variable cannot be evaluated from this equation by the Direct Solver (see also Section 8.1.3).

The solution path (SP) matrix in Figure 7.16b shows the status of each of the variables and the rules during the solution process. The input variables are marked I (for Input) in the first row of the SP matrix. The status of all the rules is initialized to U (for Unsatisfied) indicating that the constraints have not been resolved.

When the F9 key is pressed, the consecutive-substitution procedure (CSP) is triggered. The Direct Solver begins with the first equation. Of the five variables in this equation (see VEC Matrix), three are given as input—this is indicated by I in the SP matrix in Figure 7.16b. Since there are two unknowns (Ss and K), TK cannot solve the equation (neither can we, with a paper, pencil, and calculator). Therefore, the rule status remains U, and TK moves on to the second equation. Of the two variables in this equation (VEC matrix), only one is unknown (K); it is evaluated as 1.2525, and the evaluated variable is marked E (for Evaluated) in the SP matrix. The rule status

---
[1] The order can be reversed, with the columns representing equations and rows representing variables (e.g., Figure 8.4).

Eq. no. \ Var. name	S_s	K	F	C	d	D	Defl	n	G	k	Energy
(1)	x	x	x	x	x						
(2)		x		x							
(3)			x	x	x						
(4)		x	x	x			x	x	x		
(5)		x					x		x		
(6)	x								x		x

(a)

Pass no.	Eq. no.	S_s	K	F	C	d	D	Defl	n	G	k	Energy	Status
1			I	I	I					I	I		Begin
	(1)		I	I	I								U
	(2)		E		I								S
	(3)			I	I	E							S
	(4)		I	I	I				I				U
	(5)		I					E		I			S
	(6)								I				U
2		R	K	I	I	I	K	K	R	I	I	R	Continue
	(1)	E	K	I	I	I							S
	(2)		K		I								S
	(3)			I	I	K							S
	(4)		I	I	I				E	I			S
	(5)		I					K		I	I		S
	(6)	K							I			E	S
3		K	K	I	I	I	K	K	K	I	I	K	Continue
	(1)	K	K	I	I	I							S
	(2)		K		I								S
	(3)			I	I	K							S
	(4)		I	I	I				K	I			S
	(5)		I					K			I		S
	(6)	K							I		K		S

I = Input   E = Evaluated   R = Remaining   C = Contradictory   K = Known
S = Satisfied   U = Unsatisfied   O = Overdetermined

(b)

FIGURE 7.16 Workings of the Direct Solver. (a) VEC matrix. (b) The solution trace.

is changed to S (for Satisfied) indicating that the rule has been solved. This is also reflected in the Rule Sheet on the display screen—the asterisk in the Status field of that rule disappears.

In the third equation, C and d are known; therefore, D is evaluated as .375, and is marked E in the SP matrix. The rule status is changed to S. In the fourth equation, there are two unknowns (Defl and n); therefore, the rule status remains U, and TK moves on to the fifth equation. Since F and k are known, Defl is evaluated as 1 from this equation; Defl is marked E, and the rule status is changed to S. In the last equation there are two unknowns (Energy and Ss), and TK cannot solve the equation. The rule status (U) remains unchanged.

The first "pass," or cycle, of solution propagation is now complete. Since the status of at least one of the rules changed (from U to S), it means that at least one unknown variable was evaluated during this pass. The evaluated variable(s) could help the solution propagate further, i.e., assist in solving the other equations. Therefore, TK begins another cycle through the set of equations starting with the first equation. If, on the other hand, no variables were evaluated during this pass, i.e., the status of none of the rules changed, the solution process would have terminated.

Now the second pass: Three additional variables are known and three variables are remaining as indicated by the Ks and Rs, respectively, in the first row of the second pass in Figure 7.16b. Since variable K was computed in the earlier pass, Ss is the only unknown in the first equation; it is computed (48990.183), the rule status is changed to S, and TK moves on to the second equation. Though this equation was solved in the previous pass, TK checks it to make sure the equality still holds. There are instances where a variable is assigned two different values in a model; this consistency check spots such error conditions and terminates the solution process if warranted. Since the equality holds, TK moves on to the third rule and checks its consistency; it then moves to the fourth equation in which the only unknown n is evaluated as 41.594329 and the status is changed to S. The fifth equation is checked for consistency and Energy is evaluated from the last equation (52.17474), completing the second pass through the set of equations.

Since some new variables have been evaluated during this pass (indicated by the change in status of the rules), TK cycles through the set of equations one more time. However, at the beginning of this pass, all the variables are marked either I or K (first row of the third pass in Figure 7.16b). Therefore, TK checks the consistency of each of the equations during this pass. Since no errors are encountered and the status of the rules doesn't change, the solution process terminates.

Thus, using the VEC matrix in conjunction with the SP matrix, we have "played" TK and traced the workings of the Direct Solver. Note that this playing TK is not different from playing the computer for understanding the working of a program (see Section 3.2.3). As in debugging a program, this procedure is very useful for identifying problems in a TK model.

One important consequence of the CSP in TK is that the order in which the equations are entered in the Rule Sheet will, except in rare situations, not have an influence on the solution obtained.[1] This is because TK would cycle through the equations as long as at least one variable is evaluated during a pass.

Backsolving The ability to turn an equation around to evaluate any unknown variable from it is the cornerstone of TK Solver Plus and other equation-solving tools.

---

[1]There are a few exceptions to this general rule. See chapter 8 in M. Konopasek and S. Jayaraman, *The TK!Solver Book: A Guide to Problem-Solving in Science, Engineering, Business and Education,* Osborne/McGraw-Hill, Berkeley, Calif., 1984.

In TK terminology, this feature is known as **backsolving**, and it sets TK apart from spreadsheets. For example, in the second equation, though the unknown variable (K) appears on the right-hand side of the equals sign, it is evaluated! Likewise, D and n appear on the right-hand side of the respective equations and are evaluated.

Therefore, TK allows you to think in terms of relationships and concepts. Unlike programming in a conventional language, you don't have to spell out all the instructions for solving the set of equations. The equals sign in the TK rule is not an assignment operator (as in FORTRAN or BASIC) — it is a true equals sign. In other words, TK restores the mathematical meaning of the equals sign, i.e., left-hand-side expression equals the right-hand-side expression.

Check the Correctness of Results The solution in Figure 7.15b corresponds to the solution obtained using the paper-and-pencil method shown in Figure 7.3. This means the equations in Figure 7.2 have been correctly translated into TK rules and the model is sound. However, if we had made a mistake in entering one or more of the rules, the solution would have been different; for example, if we had not enclosed the denominator of the first rule in parentheses, the equation would still be syntactically correct and would be solved by TK. The resulting values, however, would not be correct. Since TK's rule syntax doesn't differ much from algebraic notation, it is a straightforward step to convert the equations into TK rules; however, semantic errors made during this translation step cannot be detected by TK.

Document the Model Since the model has been validated, the next step is to document it. We move the cursor to the Comment field of the Variable Sheet and enter the descriptions for each of the variables (Figure 7.17a). TK does not have a Document or Report Sheet for the model description. We can, however, use one of the features in the Rule Sheet of TK for entering text information. Any entry in the Rule field that follows quotation marks (") is treated by TK as a comment and not as a rule. In fact, as long as you don't mind typing a " before each line, you could use the Rule Sheet as the scratchpad of a rudimentary word processor!

Inserting a Row To enter the title of the model, we move the cursor to the Rule Sheet using the Switch command. Since there is no empty row at the top of the Rule Sheet for the title, we need to insert an empty row. With the cursor on the first row, we press the Insert key. An empty row appears and the other rows are renumbered. With the cursor in the Rule field of the empty row, we type " followed by the title * * * Helical Spring Design * * * and press the Enter key. The title is entered in the Rule field (Figure 7.17b).

The next step is to append comments to each of the rules. In many instances, a rule may have a specific name, e.g., Ohm's law, that can be appended to it. In the absence of such a name, an appropriate description can be used. For example, "Shear Stress" would be an appropriate comment for the first rule. Note that this comment neither implies that *only* shear stress is evaluated from this equation nor does it imply that shear stress *must* be evaluated from this equation. Such an assumption would run counter to the essence of TK, which is that a particular variable can be evaluated from any of the equations in which it occurs.

```
(11c) Comment: Energy Stored                                          25+/F9

--------------------VARIABLE SHEET----------------------------------------
St-Input-----Name----Output----Unit------Comment--------------------------
             Ss      48990.183 lb/in^2    Shear Stress
             K       1.2525              Wahl Factor
    10       F                 lb        Axial Load
    6        C                           Spring Index
    .0625    d                 in        Wire Diameter
             D       .375      in        Coil Diameter
             Defl    1         in        Spring Deflection
             n       41.594329           Number of Coils
    11500000 G                 lb/in^2   Modulus of Rigidity
    10       k                 lb/in     Spring Constant
             Energy  52.17474  in·lb/in^ Energy Stored
```

(a)

```
(1r) Rule: "     * * *   Helical Spring Design   * * *                25+/F9

--------------------RULE SHEET------------------------------------------
S-Rule------------------------------------------------------------------
   "    * * *    Helical Spring Design    * * *

   Ss = K*8*F*C / (pi()*d^2)            " Shear Stress
   (4*C-1)/(4*C-4) + 0.615/C = K        " Wahl Factor
   C = D/d                              " Spring Index
   Defl = 8*F*C^3*n / (d*G)             " Deflection
   k*Defl = F                           " Axial Load
   Energy = Ss^2 / (4*G)                " Energy
```

(b)

FIGURE 7.17 Documenting the model. (a) Variable descriptions in the Comment fields. (b) Model title and rule descriptions (text that follows the " is treated as a comment by TK).

Editing Field Contents To append the comments, we will use the Edit command in TK and enter the edit mode. With the cursor highlighting the first rule, we press the Function key F5 (see Figure 7.8 for key bindings). The cue (tiny blinking cursor) appears in front of the rule. We move the cue to the end of the rule using the End function key. We type " Shear Stress and press the Enter key to exit the edit mode. The comment is appended to the rule.

In the edit mode, the Left and Right Arrow keys can be used to advance the cue one character at a time (to the left or right, respectively); the Up and Down Arrow keys act like the Home and End keys on the function keyboard: they move the cue to the beginning and end of the field contents. Therefore, you must press the Enter key to exit the edit mode. We edit the remaining rules and add appropriate comments. The resulting Rule Sheet is shown in Figure 7.17b. The model is now ready for use in solving other problems.

7.4.2 Using the Model: Perform What-If and/or How-Can Analysis

Once a TK model is created, it is extremely easy to solve different types of problems without modifications to the equations in the model. We will use the helical spring model just created to illustrate this feature of TK.

What-If Analysis Let's say we want to estimate the spring parameters if the wire diameter is changed to $\frac{1}{8}$ in while the other input values remain the same as before. We move the cursor to the Input field of d and assign it 1/8. When we press the Solve key F9, the new solution shown in Figure 7.18 is obtained. The shear stress is reduced by one-half to 12,247.546 lb/in^2, the number of coils is doubled, while the spring diameter is also reduced by one-half. Thus, what-if analysis is fairly simple once we have a TK model.

How-Can Analysis The real power of TK is its ability to solve a given set of equations for different combinations of known and unknown variables without any modifications to the equations. In the first two problems that we solved, the set of known variables remained the same. We will now use the model to solve a problem with a different set of known variables.

FIGURE 7.18 What-if analysis: solving another problem.

```
(5i) Input:   .125                                                    25+/F9

---------------------VARIABLE SHEET----------------------------------------
St-Input-----Name----Output----Unit------Comment--------------------------
             Ss       12247.546 lb/in^2   Shear Stress
             K        1.2525              Wahl Factor
   10        F                  lb        Axial Load
   6         C                            Spring Index
  .125       d                  in        Wire Diameter
             D        .75       in        Coil Diameter
             Defl     1         in        Spring Deflection
             n        83.188657           Number of Coils
 11500000    G                  lb/in^2   Modulus of Rigidity
   10        k                  lb/in     Spring Constant
             Energy   3.2609213 in·lb/in^ Energy Stored
```

```
(1i) Input:   35000                                                    25+/F9
Inconsistent
--------------------VARIABLE SHEET------------------------------------------
St-Input-----Name----Output----Unit------Comment----------------------------
    35000    Ss                 lb/in^2   Shear Stress
  >          K       .89482213            Wahl Factor
    10       F                  lb        Axial Load
  > 6        C                            Spring Index
    .0625    d                  in        Wire Diameter
             D                  in        Coil Diameter
             Defl               in        Spring Deflection
             n                            Number of Coils
    11500000 G                  lb/in^2   Modulus of Rigidity
    10       k                  lb/in     Spring Constant
             Energy             in·lb/in^ Energy Stored
```

(a)

```
(4s) Status: > Inconsistent                                            25+/F9

--------------------RULE SHEET---------------------------------------------
S-Rule---------------------------------------------------------------------
  "     * * *    Helical Spring Design    * * *

    Ss = K*8*F*C / (pi()*d^2)              " Shear Stress
  > (4*C-1)/(4*C-4) + 0.615/C = K          " Wahl Factor
  * C = D/d                                " Spring Index
  * Defl = 8*F*C^3*n / (d*G)               " Deflection
  * k*Defl = F                             " Axial Load
  * Energy = Ss^2 / (4*G)                  " Energy
```

(b)

FIGURE 7.19 An overdetermined problem. (a) The variables in the overdetermined equation are marked. (b) The offending rule is marked.

Another Problem Consider the problem solved in Figure 7.17a. Estimate the axial load (F) on the spring if the shear stress is not to exceed 35,000 lb/in^2.

We assign 35,000 to S as input and press F9, the Solve key. TK beeps and displays an error message as shown in Figure 7.19. We will use the VEC matrix and SP matrix to trace the solution path so that we can isolate the problem.

Tracing the Solution Path and Error When the Solve key is pressed, TK's Direct Solver starts processing the first equation in the Rule Sheet. In this equation, all variables except K are known (see Figure 7.20, SP matrix). Therefore, the Direct Solver

Pass no.	Eq. no.	S_s	K	F	C	d	D	Defl	n	G	k	Energy	Status
1		I		I	I	I				I	I		Begin
	(1)	I	E	I	I	I							S
	(2)		C		C								O
	(3)												U
	(4)									I			U
	(5)										I		U
	(6)										I		U

I = Input   E = Evaluated   R = Remaining   C = Contradictory   K = Known
S = Satisfied   U = Unsatisfied   O = Overdetermined

FIGURE 7.20 The SP matrix for an overconstrained problem.

evaluates K and displays its value, 0.89482213, in the Output field (Figure 7.19a). The rule status is changed from U to S, and the asterisk in the Status field of the rule disappears.

The Direct Solver moves on to the second rule. Here, both C and K are known (C has been specified as input, while K has been evaluated from the first rule). Therefore, the Direct Solver checks the rule for consistency, i.e., whether the left-hand side expression is equal to the right-hand-side expression for the values of C and K. In this case, the equality does not hold. Reason: From Figure 7.17a, when C = 6, K = 1.2525; however, K has been assigned .89842 in the first equation—mutually contradictory values. The Direct Solver terminates the solution process with a beep and an overdeterminacy message. The rule status is changed to O (for Overdetermined), and the variables are marked C (for Contradictory) in the SP matrix. On the Rule Sheet, a > appears in the Status field of the second rule; the variables in the rule are also similarly marked as shown in the figure. Since the solution process is terminated, the other equations are not solved, the rule status remains U and so do the asterisks in the Status field of the rules on the Rule Sheet.

Problem Analysis We failed to remove the value of F (the axial load) from its Input field; this oversight resulted in six variables being declared as input, instead of the required five,[1] causing the error. To remove the value of F from the Input field, we move the cursor to the Status field of F; the various options (**Input, Output, List, Guess,** and **Blank**) are displayed on the prompt/error line. By typing the first letter of these options, we can assign a specific status to the variable. For example, typing

---

[1] When developing a model, it is a good idea to analyze the equations using the following simple relationship. If NIE, NV, NIV, and NOV stand, respectively, for the number of independent equations, number of variables, number of input variables, and number of output variables, then,

$$NOV = NIE = NV - NIV.$$

In this model, there are 11 variables and 6 independent equations, i.e., NOV = 6, NIE = 6, and NV = 11; therefore, only NV − NIE (= 5) variables need to be specified as input for solution by the Direct Solver. However, we assigned input values to 6 variables which overconstrained the problem and caused the error (Figure 7.19).

B blanks the variable value, while typing an I will make the variable an input variable. If the variable has a value in the Output field, it is moved to the Input field, and if no value has been previously assigned to the variable, a zero will appear in its Input field.

Since we want to blank the input value of the variable, we press <u>B</u>; the value from the Input field of F disappears. When we press the Solve key, the solution shown in Figure 7.21 is obtained. The axial load drops to 7.1443 lb, and the spring deflection decreases to 0.7144 in.

Had we carried out the detailed analysis before rushing to solve the problem, we would not have encountered this error. However, even if we failed to analyze, as in the present example, TK spots the problem and keeps us honest.

Thus, how-can analysis, or solving different types of problems, i.e., problems with different combinations of input and output variables, is easy in TK. Since TK can evaluate an unknown variable regardless of its position in the equation (left- or right-hand side), there is no need to reformulate the equations or create new models for solving a different problem each time. Moreover, since TK restores the mathematical meaning of the equal sign, it is easy for us to interpret the errors encountered during problem solving.

7.5 ITERATIVE SOLVING

Solving systems of nonlinear and transcendental equations is a major characteristic of problems in science and engineering. Such equations are neither easily amenable to symbolic (algebraic) manipulation nor can they be solved by simple substitution, i.e., using the Direct Solver in TK.

FIGURE 7.21 How-can analysis: solution of a different problem.

```
(3s) Status:                                                    25+/OK
Status options: Input  Output  List  Guess  Blank
---------------------VARIABLE SHEET----------------------------------
St-Input-----Name----Output----Unit------Comment--------------------
   35000     Ss                 lb/in^2   Shear Stress
             K        1.2525              Wahl Factor
             F        7.1442884 lb        Axial Load
     6       C                            Spring Index
    .0625    d                  in        Wire Diameter
             D        .375      in        Coil Diameter
             Defl     .71442884 in        Spring Deflection
             n        41.594329           Number of Coils
  11500000   G                  lb/in^2   Modulus of Rigidity
     10      k                  lb/in     Spring Constant
             Energy   26.630435 in·lb/in^ Energy Stored
```

Consider the following nonlinear equation dealing with fractional conversion in a chemical reactor[1]:

$$y = \frac{x}{1-x} - 5 \ln\left(0.4 \frac{1-x}{0.4 - 0.5x}\right) + 4.45977$$

where x represents the fractional conversion of species A.

If $y = 0$, solving this equation for x is not a trivial exercise by simple substitution or symbolic manipulation. One possible approach is to generate a plot of x versus y for various values of x and to identify the point where the curve crosses the X axis, i.e., $y = 0$. The value of x represents the root.

Another approach is to substitute a value for x in the right-hand-side expression and equate the value of the expression to 0 (since the left-hand side $y = 0$). If the two sides are not equal, change the value of x incrementally, until the right-hand side equals the left-hand side — that value of x represents the root. This trial-and-error or approximation method — also known as an iterative procedure — can get very tedious when you are dealing with several nonlinear equations in several unknowns.

7.5.1 Need for Iterative Solving

Consider the following problem: Design a helical spring with a coil diameter of 5 in and a spring constant of 400 lb/in. The total axial load is 1800 lb, and the allowable shear stress is 35,000 lb/in^2; assume $G = 12 \times 10^6$ lb/in^2.

The analysis of the problem shows that the values of 5 of the 11 variables are known, and we have six equations to evaluate the six unknowns. Therefore, we assign the known values to the variables and solve the model. The resulting Variable and Rule Sheets are shown in Figure 7.22. Only the last two rules in the Rule Sheet have been solved resulting in the values for Defl and Energy. Though we have assigned values to the required number of variables, viz., five, the model has not been completely solved. Let's find out why.

Figure 7.23 shows the SP matrix for this problem. Consider the first pass of the Direct Solver: Since C, K, and d are unknown, the first equation is not solved; the second equation is not solved since C and K are unknown; likewise, the third equation is not solved since C and d are unknown. In the fourth equation, there are four unknowns, Defl, C, d, and n. Defl is evaluated from the fifth equation and Energy from the last equation. During the second pass, the status of none of the rules changes, and the Direct Solver terminates the solution process.

Therefore, even if the required number of variables is known (five, in this example), the Direct Solver cannot solve the set of equations if they are not the right combination of inputs, i.e., if the unknowns lead to a set of simultaneous equations in them. This is where we need to use the other problem-solving mechanism in TK, the **Iterative Solver** (see Capsule 7.1).

---

[1]From M. Shacham, "Numerical Solution of Constrained Nonlinear Algebraic Equations," *International Journal for Numerical Methods in Engineering,* **23** (1986):1455–1481.

```
(10i) Input:   400                                                    25+/OK

---------------------VARIABLE SHEET------------------------------------------
St-Input-----Name----Output----Unit------Comment----------------------------
   35000     Ss                lb/in^2    Shear Stress
             K                            Wahl Factor
   1800      F                 lb         Axial Load
             C                            Spring Index
             d                 in         Wire Diameter
   5         D                 in         Coil Diameter
             Defl    4.5       in         Spring Deflection
             n                            Number of Coils
   12000000  G                 lb/in^2    Modulus of Rigidity
   400       k                 lb/in      Spring Constant
             Energy  25.520833 in·lb/in^  Energy Stored
```

(a)

```
(4s) Status: * Unsatisfied                                            25+/OK

---------------------RULE SHEET---------------------------------------------
S-Rule----------------------------------------------------------------------
 "      * * *    Helical Spring Design    * * *
*  Ss = K*8*F*C / (pi()*d^2)           " Shear Stress
*  (4*C-1)/(4*C-4) + 0.615/C = K       " Wahl Factor
*  C = D/d                             " Spring Index
*  Defl = 8*F*C^3*n / (d*G)            " Deflection
   k*Defl = F                          " Axial Load
   Energy = Ss^2 / (4*G)               " Energy
```

(b)

FIGURE 7.22 Need for an iterative solution. (a) Five variables specified as input, only two evaluated. (b) Only two rules are solved (no asterisks in Status field).

7.5.2 Using the Iterative Solver

In the helical spring design example (Figure 7.22), we have a set of simultaneous equations in C and d. Therefore, we can declare one of these variables as a guess variable, assign it an initial value, and see how far the solution propagates. If the model is not resolved completely, we can select another guess variable and continue the solution process.

Let's begin with d as the guess variable. From a physical or design-manufacturing perspective, this is a logical choice as well because the wire diameter will influence

Pass no.	Eq. no.	Ss	K	F	C	d	D	Defl	n	G	k	Energy	Status
1		I	I			I			I	I			Begin
	(1)	I	I										U
	(2)												U
	(3)					I							U
	(4)				I					I			U
	(5)				I			E			I		S
	(6)	I									I	E	S
2		I	R	I	R	R	I	K	R	I	I	K	Continue
	(1)	I	I										U
	(2)												U
	(3)					I							U
	(4)				I					I			U
	(5)				I		I	K			I		S
	(6)	I									I	K	S

I = Input   E = Evaluated   R = Remaining   C = Contradictory   K = Known
S = Satisfied   U = Unsatisfied   O = Overdetermined

FIGURE 7.23   SP matrix for problem in Figure 7.22.

CAPSULE 7.1

ITERATIVE SOLVING IN TK

What is iteration?

Finding the roots of a polynomial generally involves a trial-and-error process known as **iteration**. An initial guess value for the unknown root is used as the starting point to carry out a series of computations that will improve the solution until the desired solution or root is attained. This step-by-step process of approaching the solution is known as **convergence**; the measure used to determine whether a solution has been reached is known as the **convergence criterion**. These computations are carried out using some well-known algorithms. When the initial guess value does not lead to a solution but away from it, the process is said to be **diverging**.

Therefore, the success of such problem solving that requires iteration, or iterative problem-solving, depends to a large extent on the suitability of the initial guess value. This is especially true for multidimensional problems, i.e., problems requiring iterative solutions for multiple unknowns.

TK's Iterative Solver

There are several well-known algorithms or procedures for carrying out iterative computations. These include Newton's method, Bisection method, Regula Falsi method, and the Newton-Raphson method. At the heart of the Iterative Solver in TK is a modified version of the Newton-Raphson (NR) procedure.* Once the need for iterative solving is established, invoking the Iterative Solver is fairly simple and straightforward. We need to select one or more of the unknowns as guess variables (by typing a G in the Status field), assign starting or initial values to them (in the Input field), and press the Solve key, F9. We can also change the default convergence criterion (comparison tolerance) and the number of iterations (10) — both on TK's Global Sheet — if necessary.

Working of the Iterative Solver

The Iterative Solver uses the initial guess values to carry out the sequence of computations (based on the NR procedure) and attempts to satisfy the convergence criterion. It terminates the solution process when either the convergence criterion is met or the number of iterations is reached. If the convergence criterion is met, the value of the guess variable is moved to the Output field indicating that a solution has been achieved. Otherwise, the guess

*For details, see chapter 9 in Konopasek and Jayaraman, *op. cit.*

value stays in the Input field denoting a failure to converge. In this case, the last value can be used as the starting value to trigger another set of iterations. And, if the value is headed in the right direction, a few additional iterations may indeed lead to convergence. Thus, as users, we are shielded from the intricacies and problems associated with writing stand-alone programs based on the NR procedure (or other iteration algorithms) for solving sets of simultaneous nonlinear and transcendental equations.

Selecting guess variables

Selecting the right guess variable(s) for the Iterative Solver is an important task. Declaring one or more of the variables not evaluated by the Direct Solver as guesses is an obvious first step. However, there is no need to select all the unknown variables as guesses; in fact, this is not advisable since the time for one iteration step is approximately proportional to the number of guess variables. Rather, you should select one guess variable at a time, invoke the Iterative Solver, and see how far the solution propagates. Moreover, the selection of the right variable as the guess variable becomes important when assigning an initial or starting value to it. For this reason, you should select those unknowns as guesses for which you have an idea of the approximate values or solution. A systematic approach would be to select the variable that would result in the solution of the maximum number of equations. If all the unknown variables are not evaluated, another guess variable should be selected based on a similar criterion, and so on. Analysis of the VEC matrix to see how the solution will propagate will also help you in the selection of guesses.

Selecting values for guess variables

Selecting the right starting or initial values for the guess variables is as important as selecting the right guess variables. The closer the initial guess value is to the final solution, the better the chances of the Iterative Solver converging to it, and the fewer the required number of iteration steps. Where multiple roots are involved, the resulting solution depends very much on the starting value. For example, a cubic equation will require three different starting guess values for the three roots. In some instances, the guess values may not lead to a solution, but away from it.

TK is a powerful tool for solving the problem at hand; however, it cannot take our place in selecting initial values for the guess variables. We need to have an understanding of the problem domain, especially the physical significance of the various elements of the model and some idea of the range of feasible solutions, and use this knowledge to select starting values for the guess variables. This is especially critical when solving nonlinear simultaneous equations requiring two or more guess variables.

For solving sets of simultaneous linear equations, the selection of initial values is not important; even a guess value of zero will work well and the solution reached in few iteration steps.

the other parameters. Moreover, we know the range of wire diameters typically used, and this will make a good initial guess for iteration. Quite often, in problem solving, when you are playing with different combinations of inputs and outputs in a model, you can use a previous solution as a good starting point for the current solution and so on. As far as iterative solving is concerned, this is probably the best way to ensure reasonable starting values for the guess variables.

Declaring Guess Variable and Assigning Initial Value We move the cursor to the Status field of d and type G to declare d as the guess variable. The value of d from the previous problem, viz., .0625, appears in the Input field. TK has been designed to remember the previous value of the variable because it will serve as a good starting value for iteration. If no value had previously been assigned to a guess variable, a value of zero appears in the Input field. Of course, any value can be entered in the Input field. Since the axial load (F) and the coil diameter (D) are considerably higher than in the previous example, it is likely that the wire diameter (d) will also be higher. So, let's start with an initial guess of $\frac{1}{4}$ in. We type 1/4 in the Input field of d and press the Enter key. The Variable Sheet looks as shown in Figure 7.24a. We retain TK's default convergence criterion (.000001) and number of iteration steps (10).

```
(5i) Input:    .25                                                    25+/F9

--------------------VARIABLE SHEET------------------------------------
St-Input-----Name----Output----Unit------Comment---------------------
   35000       Ss                lb/in^2    Shear Stress
               K                            Wahl Factor
   1800        F                 lb         Axial Load
               C                            Spring Index
 G .25         d                 in         Wire Diameter
   5           D                 in         Coil Diameter
               Defl   4.5        in         Spring Deflection
               n                            Number of Coils
   12000000    G                 lb/in^2    Modulus of Rigidity
   400         k                 lb/in      Spring Constant
               Energy 25.520833  in·lb/in^  Energy Stored
```

(a)

```
(5i) Input:    .25                                                    25+/F9
Iterative Solver
--------------------VARIABLE SHEET------------------------------------
St-Input-----Name----Output----Unit------Comment---------------------
   35000       Ss                lb/in^2    Shear Stress
               K      1.0702237             Wahl Factor
   1800        F                 lb         Axial Load
               C      20                    Spring Index
 G .25         d                 in         Wire Diameter
   5           D                 in         Coil Diameter
               Defl   4.5        in         Spring Deflection
               n      .1171875              Number of Coils
   12000000    G                 lb/in^2    Modulus of Rigidity
   400         k                 lb/in      Spring Constant
               Energy 25.520833  in·lb/in^  Energy Stored
```

(b)

FIGURE 7.24 Iterative solving. (a) Selecting a guess variable and initial value. (b) Iterative Solver triggered. (c) Iterative Solver, second step. (d) Solution after 10 iterations.

To trigger the solution, we press the Solve key. Due to the presence of a guess variable, the Iterative Solver is invoked, and a message "Iterative Solver" appears on the prompt/error line as shown in Figure 7.24b. This is immediately followed by the message in Figure 7.24c indicating that the Iterative Solver is in the second iteration step. The phrase "Convergence" followed by the numeric value 28.06 is TK's estimate of how far it is away from converging to the final solution. Note the intermediate values of the different variables in this iteration step. These values will change at each

```
(5i) Input: .3333539798542821                                      25+/F9
Iteration: 2, Convergence: 28.06
---------------------VARIABLE SHEET---------------------------------------
St-Input-----Name----Output----Unit------Comment--------------------------
    35000    Ss                lb/in^2   Shear Stress
             K        1.0945775           Wahl Factor
    1800     F                  lb        Axial Load
             C        14.999071           Spring Index
 G  .33335398 d                 in        Wire Diameter
    5        D                  in        Coil Diameter
             Defl     4.5       in        Spring Deflection
             n        .37046214           Number of Coils
    12000000 G                 lb/in^2   Modulus of Rigidity
    400      k                  lb/in     Spring Constant
             Energy   25.520833 in·lb/in^ Energy Stored
```

(c)

```
(5i) Input:                                                        25+/OK

---------------------VARIABLE SHEET---------------------------------------
St-Input-----Name----Output----Unit------Comment--------------------------
    35000    Ss                lb/in^2   Shear Stress
             K        1.2912287           Wahl Factor
    1800     F                  lb        Axial Load
             C        5.2876669           Spring Index
             d        .94559663 in        Wire Diameter
    5        D                  in        Coil Diameter
             Defl     4.5       in        Spring Deflection
             n        23.985287           Number of Coils
    12000000 G                 lb/in^2   Modulus of Rigidity
    400      k                  lb/in     Spring Constant
             Energy   25.520833 in·lb/in^ Energy Stored
```

(d)

FIGURE 7.24 (continued).

step of the iteration and so will the convergence factor; the factor gets smaller and smaller as the Iterative Solver approaches the solution:

Iteration: 3   Convergence: 27.19
Iteration: 4   Convergence: 26.3
          ...
Iteration: 9   Convergence: 17.38

Thus, the convergence factor can serve as a helpful indicator in determining if the guess value is leading to a solution or diverging away from it. At the end of 10 iteration steps, the solution shown in Figure 7.24d is obtained.

The wire diameter is 0.94559 in, and the number of coils in the spring is 23.985. Since all the variables have been evaluated, the single guess variable has proven to be adequate. Moreover, these values are in the feasible range and make physical sense: The wire diameter is higher to support the higher axial load. Thus, we have used TK's Iterative Solver for tackling one more helical spring design problem. This example also illustrates the versatility of TK in its ability to solve different types of problems, problems with different combinations of input and output variables.

7.5.3 Other Conditions Requiring Iterative Solution

TK's strengths lie in numeric problem solving and not in the symbolic manipulation of algebraic equations. Consequently, the Direct Solver cannot evaluate, or solve for, an unknown from the equation in which it occurs more than once. This situation is denoted by the presence of # in the VEC matrix for the model. For example, in the fourth-order polynomial, $X^4 - 3X^3 + 2X^2 + X = 6$, the variable X occurs four times and so cannot be evaluated by the Direct Solver. We need to solve this equation iteratively, guessing a value for X, equating the left- and right-hand sides and modifying the guess value till the equality is satisfied. The Iterative Solver in TK greatly simplifies this process.

When to Use Iterative Solving To summarize, we need to use the Iterative Solver in TK under the following two conditions:

a The unknown variables are involved in a set of simultaneous equations.
b The unknown variable occurs more than once in the equation from which it needs to be evaluated.

Having used TK for creating and playing with a model, let's take a closer look at some of the commands in TK.

7.6 TK'S COMMAND STRUCTURE

As we discussed in Section 7.2.2, some frequently used commands in TK are displayed at the bottom of the screen. In addition to these commands invoked by typing one of the function keys (F1,...,F10), TK has a whole range of well-structured commands for carrying out a host of other functions: blanking and copying fields; inserting, deleting, and moving rows; printing and saving the sheets; resetting the sheets; and quitting TK. The command menu is displayed by typing /, the Slash key; note that the menu in a spreadsheet is also displayed generally using the Slash key.

7.6.1 The Slash Command Menu

With the cursor in the Variable Sheet, we invoke the command menu by typing /, the Slash key. TK displays the menu shown in Figure 7.25 with the cursor highlighting

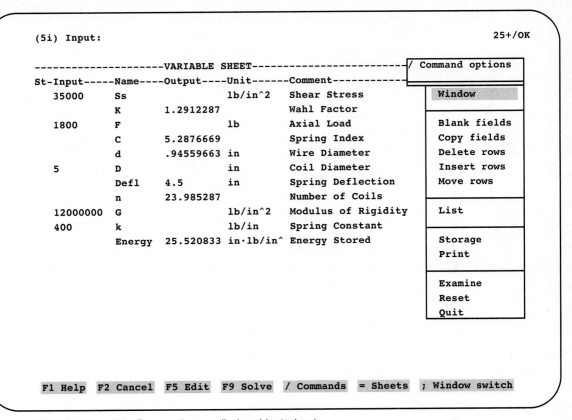

FIGURE 7.25 TK's Command menu displayed by typing /.

Window, the first option in the **Command menu.** When we press the Down Arrow key, the cursor moves down to the next command in the menu and highlights it. Similarly, the Up Arrow key can be used to move the cursor up the command menu. Since the Slash key is pressed to invoke the various commands, they are collectively referred to as **Slash commands** and the menu as the **Slash command menu.** The options in the menu are shown in Quick Reference Table 7.1.

Selecting a Command We can select a specific command in one of two ways from the menu. We can move the cursor to the command and press the Enter key. Alternatively, we can type the first character of the command, e.g., C to select the Copy command. When we select a specific command from the menu, either its submenu is displayed or prompts for executing the command are displayed on the prompt line.

For example, with the cursor highlighting Window, we press the Enter key. The resulting screen display with the Window command submenu is shown in Figure 7.26a. The **Window command** enables us to display either a single window or two windows on the screen. This means we can simultaneously view two different sheets or two

TABLE 7.1 QUICK REFERENCE TABLE FOR THE SLASH (/) COMMAND.

Editing commands	I/O
B Blank fields C Copy fields D Delete rows I Insert rows M Move rows	S Storage P Print
Changing display	**Other**
W Window	E Examine R Reset Q Quit

segments of the same sheet in the two windows. Alternatively, we can view only one sheet in a single window. Since the screen is currently displaying a single window (Figure 7.26a), selecting one of the options will display the chosen sheet in the bottom window. We type R to select the Rule option. The Rule Sheet is displayed in the bottom window as shown in Figure 7.26b. To go back to the Variable Sheet, we invoke the Slash command menu and select the Window command and the 1 Window option. A shortcut is to press the function key, F6 (see Figure 7.8).

Speeding up the Command Selection Process   Once you become proficient with TK's command structure, there is no need to go through the individual menus. You can just type / followed immediately by the first letters of the commands. For example, /WR would have resulted in the display of the Rule Sheet in the bottom window.

Cancelling a Command   There are times when you invoke the Command menu and then decide not to execute any of the commands. To quit the Command menu, you press F2, the Cancel key; TK exits the command mode, the menu disappears, and you are back in the sheet.

Invoking the Help Facility   Let's say the cursor is highlighting the Copy command in the Slash command menu and we want to find out how the command should be used. We press F1, the Help key (see Function key layout in Figure 7.8). TK displays the help screen shown in Figure 7.27a — information on the usage of the Copy command. This type of help facility is said to be **context-sensitive** because the help information displayed by TK pertains to the current context, viz., the cursor is highlighting the Copy command in the menu. Similarly, if the cursor were highlighting the Move command, the Help screen for the Move command would be displayed. The main advantages of a context-sensitive Help facility are that it is fast and provides information that will be useful to the context; you don't have to wade through the entire Help system to obtain the desired information.

Navigating through the Help System   At the top of the Help screen in the figure, the various options in the Help menu are displayed. The **PgUp** and **PgDn** options scroll the screen display (up or down) a page at a time, while **Index** displays the Help index. The Index can be used for selecting help on a particular topic or command. At

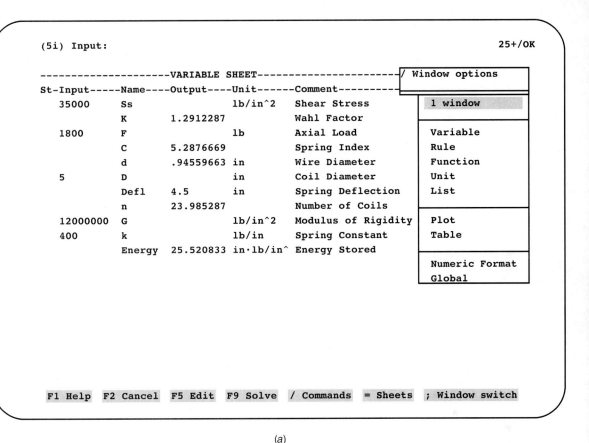

(a)

FIGURE 7.26 The /Window command. (a) The options under the Window command. (b) The Rule Window option: Rule Sheet in the bottom window.

the bottom of the screen, a list of related topics is displayed. By selecting one of these topics, additional help on the chosen topic can be obtained. This facility comes in handy if you are not a regular user of TK and want to be refreshed on its usage. For example, when we select the Editing commands option from the bottom of Figure 7.27a, the screen shown in Figure 7.27b is displayed. We can quit the Help mode either by selecting the **Quit** option from the top of the Help screen or press F2 to cancel the Help command.

Thus the Help facility in TK can be used not only as an assistant for navigating through the program but also for learning about its various commands and features.

7.6.2 Printing the Sheets

As shown in Figure 7.10, the next step in problem solving with TK is to present the results in the desired format. Since we have not generated any graphs or tables up to this point, the appropriate output would be the printed Variable and Rule Sheets. We can print the sheets using the Print command in TK.

```
(1r) Rule: Ss = K*8*F*C / (pi()*d^2)              " Shear Stress              25+/OK

---------------------VARIABLE SHEET-------------------------------------------
St-Input-----Name----Output----Unit-----Comment------------------------------
    35000     Ss                lb/in^2  Shear Stress
              K       1.2912287          Wahl Factor
    1800      F                 lb       Axial Load
              C       5.2876669          Spring Index
              d       .94559663 in       Wire Diameter
    5         D                 in       Coil Diameter
              Defl    4.5       in       Spring Deflection
              n       23.985287          Number of Coils
    12000000  G                 lb/in^2  Modulus of Rigidity
---------------------RULE SHEET----------------------------------------------
S-Rule-----------------------------------------------------------------------
  Ss = K*8*F*C / (pi()*d^2)           " Shear Stress
  (4*C-1)/(4*C-4) + 0.615/C = K       " Wahl Factor
  C = D/d                             " Spring Index
  Defl = 8*F*C^3*n / (d*G)            " Deflection
  k*Defl = F                          " Axial Load
  Energy = Ss^2 / (4*G)               " Energy


F1 Help  F2 Cancel  F5 Edit  F9 Solve  / Commands  = Sheets  ; Window switch
```

(b)

FIGURE 7.26 (continued).

Using the Print Command TK lets you print one sheet at a time, and the output can be sent to a printer or to a text file. The printing can be controlled by the printer settings on TK's Global Sheet, i.e., page dimensions, page breaks, and so on. In most cases, the default settings will work fine. The address of the printer port (e.g., LPT1) should be entered in the Global Sheet to route the output to the printer.

With the cursor highlighting the first row of the Variable Sheet, we invoke the Slash command menu by typing /. To select the Print option, we type P. The resulting screen display is shown in Figure 7.28*a*. The first row is highlighted, and, on the prompt/error line, TK prompts us for the last line to be printed. It also displays the address or current location of the cursor, viz., 1s, along with the prompt.

We can select the range of rows to be printed by highlighting it using the Arrow keys. As the cursor moves, the rows are highlighted and the address of the cursor changes in the prompt/error line. Since we want to print the entire Variable Sheet, we can move the cursor to the last line by just pressing the End key. Figure 7.28*b* shows the resulting screen display with all the rows highlighted and the new address in the

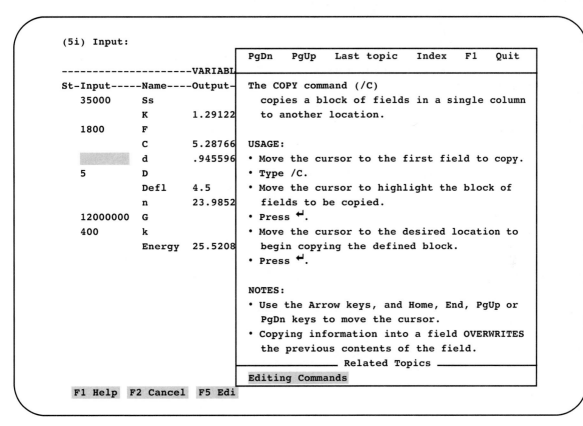

(a)

FIGURE 7.27 Context-sensitive Help facility. (a) Cursor highlights Copy option. Help key is pressed. (b) More help on associated editing commands.

prompt/error line. When we press the enter key, the highlighted rows are printed, and the output is shown in Figure 7.28c.

Generating a Text File Instead of sending the output to the printer, we can route it to a text file. To do this, we need to enter the name of the file (including the path, if necessary) in the Printer Output field of the Global Sheet. The resulting output can be imported into a word processing program. In fact, most of the figures in Chapters 7, 8, and 9 were produced using this technique. TK adds .PRF as the extension to the specified filename. To append more than one sheet to the same text file (e.g., Rule Sheet to the Variable Sheet), a + should be suffixed to the filename in the Global Sheet.

7.6.3 Saving the Model

The final step in problem solving with TK is to save the model for subsequent retrieval and use (Figure 7.10). The Storage command in TK can be used to save the model. Alternatively, the sheets (with a few exceptions) can be saved individually and loaded

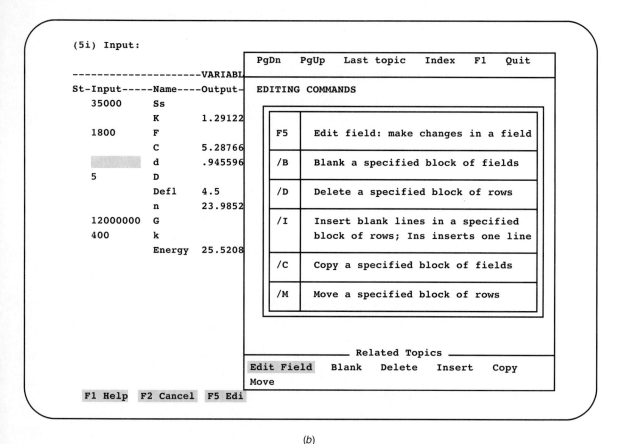

FIGURE 7.27 (*continued*).

into other models. For example, you could define the unit conversions you normally use and save the Unit Sheet by itself. You can then load this Unit Sheet into other models. This way, you don't have to reenter the unit conversions in every model you create. Not only does this save time, you also minimize the risk of making errors while defining such conversions.

The Storage command can be invoked from any sheet in TK. We will illustrate the command by saving the helical spring design model.

Using the Storage Command We invoke the Slash Command menu by typing /. TK displays the top-level Command menu shown earlier in Figure 7.25. To select the Storage command, we type S̲. TK displays the options in the Storage command as shown in Figure 7.29a. As with any of the Slash commands, a specific option can be selected by typing the first character of the option or by moving the cursor to it and pressing the Enter key.

Storage Command Options As shown in the figure, the three major options in the Storage command are: **Load** a model, **Save** the entire model or select sheets in the

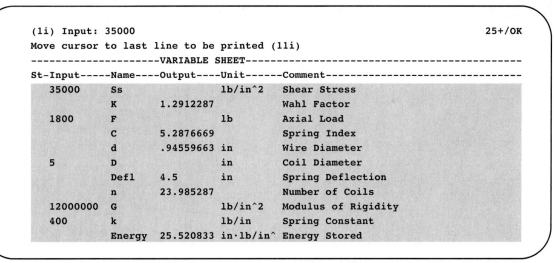

FIGURE 7.28 The Print command (/P). (a) Prompt/error line displays current location of cursor. (b) Cursor at the last line to be printed. (c) Printer output.

model, and **Delete** a file. The Load File option is used for loading a previously stored TK model or sheet. You select this option by typing L from the Slash Command menu; TK displays the prompt for a filename. When you type the name of an existing file and press the Enter key, TK will load that model or sheet. The new information in the file is appended to the existing contents, if any, in TK. For example, if you are

```
-------------------VARIABLE SHEET----------------------------------------
St-Input-----Name----Output----Unit------Comment-----------------------------
   35000      Ss                 lb/in^2    Shear Stress
              K       1.2912287             Wahl Factor
   1800       F                  lb         Axial Load
              C       5.2876669             Spring Index
              d       .94559663  in         Wire Diameter
   5          D                  in         Coil Diameter
              Defl    4.5        in         Spring Deflection
              n       23.985287             Number of Coils
   12000000   G                  lb/in^2    Modulus of Rigidity
   400        k                  lb/in      Spring Constant
              Energy  25.520833  in·lb/in^  Energy Stored
```

(c)

FIGURE 7.28 (continued).

playing with a model and use the Storage command to load another model, the current contents (variables, rules, etc.) are not lost. This feature greatly helps you in merging models and/or in building models in stages. Note that this is in contrast to the Load File command in a spreadsheet, where typically the existing contents are lost when you load a new file.

The Save Model option is used for saving the entire model. This is the most frequently used of the different Save options. To save the helical spring design model, we type S in response to the Storage Command options. TK displays the prompt for the file name as shown in Figure 7.29b. We enter the name of the file, including the path name, c:\wp\book\spiter and press the Enter key. TK adds the extension .tk to the filename and saves the model. This model can be loaded at a subsequent time using the Load option of the Storage command.

The Variable Save option of the Storage command can be used to save the contents of the Variable Sheet alone on to a disk. Likewise, the Function, Unit, and other Sheet Save options can be used to save the individual sheets for use in other models. Thus, the Storage command gives you the flexibility to build and use models in a modular fashion.

The Delete File option of the Storage command can be used for deleting a file from the disk. It is invoked by typing D from the Slash Command menu. When you specify a filename in response to TK's prompt for a filename, TK seeks confirmation prior to erasing the file. This minimizes accidental deletions of models and files.

Miscellaneous Options The remaining options of the Storage command, shown in Figure 7.29a, are used for exchanging information with other programs. For example, when a .WKS file is created by TK, it can be used by Lotus's spreadsheet program 1-2-3. The #DIF option can be used for exchanging data with programs conforming to the DIF standard (an early standard) defined by Software Arts. The options under the Storage command are shown in Quick Reference Table 7.2.

```
(1i) Input: 35000                                              25+/OK

--------------------VARIABLE SHEET------------------ ┌─Storage options──┐
St-Input-----Name----Output----Unit-----Comment----  │ Load file        │
   35000     Ss                lb/in^2  Shear Stress │                  │
             K       1.2912287          Wahl Factor  ├──────────────────┤
   1800      F                 lb       Axial Load   │ Save model       │
             C       5.2876669          Spring Index │ Variable save    │
             d       .94559663 in       Wire Diameter│ Function save    │
   5         D                 in       Coil Diameter│ Unit save        │
             Defl    4.5       in       Spring Deflecti│ Numeric format save │
             n       23.985287          Number of Coils│ Plot save      │
   12000000  G                 lb/in^2  Modulus of Rigi│ Table save     │
   400       k                 lb/in    Spring Constant├────────────────┤
             Energy  25.520833 in·lb/in^ Energy Stored │ # DIF storage  │
                                                      │ WKS storage     │
                                                      │ ASCII storage   │
                                                      │ Configuration save│
                                                      ├──────────────────┤
                                                      │ Delete file      │
                                                      └──────────────────┘
```

(a)

```
(1i) Input: 35000                                              25+/OK
Save; filename: \wp\book\spiter
--------------------VARIABLE SHEET----------------------------------------
St-Input-----Name----Output----Unit-----Comment--------------------------
   35000     Ss                lb/in^2  Shear Stress
             K       1.2912287          Wahl Factor
   1800      F                 lb       Axial Load
             C       5.2876669          Spring Index
             d       .94559663 in       Wire Diameter
   5         D                 in       Coil Diameter
             Defl    4.5       in       Spring Deflection
             n       23.985287          Number of Coils
   12000000  G                 lb/in^2  Modulus of Rigidity
   400       k                 lb/in    Spring Constant
             Energy  25.520833 in·lb/in^ Energy Stored
```

(b)

FIGURE 7.29 Saving a model. (a) Options under the Storage command (/S). (b) The Save option: prompt for filename.

TABLE 7.2 QUICK REFERENCE TABLE FOR STORAGE (/S) COMMAND.

File options	Sheet options	Data transfer
L–Load file	**F**–Functions save	**W**–WKS storage
S–Save model	**U**–Units save	**A**–ASCII storage
D–Delete file	**N**–Numeric Format save	**C**–TK.CFG save
	P–Plot save	**#**–DIF save
	T–Table save	

7.6.4 TK Sheets and the Select Command

As we saw in Figure 7.5, TK has a set of nine sheets. The **Select command** can be used for selecting and displaying a specific sheet in the current window.

Using the Select Command With the cursor in the Variable Sheet, we invoke the Select command by pressing =. The menu with the various sheet options is displayed as shown in Figure 7.30. As with the Slash commands, to select a particular option, we can move the cursor to that option and press the Enter key or type the first letter of the option. For example, to display the Rule Sheet, we type R. The Variable Sheet on the screen is replaced by the Rule Sheet (see Figure 7.17*b*).

Subsheets in TK Several of the sheets in TK have associated subsheets as shown in Figure 7.31. For example, a Variable Subsheet is associated with each variable on the Variable Sheet and provides additional information about the variable. A subsheet can be accessed from its parent sheet by pressing >, the **Dive command.** The sheet display in the window is replaced by the subsheet. Conversely, to return to a sheet from its subsheet, you must use the **Return** or **Surface command,** press <.

FIGURE 7.30 The Select Command options (command invoked by typing =).

```
(1i) Input: 35000                                            25+/OK

---------------------VARIABLE SHEET-------------------- = Sheet options
St-Input-----Name----Output----Unit------Comment--------
    35000    Ss                lb/in^2   Shear Stress      Variable
             K       1.2912287           Wahl Factor       Rule
    1800     F                 lb        Axial Load        Function
             C       5.2876669           Spring Index      Unit
             d       .94559663 in        Wire Diameter     List
    5        D                 in        Coil Diameter
             Defl    4.5       in        Spring Deflection Plot
             n       23.985287           Number of Coils   Table
    12000000 G                 lb/in^2   Modulus of Rigidity
    400      k                 lb/in     Spring Constant   Numeric Format
             Energy  25.520833 in·lb/in^ Energy Stored     Global
```

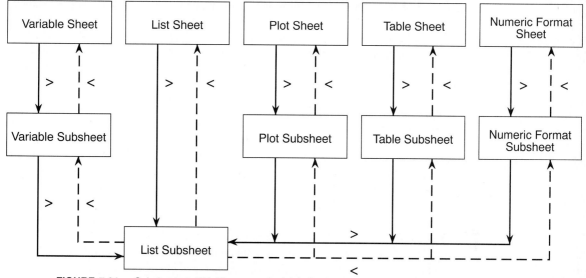

FIGURE 7.31 Subsheets in TK. Dive command (>): Go down to a subsheet. Return command (<): Come up from a subsheet.

7.6.5 Resetting and Exiting TK

When you finish playing with a model, it is not necessary to exit and reload TK in order to work with a different model. The **Reset option** of the Slash command lets you clean up the environment without exiting TK. When you select the Reset option, TK provides three choices: reset the variable values (/RV), reset the sheet on which the cursor is located when the command is invoked (/RS), or reset the entire model, i.e., clean up the complete environment (/RA). Before executing the command, TK seeks confirmation for all the three options.

The Quit Command The **Quit** option of the Slash Command menu lets you exit TK and return to the operating system. However, before invoking the Quit option, you should save your work. Otherwise, it will be lost. Since we have already saved the model, we press / and choose the Quit option from the menu. TK seeks confirmation to exit the program. This ensures that you don't accidentally leave TK and lose your work. When we respond with a Y to confirm, the TK screen disappears, and we are returned to the operating system.

7.7 A TOUCH OF HISTORY

TK's origins go back to the work of Milos Konopasek in the late 1960s in Manchester, United Kingdom, where he was developing, in today's terminology, an expert system for the textile professional. Faced with the daunting task of increasing the computer's role (both quantitatively and qualitatively) in the problem-solving process, Konopasek developed the "Question Answering System," or QAS, for short in 1972. QAS let the user communicate with the computer not through programs in conventional languages

(FORTRAN/ALGOL at that time), but in the engineer's natural language of mathematical relationships (algebraic equations) and related databases. QAS restored the mathematical meaning of the equals (=) sign in an algebraic equation, a concept lost to programmers in FORTRAN and other conventional languages where the equals sign was an assignment operator. It was initially implemented on PDP-10 and other mainframe systems and much later on microcomputers.[1]

Software Arts and TK!Solver In 1980, the author of this book had started working as a graduate student with Konopasek and was excited by QAS and how it helped an engineer take advantage of the power of the computer for solving complex problems. At this time, VisiCalc, the first electronic spreadsheet for personal computers, was beginning to revolutionize the personal computer industry. It became clear that bringing QAS to the world of personal computers would advance the spreadsheet concept a step further; Software Arts, the creator of VisiCalc, was the ideal medium to accomplish this goal. Thus began the association between Konopasek, the author, and Software Arts, which led to the development of TK!Solver, announced in May 1982. In addition to Milos Konopasek, the author, Dan Bricklin, and Bob Frankston, the other major participants during the design and development phase were Seth Steinberg, Jim O'Dell, David Reed, Diane Curtis, and Melinda Mayer and later Fred Helenius.

In 1985, Software Arts was acquired by Lotus, and the rights to TK were subsequently acquired by Universal Technical Systems. Problem solving in science and engineering involves both declarative and procedural computations.[2] So, a major step in the evolution of TK was inclusion of the procedural programming capability in the form of user-defined procedure functions in TK Solver Plus, released in 1987.

SUMMARY

We introduced the basics of an equation solver and developed a structured methodology for using equation solvers in quantitative problem solving. We used this methodology to solve problems related to the design of helical springs. This methodology is generic and can be easily adopted for problems in other fields. We explored the Direct and Iterative Solvers, the two major problem-solving mechanisms in TK. The helical spring model shows that once you develop a model, TK lets you concentrate on the creative aspects of the problem without worrying about modifying the equations in the model or writing programs every time you want to explore a different aspect.

We then looked at TK's menu-based command structure and tried a few of the commands. Finally, we provided a glimpse of the evolution of equation solvers.

At this stage, you should be comfortable using an equation solver such as TK. In the next chapter, we will explore other features and capabilities of equation solvers with appropriate examples. Following that, we will present several application examples as case studies to further illustrate the uses of equation solvers.

[1] For a historical perspective of TK, see M. Konopasek and S. Jayaraman, "Expert Systems for Personal Computers: The TK!Solver Approach" in *Insights into Personal Computers,* A. Gupta and H. Toong (eds.), IEEE Press, New York, 1985. The extensive bibliography provides references to early papers on QAS/TK.

[2] M. Konopasek and S. Jayaraman, "Constraint and Declarative Languages for Engineering Applications: The TK!Solver Contribution," *Proceedings of the IEEE,* **73** (no. 12) (December 1985):1791–1806.

CHAPTER 8

EQUATION SOLVING AND ENGINEERING DESIGN AND ANALYSIS OF BEAMS

In this chapter, we continue the exploration of other features in TK Solver Plus that render it a very powerful tool for engineering problem solving. In creating a model for the engineering design of beams, we look at defining conversions between units of measurement, List Solving, plotting, and tabular display of results. We then discuss the ability to define relationships between sets of values with the user-defined function feature in TK. Finally, we cover conditional rules (IF-THEN construct) in TK along with some important built-in functions. By the end of this chapter, you should have gained a good appreciation for the power of an equation solver as a problem-solving tool and be fairly conversant with its major features.

8.1 ENGINEERING DESIGN

Engineering is often defined as the application of scientific principles and knowledge for practical purposes. As engineers, we are constantly engaged in engineering design, whether it be the design of a bridge, a stadium dome, a machine, or even a process. Engineering design, the essence of engineering, is a process that typically goes through several phases: requirements analysis and problem definition, information gathering, synthesis, analysis, evaluation, and refinement-optimization of the design.

8.1.1 Characteristics of the Design Process

The design process is creative and requires the knowledge and combined application of mathematical relationships and databases of subject-specific information (e.g., stress-strain behavior of structural materials)—within specific constraints (resources, dimensions, and/or cost)—to generate concrete or feasible solutions to given problems.

The synthesis-analysis-evaluation-refinement cycle is iterative in nature: We change the values of one or more variables and assess the impact on the others and repeat this process until the requirements are met under the specified constraints. To interpret and understand the relationships, we normally create tables and plots of the results.

Units of measurement are an integral part of all mathematical relationships. As engineers, we are constantly dealing with different sets of units, British and SI, for example, and need to quickly convert from one unit to another. This is especially true in engineering design where the product is likely to be designed to meet the dimensions specified in one set of units of measurement that may differ from the units adopted in manufacturing.

Additionally, in the design process, we make use of empirical values generated in the past from experiments and observations.

Problem-Solving Tool Requirements From the characteristics of the design process, it becomes clear that the problem-solving tool should provide the following features:

a The flexibility to handle different combinations of knowns and unknowns
b The ability to explore different sets of values for specific combinations of knowns and unknowns
c The ability to "shoot" for one or more variable values, generally using iteration
d The facility to define and use conversions between units of measurement
e The facility to create and use databases of subject-specific information and to define relationships between sets of empirical values
f The ability to generate tabular and graphical outputs

Here is where TK's prowess beyond being just an equation solver comes in. As we saw in Section 7.4, TK's Direct and Iterative Solvers can handle sets of equations for different combinations of known and unknown variables with ease. In addition, TK's Unit Sheet can be used to define conversions between units of measurement, while the user-function feature can be used for creating and utilizing databases. Finally, TK's Plot and Table Sheets can handle graphical and tabular outputs. In other words, TK provides the major features desired in an engineering design tool. We will utilize some of these features while developing a model for the engineering design of beams.

8.1.2 Engineering Design of Beams

A **beam** is defined as a structural member that is reasonably long compared with its lateral dimensions when suitably supported, and subjected to forces or couples that lie in a plane containing the longitudinal axis of the member. A beam that is freely supported at both ends is referred to as a **simply supported beam** (see Figure 8.1). This means that the end supports are capable of exerting only forces upon the beam and are not capable of exerting any moments. A beam with a fixed or restrained end, usually built into a wall, is referred to as a **cantilever beam.**

A beam is typically subjected to a load in one of three ways: a concentrated load at a point, a uniformly distributed load, or an applied couple. The response of the beam

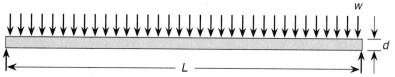

FIGURE 8.1 A simply supported beam: uniformly loaded.

will vary with the type and magnitude of loading. Engineering design of beams consists of selecting the beam of the right dimensions (e.g., length, width, and depth) and of the right material (e.g., wood, steel, or aluminum) to support a given load without exceeding the safe limits of deflection, tensile, and bending stresses.

Example Consider the following problem: A 5.5- by 11.5-in simply supported rectangular beam, 12 ft long, is subjected to a uniformly distributed load of 50 lb/in. If the elastic modulus of the beam material is 1.6×10^6 lb/in^2, estimate the deflection, bending moment, and shearing force at a distance of 3 ft from the end of the beam. Also, calculate the maximum for each of three variables.

8.1.3 Problem Analysis and Model Design

Following our problem-solving methodology discussed in Chapter 2, the first step is to develop the mathematical model. Figure 8.2 shows the mathematical relationships for a simply supported beam subjected to a uniformly distributed load along its length. The equations can be derived from first principles or obtained from any textbook on strength of materials.[1]

There are 19 variables and 13 equations in the model, i.e., $NV = 19$, $NIE = 13$. Therefore, from the relationship discussed in Section 7.4.2, viz., $NOV = NIE = NV - NIV$, we should know the values of six variables (NIV) for solving the model. From the problem statement, we know that six variables have been specified; therefore, a solution can be attempted. However, to determine if they are the right set of six variables, we can create a VEC matrix for the model (see Section 7.4.1) and analyze it. An alternate technique is to draw and analyze a Relationship, or R graph, for the model.

Relationship Graph or R Graph In a Relationship or R graph, each equation or rule is represented by a polygon with nodes for each of the variables. Therefore, if there are only two nodes, the polygon reduces to a straight line connecting the two nodes. An example of this is the fourth rule in the model ($c = d/2$) which is shown in Figure 8.3a. If a particular variable cannot be evaluated from a particular equation (by the Direct Solver), the nonresolvability is denoted by a mark around the node. In Figure 8.3b, the mark around node x indicates that x cannot be evaluated from the corresponding (seventh) equation (by the Direct Solver) since it appears more than once.

---

[1]See S. Timoshenko, *Elements of Strength of Materials,* VNR, New York, 1968.

$$I = \frac{bd^3}{12} \tag{1}$$

$$S = \frac{bd^2}{6} \tag{2}$$

$$Area = bd \tag{3}$$

$$c = \frac{d}{2} \tag{4}$$

$$y\_ul = \frac{wx(L^3 + x^3 - 2Lx^2)}{24EI} \tag{5}$$

$$y_{max}\_ul = \frac{5wL^4}{384EI} \tag{6}$$

$$M_x = \frac{wx(L - x)}{2} \tag{7}$$

$$M_{max} = \frac{wL^2}{8} \tag{8}$$

$$S_{bx} = \frac{M_x c}{I} \tag{9}$$

$$S_{b\_max} = \frac{M_{max} c}{I} \tag{10}$$

$$V_x = w\left(\frac{L}{2} - x\right) \tag{11}$$

$$V_{max} = \frac{wL}{2} \tag{12}$$

$$S_{sh\_max} = \frac{3V_{max}}{2 Area} \tag{13}$$

b	= beam width (in)
d	= beam depth (in)
L	= beam length (in)
w	= uniform load on beam (lb/ft)
E	= elastic modulus of beam material (lb/in^2)
x	= beam end to desired point (in)
y_ul	= deflection at x (in)
y_{max}_ul	= maximum deflection (in)
M_x	= bending moment at x (in · lb)
M_{max}	= maximum bending moment (in · lb)
S_{bx}	= bending stress at x (lb/in^2)
S_{b_max}	= maximum bending stress (lb/in^2)
V_x	= shearing force at x (lb)
V_{max}	= maximum shearing force (lb)
S_{sh_max}	= maximum shearing stress (lb/in^2)
c	= maximum distance to neutral axis (in)
I	= moment of inertia of beam section (in^4)
S	= section modulus (in^3)
$Area$	= area of cross-section (in^2)

FIGURE 8.2 Mathematical relationships: simply supported beam.

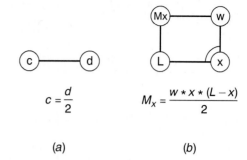

FIGURE 8.3 The Relationship or R graph.
(a) Equation in two variables.
(b) Mark around node *x* indicates nonresolvability of *x* from that equation.

The VEC matrix for the model is shown in Figure 8.4. Note the # for *x* denoting its nonresolvability by the Direct Solver in the fifth and seventh equations. When we try to draw the R graph for the beam model, we find that topologically, it is hard to include all the relationships in a single diagram. Therefore, for large models, we can either draw R graphs for parts of the model or just use the VEC matrix for the analysis.

8.2 CREATING THE MODEL

As we saw in 7.3.1, the first major step in creating the model is to enter the equations in the Rule Sheet. As each rule is parsed and checked for integrity, TK automatically adds the new variables to the Variable Sheet in the order in which they appear in the rules. At times, the automatic ordering of variables in the Variable Sheet may not suit your needs. During the course of problem solving when you are changing the values of several variables and assessing the impact on others, you would not have to search for the variables (either input or output) in the Variable Sheet if they are grouped logically. This is especially important when you have a large number of variables in the model. In the present example, a logical ordering of variable names would be those pertaining to the beam dimensions, followed by deflection, moment, shearing force, and so on. Another logical grouping would be the "usual" input variables followed by the output variables. This will again improve productivity during problem solving. We can use the **Append Variable Names feature** to turn off the automatic entry of variable names and enter them manually in the desired order.

Entering Variable Names and Equations We load TK and enter the variable names in the Name field of the Variable Sheet, beginning with the beam dimensions. Following this, we enter the equations in the Rule Sheet. Figure 8.5*a* shows the completed Rule Sheet for the model, including comments for the rules. Similarly, we use the Comment field of the Variable Sheet for variable descriptions.

Enter Associated Information We enter the units of measurement associated with each of the variables (from Figure 8.2) in the corresponding Unit field on the Variable Sheet (Figure 8.5*b*). Notice that the unit for beam length L is inches (*in*), while the value is specified in feet in the problem. The same applies to x. We have

Equation no. / Variable name	1	2	3	4	5	6	7	8	9	10	11	12	13
b	x	x	x										
d	x	x	x	x									
L					#	x	x	x			x	x	
w						x	x	x	x		x	x	
E						x	x						
x						#		#			x		
y_ul						x							
y_{max_ul}							x						
M_x								x		x			
M_{max}									x	x			
S_{bx}										x			
S_{b_max}										x			
V_x											x		
V_{max}												x	x
S_{sh_max}													x
c				x						x	x		
I	x				x	x				x	x		
S		x											
Area				x									x

FIGURE 8.4 VEC matrix for the beam model. *Note:* As compared to Figure 7.16a, variable names correspond to rows and equation numbers to columns.

two options under the circumstances: One is to convert the feet into inches ourselves and assign the converted values as inputs. The second option is to define unit conversions on the Unit Sheet and let TK perform the conversions as and when necessary. The latter has the advantage that once you define a conversion, it is there and you don't have to do the arithmetic again just because the value (e.g., beam length) changes in the problem. Moreover, you can create and save a library of conversions that you can load into any model.

8.2 CREATING THE MODEL

```
----------------------RULE SHEET--------------------------------------------
S-Rule----------------------------------------------------------------------
* I = b*d^3/12                              " Moment of Inertia
* S = b*d^2/6                               " Section Modulus
* Area = b*d                                " Cross-sectional Area
* c = d/2                                   " Max Distance to Neutral Axis
* y_ul = w*x*(L^3 + x^3 - 2*L*x^2) / (24*E*I) " Beam Deflection at x
* ymax_ul = 5*w*L^4/(384*E*I)               " Max Beam Deflection (at center)
* Mx =   w*x/2*(L-x)                        " Bending Moment at x
* Mmax =  w*L^2/8                           " Max Bending Moment
* Sbx = Mx*c/I                              " Bending Stress (max) at x
* Sb_max = Mmax*c/I                         " Max Bending Stress in beam
* Vx = w * (L/2 - x)                        " Shearing Force at x
* Vmax = w*L/2                              " Max Shearing Force (at supports)
* Ssh_max = 3*Vmax / (2*Area)               " Max Shearing Stress in beam
```

(a)

```
----------------------VARIABLE SHEET---------------------------------------
St-Input-----Name----Output----Unit------Comment---------------------------
             b                  in        beam width
             d                  in        beam depth (height)
             L                  in        beam length
             w                  lb/in     uniform load on beam
             E                  lb/in^2   elastic modulus of beam material
             x                  in        beam end to desired point
             y_ul               in        deflection at desired point (x)
             ymax_ul            in        maximum deflection
             Mx                 in·lb     bending moment at x
             Mmax               in·lb     maximum bending moment
             Sbx                lb/in^2   bending stress at x
             Sb_max             lb/in^2   maximum bending stress
             Vx                 lb        shearing force at x
             Vmax               lb        maximum shearing force
             Ssh_max            lb/in^2   maximum shearing stress
             c                  in        maximum distance to neutral axis
             I                  in^4      moment of inertia of beam section
             S                  in^3      section modulus
             Area               in^2      area of cross-section
```

(b)

FIGURE 8.5 Simply supported beam model. (a) Rule Sheet. (b) Variable Sheet.

8.2.1 Defining Unit Conversions

There are two units associated with each variable in TK. They are the **Calculation unit** and **Display unit.** The first time you enter a unit in the Unit field of the Variable Sheet, it is treated as the Calculation unit. The unit appearing on the Variable Sheet is the Display unit, and the variable values are displayed in this unit. Therefore, the first time you associate a unit with the variable, it is treated both as the Calculation unit and Display unit. For example, when we entered in[1] for the first time in the Unit field of L, TK assumed the Calculation unit to be inches (*in*); and, because *in* is displayed on the Variable Sheet, it is also the Display unit. Since we want to assign the variable value in feet, the Display unit will be *ft* (or feet; no restrictions are imposed by TK on choice of unit names). We should therefore define a conversion between the two units on the Unit Sheet. Otherwise, TK cannot convert the value and will display a ? next to the value.

The Unit Sheet We invoke the Sheet Select command by typing = and select the Unit Sheet option from the menu. The Unit Sheet is displayed as shown in Figure 8.6*a*. The field names in the sheet are mnemonic and should be read as follows:

To convert **from** *unit_name* **to** *unit_name*, **multiply by** *factor* and **add offset** *factor*.

The Comment field can be used for documenting the unit conversion. For example, the conversion between feet (*ft*) and inches (*in*) would be read as follows:

From ft **to** in, **multiply by** 12 and **add offset** zero.

This conversion definition is shown in the first row of the Unit sheet in Figure 8.6*b*. The Add Offset field is used to adjust the zeros of scales, e.g., Celsius-Fahrenheit, Celsius-Kelvin, etc. The value in this field is added after the multiplication is carried out. If the field is left blank, as in the *ft-in* example, the value is assumed to be zero. For example, the entry in the Unit Sheet for converting from Celsius to Fahrenheit would be as follows:

From oC **to** oF, **multiply by** 1.8 and **add offset** 32.

The Multiply By field acts like a calculator. This means that if you don't remember the exact conversion factor, you can carry out the necessary arithmetic operations in the field to arrive at the conversion factor. Let's say we are defining the conversion between kilograms (*kg*) and pounds (*lb*) but don't remember the factor. We know that there are 1000 g in 1 *kg* and 1 *lb* is equal to 453.6 g. We enter 1000/453.6 in the Multiply By field for the kilogram-to-pound conversion, and TK responds with the factor shown in the figure. In fact, the calculator feature can be used to learn about unit conversions from first principles. The Add Offset field also acts like a calculator.

[1] All entries that we type at the computer are shown underlined in text.

```
---------------------UNIT SHEET-------------------------------------------
From------To--------Multiply By---Add Offset----Comment------------------
```
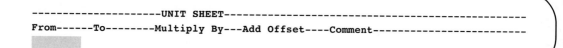

(a)

```
---------------------UNIT SHEET-------------------------------------------
From------To--------Multiply By---Add Offset----Comment------------------
ft         in        12
yd         ft        3
in         m         .0254
m          cm        100
cm         mm        10
kg         lb        2.2045855
lb/in      lb/ft     12
lb/in      kg/m      17.8954903364
ft^3       in^3      1727.99999999
in^3       cm^3      16.387064
m^3        cm^3      1000000
ft^4       in^4      20736
in^4       cm^4      41.62314256
m^4        cm^4      1E8
in·lb      ft·lb     .083333333333
in·lb      N·m       .113
lb/in^2    Pa        6894.75729307
```

(b)

FIGURE 8.6 Defining unit conversions. (a) The Unit Sheet. (b) Unit conversions.

Inverse Unit Conversions When we define the conversion between feet (*ft*) and inches (*in*) (as shown in the first row of the figure), the inverse conversion, i.e., inches (*in*) to feet (*ft*), is automatically "understood" by TK. Therefore, once a conversion between two units is defined, there is no need to define its inverse. The entries on the Unit Sheet could be called "rules" for conversion.

Indirect Unit Conversions TK can also perform conversions between two units defined in terms of a common third unit, a feature referred to as **indirect unit conversion**. As seen in the figure, the units inches (*in*) and yards (*yd*) have been defined in terms of the common unit feet (*ft*). So conversion between all three units can be performed. Moreover, TK can continue such indirect conversions in sequence as long as there is a common unit in the chain of conversions. For example, from the first three definitions on the Unit Sheet (the first three rows), TK can convert between yards (*yd*) and meters (*m*) [**sequence:** yards (*yd*) to inches (*in*) through feet (*ft*), and then to meters (*m*) through inches (*in*)]. This chaining is possible due to the common unit at each stage of the conversion process, viz., feet (*ft*) and inches (*in*), respectively. Likewise, there is no need to define a separate conversion between millimeters (*mm*) and yards (*yd*): TK can convert from millimeters (*mm*) to meters (*m*) through centimeters (*cm*), and then to feet (*ft*) through inches (*in*), and finally to yards (*yd*) through feet (*ft*). On the other hand, if we define two conversions, one *meter* to *feet* and another *hour* to *minutes*, TK does not automatically convert from meters per hour (*m/h*) to feet per minute (*ft/min*) or vice versa.

Integrity of Computations and Dimensional Analysis TK can convert only between units of measurement defined on the Unit Sheet. TK does not carry out any dimensional analysis—the systematic process of checking to ensure that the various units in an equation are multiplied, divided, added, and subtracted correctly so that the integrity of computations is maintained and resulting values are in the right set of units.

Example Consider the following simple equation:

$$\text{Force} = \text{mass} * \text{acceleration}$$

If the mass is specified in kilograms, the acceleration in meters per second squared, then the unit of force on the left-hand side is newtons. So all computations should be performed in these three units to ensure the integrity of the result. This means that the Calculation units for the three variables mass, acceleration, and force in a TK model should be kilograms, meters per second squared, and newtons, respectively.

If, in a problem, the mass is given in grams, the acceleration in meters per second squared, we should convert the value of mass from grams to kilograms and then perform the computation to evaluate force. Otherwise, the resulting force will not be in newtons. Just as you would assume responsibility for ensuring dimensional integrity when solving a problem with paper and pencil, it is up to you to do the same when using TK.

Explicit Conversion Definitions In Figure 8.6*b*, even though we have defined the individual conversions between kilograms (*kg*) and pounds (*lb*), and inches (*in*) and meters (*m*), we have explicitly defined the conversion between pounds per inch (*lb/in*) and kilograms per meter (*kg/m*). This is because TK cannot recognize such dimensional relationships. For the same reason, we have separate conversions for inches (*in*) to centimeters (*cm*) and cubic inches (in^3) to cubic centimeters (cm^3). TK treats the unit names as strings and carries out the specified arithmetic computation to go from one unit to the other. Consequently, the unit names are **case-sensitive**. Thus, *Lb* is different from *lb*, and *Kg*, *KG*, *kG*, and *kg* are all different. In fact, this is one of the most frequently encountered errors while using units in TK. When TK cannot convert between the Calculation and Display units of a variable, it will display a ? next to the value in the Variable Sheet—a helpful warning that TK does not know what to do and that something is amiss with the units.

Conversion Factors TK has no built-in knowledge of conversion factors. For example, it does not know that 1 ft equals 12 in. Consequently, if you erroneously enter that 1 ft equals 120 in, TK will not protest. The lack of built-in knowledge of converion factors could be construed as a drawback in TK. On the other hand, the unit conversion feature provides a flexible framework that could be individually molded to meet the requirements. Likewise, there are no "standards" for unit names. Thus, the unit kilogram could be represented as *k*, *kg*, *kilogram*, *kilo*, *kgs*, *kilograms*, or *kilos*. However, it is to your benefit to adopt a fixed set of unit names to avoid any unit-related errors during problem solving with TK.

Units Library As an engineer, you will be dealing with different sets of units and even units related to different fields. For example, while solving a design problem, you may be concerned with units related to both electrical engineering and mechanical engineering. Since it is not possible to remember all the conversions you would need and it may take time to derive the conversion factors each and every time, you will frequently refer to handbooks or textbooks for the factors. This task can be daunting and time-consuming.

Instead, you can use the various sources and create your own library of unit conversions in TK. You can then use the Save Units option of the Storage command to save the Unit Sheet (/SU). You can create a single model with all the unit conversions you normally use or several models, each dealing with units in a different field. Then, you can use the Load option of the Storage command and load the Unit Sheet into any model you are creating or working with. Thus, the Unit Sheet and the Unit Conversion feature in TK can greatly enhance your problem-solving productivity.

The Units TK!SolverPack by Steven Ross, published by McGraw-Hill, is an excellent compendium of units that can be readily imported into TK models. Ross has endeavored to assemble units from various disciplines, including those used in times past, and as the single largest collection of units and unit conversions, in a readily usable form, it is unique.

8.2.2 Assigning Values and Solving the Model

After entering all the unit conversions on the Unit Sheet, we invoke the Storage command (/S) and use the Unit option to save the Unit Sheet under the name beamunit. We can subsequently load this sheet into other models, as and when necessary.

We display the Variable Sheet in the window. Since the value of L and x are specified in feet, we move the cursor to their respective Unit fields and type ft, which becomes the Display unit. We assign the various input values and press F9, the Solve key. The resulting solution, obtained by the Direct Solver, is shown in Figure 8.7. All the unknown variables are evaluated, and the asterisks from the Status field of the rules disappear. The solution path is fairly easy to trace from the VEC matrix shown in Figure 8.4.

The deflection y_ul at the desired point (3 ft), is less than the maximum deflection y_{max}_ul of 0.251 in. The shearing force is one-half of the maximum shearing force, and the bending moment at that point is also smaller than the maximum bending moment. This response of the beam and the various values "appear" reasonable from a physical or engineering standpoint. Therefore the solution can be assumed to be correct. Note that the values of L and x are specified (and displayed) in feet. TK uses the con-

```
-------------------VARIABLE SHEET---------------------------------------
St-Input-----Name----Output----Unit------Comment-------------------------
    5.5        b                in       beam width
   11.5        d                in       beam depth (height)
   12          L                ft       beam length
   50          w                lb/in    uniform load on beam
 1600000       E                lb/in^2  elastic modulus of beam material
    3          x                ft       beam end to desired point
               y_ul    .17883342 in      deflection at desired point (x)
               ymax_ul .25099427 in      maximum deflection
               Mx      97200     in·lb   bending moment at x
               Mmax    129600    in·lb   maximum bending moment
               Sbx     801.78725 lb/in^2 bending stress at x
               Sb_max  1069.0497 lb/in^2 maximum bending stress
               Vx      1800      lb      shearing force at x
               Vmax    3600      lb      maximum shearing force
               Ssh_max 85.375494 lb/in^2 maximum shearing stress
               c       5.75      in      maximum distance to neutral axis
               I       697.06771 in^4    moment of inertia of beam section
               S       121.22917 in^3    section modulus
               Area    63.25     in^2    area of cross-section
```

FIGURE 8.7 Variable Sheet with problem solution.

versions defined in the Unit Sheet to convert these values into inches, the Calculation unit, and performs the various calculations.

In models with several equations where it will be time-consuming to manually solve each of the equations and verify the solution, you can ensure the correctness of the solution by picking one of the rules and evaluating the unknown in it using a calculator. If this matches the solution obtained by TK, the model can be assumed to be okay. However, this is not a guarantee against errors in other equations or computations. In the present model, the fifth rule would be a good candidate test equation, because it is a lengthy equation.

A better approach to debugging the model is to use TK's built-in function **debug (var1, var2, ..., varn)**, where **var1, var2, ..., varn** are variable names, to display values of desired variables during the solution process (see Section 9.4.2 for more).

Saving the Model Since the model is working well, we use the Storage command to save the model under the name beamul.

How-Can Analysis Design for performance is yet another aspect of engineering design when you are trying to design a system to meet certain performance requirements. In the present example, let's say we want to find out the load the beam can support (in pounds per foot, lb/ft) so that the maximum deflection is 0.3999 in. This is a simple case of how-can analysis—the combination of known and unknown variables is different from the previous problem.

```
----------------------VARIABLE SHEET------------------------------------------
St-Input-----Name----Output----Unit------Comment-------------------------------
    5.5       b                  in       beam width
   11.5       d                  in       beam depth (height)
   12         L                  ft       beam length
              w       955.95808  lb/ft    uniform load on beam
 1600000      E                  lb/in^2  elastic modulus of beam material
    3         x                  ft       beam end to desired point
              y_ul    .28492875  in       deflection at desired point (x)
    .3999     ymax_ul            in       maximum deflection
              Mx      12905.434  ft·lb    bending moment at x
              Mmax    17207.245  ft·lb    maximum bending moment
              Sbx     1277.4583  lb/in^2  bending stress at x
              Sb_max  1703.2778  lb/in^2  maximum bending stress
              Vx      2867.8742  lb       shearing force at x
              Vmax    5735.7485  lb       maximum shearing force
              Ssh_max 136.02566  lb/in^2  maximum shearing stress
              c       5.75       in       maximum distance to neutral axis
              I       697.06771  in^4     moment of inertia of beam section
              S       121.22917  in^3     section modulus
              Area    63.25      in^2     area of cross-section
```

FIGURE 8.8 Estimation of load on beam for a given maximum deflection.

We blank the value of *w* from its Input field and change its unit from pounds per inch (lb/in) to pounds per foot (lb/ft). We assign .3999 to ymax_ul and solve the model. The resulting solution is shown in Figure 8.8. The beam can support a load of 955.958 lb/ft for the specified maximum deflection. The bending moment at x and the maximum bending moment are displayed in ft·lb, instead of in·lb, the Calculation unit. Note that the maximum bending and shearing stresses are higher than those in Figure 8.7.

Combining Systems of Units of Measurement Once the Calculation units of the variables are defined in a consistent set of units (British, in the present model), you can display the values in other systems, e.g., SI, or combinations of systems. Of course, the necessary conversions should have been defined on the Unit sheet.

Example Consider the following example: An 8-m long simply supported rectangular beam 10 × 15 cm supports a uniformly distributed load of 60 kg/m. If the modulus of elasticity is 1.4E10 Pa, calculate the deflection in inches at a distance of 2 m from the end, the maximum deflection in meters, and the bending moments in N·m. Also calculate the other parameters.

In this problem, the variable values are specified in units that differ from their respective Calculation units defined in the TK model. To use these values in solving the problem, we change the Display unit of the variables to the appropriate units. We assign the given values and solve the model. The resulting solution is shown in

```
---------------------VARIABLE SHEET-----------------------------------------
St-Input-----Name----Output----Unit------Comment-------------------------------
    10       b                  cm       beam width
    15       d                  cm       beam depth (height)
    8        L                  m        beam length
    60       w                  kg/m     uniform load on beam
    1.4E10   E                  Pa       elastic modulus of beam material
    2        x                  m        beam end to desired point
             y_ul    2.2309491  in       deflection at desired point (x)
             ymax_ul .07953138  m        maximum deflection
             Mx      3523.4646  N·m      bending moment at x
             Mmax    4697.9528  N·m      maximum bending moment
             Sbx     1362.5779  lb/in^2  bending stress at x
             Sb_max  1816.7706  lb/in^2  maximum bending stress
             Vx      264        lb       shearing force at x
             Vmax    528        lb       maximum shearing force
             Ssh_max 34.064448  lb/in^2  maximum shearing stress
             c       2.9527559  in       maximum distance to neutral axis
             I       67.570583  in^4     moment of inertia of beam section
             S       22.883904  in^3     section modulus
             Area    23.250047  in^2     area of cross-section
```

FIGURE 8.9 Flexibility to handle different units of measurement.

Figure 8.9. The bending moment is displayed in N·m, while the bending and shearing stresses are displayed in pounds per square inch (lb/in^2). So, as long as relationships between units are defined on the Unit Sheet, values are converted to the appropriate Calculation units and used during the solution of the model. Thus, it is possible to assign and display values in different units of measurement.

Absence of Conversion Since the depth and width are given in centimeters (cm), it would only be appropriate to display the area of cross-section (Area) in square centimeters (cm^2) or square meters (m^2). So we move the cursor to the Unit field of Area in Figure 8.9 and enter m^2. A "?" appears in front of the value as shown in Figure 8.10. **The reason:** No unit conversion, either direct or indirect, is defined in the Unit Sheet between the Calculation and Display units of Area, square inches (in^2) and square meters (m^2), respectively (see Figure 8.6*b*). Consequently, TK displays the question mark indicating that it is "puzzled" by the new unit name — a helpful hint that something is wrong with the values displayed in the Variable Sheet.

We display the Unit Sheet and enter the last two conversions shown in Figure 8.11. As soon as the second conversion is defined, the ? disappears from the value of Area, and the converted value (in square meters, m^2) is displayed in the field as shown in the figure. We save the model again under the same name, beamul.

These examples illustrate TK's treatment of equations as multi-directional mathematical relationships between the variables, thus providing the flexibility to handle

```
(19u) Unit: m^2                                                        25+/OK

--------------------VARIABLE SHEET--------------------------------------------
St-Input-----Name----Output----Unit------Comment-----------------------------
    10         b                 cm       beam width
    15         d                 cm       beam depth (height)
     8         L                 m        beam length
    60         w                 kg/m     uniform load on beam
   1.4E10      E                 Pa       elastic modulus of beam material
     2         x                 m        beam end to desired point
               y_ul     2.2309491  in     deflection at desired point (x)
               ymax_ul  .07953138  m      maximum deflection
               Mx       3523.4646  N·m    bending moment at x
               Mmax     4697.9528  N·m    maximum bending moment
               Sbx      1362.5779  lb/in^2 bending stress at x
               Sb_max   1816.7706  lb/in^2 maximum bending stress
               Vx       264        lb     shearing force at x
               Vmax     528        lb     maximum shearing force
               Ssh_max  34.064448  lb/in^2 maximum shearing stress
               c        2.9527559  in     maximum distance to neutral axis
               I        67.570583  in^4   moment of inertia of beam section
               S        22.883904  in^3   section modulus
               Area     ?23.25005  m^2    area of cross-section
```

FIGURE 8.10 TK unable to convert from square inches (in^2) to square meters (m^2): displays a ? in the Output field of Area.

different combinations of known and unknown variables without any additional programming effort on our part. The TK framework also lets us handle different sets of units of measurement with ease. Therefore, it is ideally suited for use in engineering design.

8.3 SHEARING FORCE AND BENDING MOMENT DIAGRAMS

Another important aspect of engineering beam design is computing the shearing force and bending moment at various points along the length of the loaded beam and representing them in **shearing force** and **bending moment** diagrams, respectively. In all the problems solved until now, we computed these two parameters at a given distance (x) from the end of the beam (e.g., Figures 8.7, 8.8, 8.11). To compute the shearing force and bending moment at various points along the length of the beam, we need to solve the model for each value of x. The more the number of intermediate points, the more accurate will be the graphical representation. Note that the values of x will range from 0 (one end of the beam) to L, the length of the beam (the other end of the beam). What we have here is a repetitive solution of the set of equations for various values of x. TK provides a facility known as **List Solving** that can be used for such repetitive computations.

```
(18m) Multiply By: 6.4516                                          25+/OK

--------------------UNIT SHEET----------------------------------------------
From------To--------Multiply By---Add Offset----Comment--------------------
m^3        cm^3       1000000
ft^4       in^4       20736
in^4       cm^4       41.62314256
m^4        cm^4       1E8
in·lb      ft·lb      .083333333333
in·lb      N·m        .113
lb/in^2    Pa         6894.75729307
m^2        cm^2       10000
in^2       cm^2       6.4516
--------------------VARIABLE SHEET------------------------------------------
St-Input-----Name----Output----Unit------Comment---------------------------
             Sbx       1362.5779 lb/in^2  bending stress at x
             Sb_max    1816.7706 lb/in^2  maximum bending stress
             Vx        264       lb       shearing force at x
             Vmax      528       lb       maximum shearing force
             Ssh_max   34.064448 lb/in^2  maximum shearing stress
             c         2.9527559 in       maximum distance to neutral axis
             I         67.570583 in^4     moment of inertia of beam section
             S         22.883904 in^3     section modulus
             Area      .015      m^2      area of cross-section
```

FIGURE 8.11 Conversion defined on Unit Sheet; ? disappears and value is converted.

8.3.1 Repetitive Computations and List Solving

In science and engineering problem solving, we frequently examine the effect of one or more variable values on others in the model. The sequence of computations remains the same in each instance — only the values change. The effect of changing the current in a circuit on the resistance in the circuit, the effect of temperature on the viscosity of a solution, or the effect of interest rate on the monthly payment on a loan are all good examples of repetitive computations. TK's List Solving feature comes in handy for each of these and similar examples. Current, temperature, and interest rate are known as **independent variables,** while resistance, viscosity, and monthly payment are termed **dependent variables** in each of the examples, respectively.

What Is List Solving? TK lets you create lists of values for the independent variables whose effect you are trying to investigate. To store the computed values of the dependent variables, you can associate lists with the dependent variables. When you invoke the List Solving feature, the model is solved for each of the values or instances in the input list. Depending on the problem formulation, TK uses the Direct or Iterative Solver for solving the model. Of course, if an error is encountered in any instance, i.e., in any of the steps, TK terminates the solution for that instance and moves on to the next value in the list.

Declaring a list is analogous to declaring and dimensioning an array variable in a conventional programming language; the exception, however, is that the lists are limited to a single dimension. You cannot have multidimensional arrays in TK. Moreover, the size of the list is defined by the number of values entered in the list—there is no need to declare the dimension separately. List Solving is especially useful when a large number of equations is involved in the cause-effect relationship between the independent and dependent variables. Additionally, the computed values can be easily displayed in the form of tables and graphs. We will use the beam model to illustrate List Solving in TK.

Example Consider the following example: A 6- by 10-in simply supported rectangular beam, 12 ft long, supports a uniformly distributed load of 650 lb/ft. If the modulus of elasticity is 1.6E6 lb/in^2, compute the shearing force and bending moment along the length of the beam; display the results in a table and draw the shearing force and bending moment diagrams.

Steps in List Solving The first step in List Solving is to create a model and ensure its correct working for one set of variable values. This step will also help you identify the type of solver (Direct or Iterative) you would need to solve the problem, and accordingly formulate the problem. Once the model is created and debugged, the remaining steps are simple and straightforward. Since we have already created and tested the model, we begin the process by loading the model, *beamul*.

Declaring and Defining Lists In the Variable Sheet, we assign the given values to the different variables. As we mentioned earlier, since x represents the point or distance along the length of beam, we need to compute the shearing force and bending moment for different values of x. Therefore, a list of values should be associated with x. To do this, we move the cursor to the Status field of x and type L, for List. TK associates a list of the same name, viz., x with the variable x. This is denoted by the appearance of an L in the Status field of x as shown in Figure 8.12. At the same time, the list name x appears in the List Sheet, the repository for all list names in a model. For each list declared, TK also creates a List Subsheet in which the different values of the variable are stored.

You can **disassociate** a list from a variable by typing another L in the Status field of the variable with the associated list. The L disappears, and the list is no longer associated with the variable. However, the list name and the values (if any) remain on the List Sheet and Subsheet, respectively.

Since we need to compute the bending moment and shearing force along the length of the beam, the value of x will vary from 0 to 12 ft, the beam length. As for the number of values, an interval of 1 ft would be adequate, because it would give us 13 points, sufficient for obtaining a smooth plot. To enter these values for x, we need to access its List Subsheet. There are two ways of doing this (see Figure 7.31). We could display the List Sheet and, with the cursor highlighting x, invoke the **Dive command** by pressing >. Another approach is to dive from the Variable Sheet into the Variable Subsheet and dive again into the List Subsheet. We will select the latter option.

Diving into the Variable Subsheet With the cursor in the row of variable x, we invoke the Dive command by pressing >. The Variable Sheet display is replaced by

```
(6s) Status: L                                                    25+/OK

--------------------VARIABLE SHEET------------------------------------------
St-Input-----Name----Output----Unit------Comment-----------------------------
     6        b                 in         beam width
    10        d                 in         beam depth (height)
    12        L                 ft         beam length
   650        w                 lb/ft      uniform load on beam
 1600000      E                 lb/in^2    elastic modulus of beam material
L             x                 ft         beam end to desired point (x)
              y_ul              in         deflection at desired point (x)
              ymax_ul           in         maximum deflection
              Mx                in·lb      bending moment at x
              Mmax              in·lb      maximum bending moment
              Sbx               lb/in^2    bending stress at x
              Sb_max            lb/in^2    maximum bending stress
              Vx                lb         shearing force at x
              Vmax              lb         maximum shearing force
              Ssh_max           lb/in^2    maximum shearing stress
              c                 in         maximum distance to neutral axis
              I                 in^4       moment of inertia of beam section
              S                 in^3       section modulus
              Area              in^2       area of cross-section
```

FIGURE 8.12 Associating a list with x.

```
(s) Status: L                                                     25+/OK
Status options: Input  Output  List  Guess  Blank
--------------------VARIABLE: x---------------------------------------------
Status:                         L
First Guess:
Associated List:                x
Input Value:
Output Value:
Numeric Format:
Display Unit:                   ft
Calculation Unit:               in
Comment:                        beam end to desired point
```

FIGURE 8.13 Variable Subsheet for x (Dive command).

the **Variable Subsheet** shown in Figure 8.13. At the top of the sheet, the variable name is displayed. The Status field shows the L denoting its List status. The **First Guess field** is used for entering an initial guess value for the variable. If a value is present in the field, TK automatically triggers the Iterative Solver when the Direct

Solver fails. The **Associated List field** contains the name of the list associated with the variable. In this example, it is same as the variable name, x. This, however, is not necessary. You can associate a list of a different name with the variable; this feature enables you to associate a different list with the variable every time you solve the model.

The Input, Output, and Comment fields correspond to the display on the Variable Sheet. The **Numeric Format field** is used for associating a specific display format with the variable (see Section 8.4 for more). Both the Calculation and Display units of the variable are displayed in the Subsheet. Thus the Variable Subsheet provides detailed information about a specific variable.

Diving into the List Subsheet From the Variable Subsheet, we use the Dive command and display the List Subsheet. The various values of x should be entered in the **Value field** of the List Subsheet shown in Figure 8.14a. As a value is entered in the Value field, TK automatically displays the element number in the **Element field.** The Calculation and Display units on the List Subsheet are the same as that of the variable x. **Reason:** When the list name is the same as the variable name, the list "inherits" the units from the variable. Since we need to enter values from 0 to 12 ft in increments of 1 ft, we can go to the Value field and enter the values one by one. This can be tedious when you have a large number of values in the list. TK provides the **Fill List feature** that simplifies the task of entering the values in the list.

The Fill List Feature We invoke the Fill List feature by typing !, the exclamation key. TK displays the menu with several options shown in Figure 8.14a. The **Add Step option** asks you to specify an initial value, an increment or step value, and a final value. TK computes the second value as the algebraic sum of the first value and the increment, the third value as the algebraic sum of the second and the increment, and so on, till all the values have been created and displayed in the Value field. In other words, the current value is the algebraic sum of the previous value and the increment. See Capsule 8.1 for other options of the Fill List command.

Since we know the first and last values and the increment, we select the Add Step option by pressing the first character for the option name, A. TK prompts us for the first value, and we enter 0. This is followed by prompts for the step value (response = 1) and the final value (response = 12). Once we specify the final value, TK calculates the intermediate values and displays them in the List Subsheet as shown in Figure 8.14b. We now have the points along the beam at which the bending moment and shearing force need to be calculated.

Storing Calculated Values To store the calculated values of bending moment, shearing force, deflection, etc., at each point along the length of the beam, we need to associate lists with each of the variables. We do this by displaying the Variable Sheet and typing L in the Status fields of the desired variables (see Figure 8.15). Note that the associated list name will be the same as the corresponding variable name.

Triggering List Solving In Figure 8.12, there is no value in the Input field of x; this means the values in the associated list will not be used as input in the course of List Solving. But these values need to be used as input. Therefore, we move the cursor to the Status field of x and type I, for Input. A default value of zero appears in the Input field, making x an input variable. The actual value in the field does not have

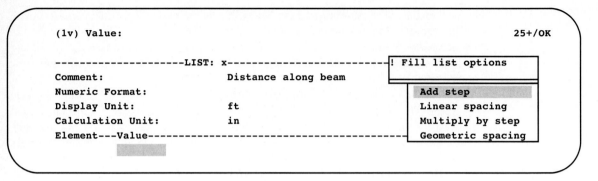

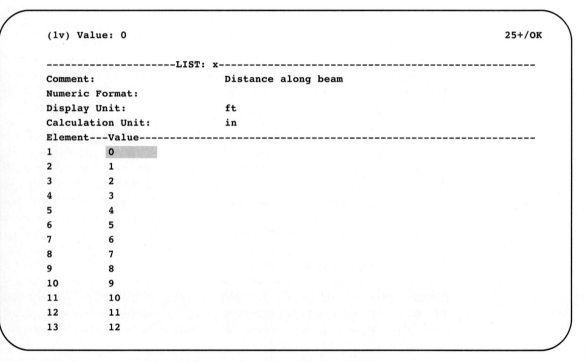

FIGURE 8.14 Creating a list of values. (*a*) Fill list options triggered by pressing !. (*b*) List of values created using the Add Step option.

any effect on the results obtained during List Solving. **Reason:** During List Solving, TK uses only the values from the associated list.

We trigger List Solving by pressing the function key F10. TK brings in the first value of x from the List Subsheet, computes the unknown variables, and stores them in the corresponding lists. TK then moves on to the next element in the list and solves

> **CAPSULE 8.1**
>
> OTHER OPTIONS OF FILL LIST COMMAND
>
> As shown in Figure 8.14a, the Fill List command has three other options, in addition to the Add Step option used in the present example. The **Linear Spacing option** prompts you for an initial value, number of values, and final value. TK assumes a linear relationship between the values and computes the increment or step size as:
>
> $$= \frac{\text{Final value} \sim \text{initial value}}{\text{Number of values} - 1}$$
>
> The current value is then computed as the algebraic sum of the previous value and the increment. For example, if the first value is 4, the number of values is 13, and the last value is 40, then:
>
> $$\text{Increment} = \frac{40 \sim 4}{13 - 1} = 3$$
>
> and the values are 4, 7, 10, 13, 16, ..., 40.
>
> In the **Multiply by Step option,** you specify the first value, the increment (step value), and the final value. TK computes the current value by multiplying the previous value by the specified increment. The resulting list of values is a geometric series. For example, if the first value is 5, the step value is 2, and the final value is 200, then TK enters the values as 5, 10, 20, 40, 80, and 160. Note that TK stops with 160, which is less than 200, the final value specified. This is because the next value in the series viz., 320 (= 160 * 2) is greater than the final value of 200.
>
> When the exact increment or step value is not known and you want to create a set of values in a geometric series, you can use the **Geometric Spacing option** instead of the Multiply by Step option. The Geometric Spacing option lets you specify the first and last values and the number of values and computes the intermediate values such that the ratio of consecutive values is constant.
>
> While executing the List Fill options, TK will spot such inconsistencies as specifying a negative increment when the final value specified is greater than the initial value.

the model again. The message on the prompt/error line denotes the instance or element number being processed, viz., 6 in Figure 8.15. Once all the elements in the list are processed, the prompt disappears and the original display is restored. If the Variable Sheet is displayed during List Solving, the variable values are updated after every solution. See Capsule 8.2 for some special aspects of List Solving.

Viewing the Results To view the computed values of the different variables, we can dive into the corresponding List Subsheet from the Variable Sheet. For instance, Figure 8.16a shows the values of deflection while Figure 8.16b shows the corresponding bending moments along the length of the beam. However, the distance along the beam is not seen in either figure. Moreover, these values are being viewed separately. Similarly, to view the other computed values sheet by sheet would not provide the total picture associated with the problem and the solution. For example, to find out the value of the bending moment and the shearing force for a specific shearing force, you would have to go through several sheets in succession. A better approach would be to display the various values in a table. We will use TK's Table feature for this purpose.

8.3.2 Tables and Graphs

Organization of the generated results is an important step in the problem-solving process. Tables and graphs are two effective means of organizing information. The

```
(6s) Status: L                                                        25+/OK
Solving for element 6, Direct Solver
---------------------VARIABLE SHEET----------------------------------------
St-Input-----Name----Output----Unit------Comment----------------------------
        6         b                in          beam width
       10         d                in          beam depth (height)
       12         L                ft          beam length
      650         w                lb/ft       uniform load on beam
  1600000         E                lb/in^2     elastic modulus of beam material
L       4         x                ft          beam end to desired point
L                 y_ul    .329472  in          deflection at desired point (x)
                  ymax_ul .37908   in          maximum deflection
L                 Mx      124800   in·lb       bending moment at x
                  Mmax    140400   in·lb       maximum bending moment
L                 Sbx     1248     lb/in^2     bending stress at x
                  Sb_max  1404     lb/in^2     maximum bending stress
L                 Vx      1300     lb          shearing force at x
                  Vmax    3900     lb          maximum shearing force
                  Ssh_max 97.5     lb/in^2     maximum shearing stress
                  c       5        in          maximum distance to neutral axis
                  I       500      in^4        moment of inertia of beam section
                  S       100      in^3        section modulus
                  Area    60       in^2        area of cross-section
```

FIGURE 8.15 List Solving in progress.

purpose of creating tables and graphs is to summarize and highlight important aspects of the results. Tables and graphs enable us to "discover" or "spot" trends in the results and to draw conclusions on the relationships between the variables. A graphical representation of data is generally easier to comprehend than a table of data—a picture is worth a thousand words. The very purpose of generating huge amounts of results is lost if the data are not presented in a meaningful format. In the present example, viewing the various List Subsheets by themselves does not highlight the relationship between the distance along the beam (the independent variable) and the bending moment (dependent variable).

The Table Feature TK's Table feature lets us explore the results of List Solving in the desired form. The Table Sheet is a framework in which we can specify lists that need to be displayed in the form of a table. We can also specify the format of the output. The output can be sent to a printer or to a text file, and the latter can be included as part of a document in a word processor.

Defining Table Contents The first step in displaying the values in the form of a table is to define the contents of the table. We invoke the Select command (=) and choose the Table Sheet option. The Table Sheet is displayed (see Figure 8.17a). TK lets us assign a name (of our choice) to the table being created, and this name is

CAPSULE 8.2

SPECIAL ASPECTS OF LIST SOLVING

Blank elements in a list

Consider two lists, speed and time, shown in Table 8.1, used as inputs for computing the distance traveled (*distance = speed * time*). When List Solving is invoked, TK computes *distance* for each instance of the values, until it reaches the fifth element. At this point, the problem is underdefined because only *speed* is known in the equation, and there are two unknowns. Therefore, TK cannot solve the equation and moves on to the next element. Here *speed* and *time* are both given, TK computes *distance* and moves on to the next element. Since there are no more values, it terminates. Therefore, when using List Solving, you should be careful to ensure that there are no unintentional blank elements in the input lists.

List Solving versus loop construct

Let's consider a modification of the problem. For each value of time, we want to compute the distance traveled at various speeds, i.e., distance traveled in 6 h at speeds ranging from 35 to 60 mi/h, distance in 5 h at different speeds, and so on. This computation is similar to a FOR-NEXT or DO loop in a procedural programming language.

TABLE 8.1 SPEED AND TIME LISTS

Speed, mi/h	Time, h
35	6
40	5
45	4
50	3
55	
60	1

Note the blank in the time list.

TK's List Solving feature is different from the traditional Loop construct. For each instance of the solution process, the solver accesses the corresponding elements from the various lists, performs the computations and stores the results appropriately. Therefore, the current problem cannot be solved using TK's List Solving feature. However, TK does provide a traditional loop (FOR-NEXT) construct (more on this and TK's other procedural programming capabilities in Chapter 9).

entered in the **Name field** on the Table Sheet. A suitable title for the table can be entered in the **Title field** of the Sheet. Associated with each row on the Table Sheet is a Table Subsheet, on which the list names and related information are defined. This relationship is similar to the Variable Sheet-Subsheet link—detailed information is present in the Subsheet.

Since the example deals with beam design, we enter Beam1 as the name of the table in the Name field. The selected name should be useful in identifying the contents of the table. We also enter an appropriate title as shown in Figure 8.17a. We access the Table Subsheet using the Dive command (>). The **List, Width,** and **Heading** fields of the Table Subsheet shown in Figure 8.17b are used frequently. The List field is used for entering the names of lists to be included in the table, while the Width field is used for the width of the column display (number of characters) for that list (the default width is 10). The Heading field is used for entering the appropriate column title for the list. If this field is left blank, the list name is used as the column heading. The **First Element** and **Last Element fields** are used for specifying which portion of the lists are to be included in the display. By default all values in the lists are displayed.

We define the table contents as shown in Figure 8.17b. Once the table is set up, we can view the information in one of several ways: as an **interactive table** on the screen, as a display on the screen, or as printed output. If the values are displayed in an interactive table, we can go and modify them much like a spreadsheet. Hence the term "interactive table." We use the Dive command (>) to select the Interactive Table

```
---------------------LIST: y_ul------------------------------------------
Comment:                    Deflection along beam
Numeric Format:
Display Unit:               in
Calculation Unit:           in
Element---Value----------------------------------------------------------
1          0
2          .0997425
3          .19188
4          .2700945
5          .329472
6          .3665025
7          .37908
8          .3665025
9          .329472
10         .2700945
11         .19188
12         .0997425
13         0
```

(a)

```
(7v) Value: 140400                                              25+/F9

---------------------LIST: Mx-------------------------------------------
Comment:                    Bending moment along beam
Numeric Format:
Display Unit:               in·lb
Calculation Unit:           in·lb
Element---Value---------------------------------------------------------
1          0
2          42900
3          78000
4          105300
5          124800
6          136500
7          140400
8          136500
9          124800
10         105300
11         78000
12         42900
13         0
```

(b)

FIGURE 8.16 Values created during List Solving. (*a*) Deflection along length of beam. (*b*) Bending moment along length of beam.

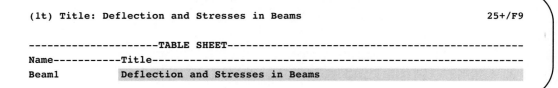

```
(1t) Title: Deflection and Stresses in Beams                          25+/F9

--------------------TABLE SHEET--------------------------------------------
Name-----------Title------------------------------------------------------
Beam1          Deflection and Stresses in Beams
```

(a)

```
--------------------TABLE: Beam1------------------------------------------
Screen or Printer:          Screen
Title:                      Deflection and Stresses in Beams
Vertical or Horizontal:     Vertical
Row Separator:
Column Separator:
First Element:              1
Last Element:
List------Numeric Format---Width---Heading--------------------------------
x                           10
y_ul                        10      deflection
Mx                          10      Mx
Vx                          10      sh_force
Sbx                         10      bending_st
```

(b)

```
--------------------TABLE: Beam1------------------------------------------
Title:     Deflections and Stresses in the beam
Element-x----------deflection-Mx---------sh_force---bending_st------------
1        0         0           0          3900       0
2        1         .0997425    42900      3250       429
3        2         .19188      78000      2600       780
4        3         .2700945    105300     1950       1053
5        4         .329472     124800     1300       1248
6        5         .3665025    136500     650        1365
7        6         .37908      140400     0          1404
8        7         .3665025    136500     -650       1365
9        8         .329472     124800     -1300      1248
10       9         .2700945    105300     -1950      1053
11       10        .19188      78000      -2600      780
12       11        .0997425    42900      -3250      429
13       12        0           0          -3900      0
```

(c)

FIGURE 8.17 Creating tabular output. (*a*) The Table Sheet. (*b*) Declaring the lists in the table. (*c*) Interactive table: values generated by List Solving. (*d*) Hard copy of table of values.

Deflection and Stresses in Beams

x	deflection	Mx	sh_force	bending_st
0	0	0	3900	0
1	.0997425	42900	3250	429
2	.19188	78000	2600	780
3	.2700945	105300	1950	1053
4	.329472	124800	1300	1248
5	.3665025	136500	650	1365
6	.37908	140400	0	1404
7	.3665025	136500	−650	1365
8	.329472	124800	−1300	1248
9	.2700945	105300	−1950	1053
10	.19188	78000	−2600	780
11	.0997425	42900	−3250	429
12	0	0	−3900	0

FIGURE 8.17 (continued). (d)

option. The resulting screen display shown in Figure 8.17c resembles a spreadsheet screen display.

In the figure, we see that beam deflection, bending moment, and bending stress gradually increase from the two ends and reach a maximum at the center. The shearing force is zero at the center. These results are what you would expect from a uniformly loaded simply supported beam. By being able to view all the values at once, and in a structured format, we can gain an appreciation for the interrelationships and for the total behavior of the system.

Function Key and Table Feature   Once a table is set up, i.e., the contents are defined, the table can be displayed by pressing F8, the Function key. Depending on the output device specified, the table is displayed on the screen, sent to a printer or to a text file. This means that you don't have to set up the table every time you use List Solving.

If you use the Function key F8 to display the values on the screen, the table would not be interactive — you cannot change any of the values. To return to the Table Subsheet display, you can press any key. If the Printer option is selected in the Table Subsheet, the output is sent to the printer as shown in Figure 8.17d. Note the special row and column separators (-, |, respectively) in the figure.

Thus the Table feature in TK provides an elegant framework for viewing the results of computations.

Multiple Tables   TK also lets us define multiple tables, each with a set of lists. For example, instead of the single table created here (Beam1), we could have created two tables, one with the values of x, y_ul, and Mx, and the other with the values of x, Vx, and Sbx. By creating two separate tables, we can focus on specific facets of the solution instead of looking at all the results at once.

Multiple tables can be created by declaring table names on separate rows of the Table Sheet and subsequently defining the lists in the corresponding subsheets. We can thus create and save different views of the results.

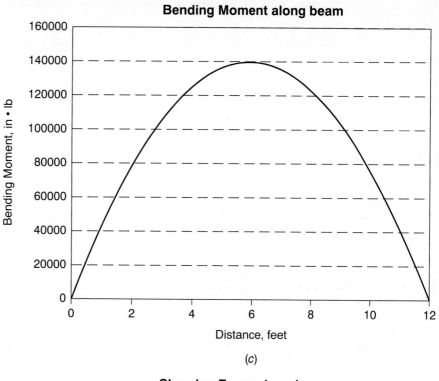

(c)

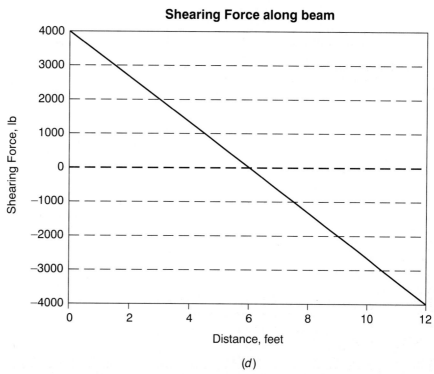

(d)

FIGURE 8.18 (*continued*).

```
(7o) Output: .1778650746803948                                    25+/OK

---------------------VARIABLE SHEET------------------------------------------
St-Input-----Name----Output----Unit-----Comment--------------------------------
    5.5        b                in        beam width
   11.5        d                in        beam depth (height)
   12          L                ft        beam length
   70          w                lb/in     uniform load on beam
 1600000       E                lb/in^2   elastic modulus of beam material
    2          x                ft        beam end to desired point
               y_ul    .17786507  in      deflection at desired point (x)
               ymax_ul .35139198  in      maximum deflection
               Mx      100800     in·lb   bending moment at x
               Mmax    181440     in·lb   maximum bending moment
               Sbx     831.48307  lb/in^2 bending stress at x
               Sb_max  1496.6695  lb/in^2 maximum bending stress
               Vx      3360       lb      shearing force at x
               Vmax    5040       lb      maximum shearing force
               Ssh_max 119.52569  lb/in^2 maximum shearing stress
               c       5.75       in      maximum distance to neutral axis
               I       697.06771  in^4    moment of inertia of beam section
               S       121.22917  in^3    section modulus
               Area    63.25      in^2    area of cross-section
```

FIGURE 8.19 Variable values are unformatted.

similar to the Table or Plot Sheet-Subsheet combination, and it is used for declaring and defining specific display formats for variable values, e.g., displaying values with up to four decimal places, in scientific notation, and with punctuation marks. We can assign names to the formats, again similar to the Plot/Table names, in the **Name field** of the Numeric Format Sheet. A helpful description of the format can be entered in

FIGURE 8.20 Declaring numeric formats. (a) Numeric Format Sheet: Format names declared. (b) Numeric Format Subsheet for f2. (c) Numeric Format Subsheet for fs.

```
(3n) Name: fs                                                     25+/F9

---------------------NUMERIC FORMAT SHEET-----------------------------------
Name----------Comment--------------------------------------------------------
  f2           format to two decimal places
  f4           format to four decimal places
  fs           scientific notation
```

(a)

8.4 CONTROLLING THE DISPLAY: NUMERIC FORMATTING

```
     (n) Numeric Notation: Decimal                                        25+/OK
     Notation options: Scientific  Decimal  Either
     --------------------NUMERIC FORMAT: f2----------------------------------
     Comment:                       format to two decimal places
     Numeric Notation:              Decimal
     Significant Digits:            15
     Decimal Places:                2
     Padding:                       None
     Decimal Point Symbol:          .
     Digit Grouping Symbol:         ,
     Zero Representation:           0
     +/- Notation:                  - Only
     Prefix:
     Suffix:
     Justification:                 Left
     Left Margin Width:             0
     Right Margin Width:            0
```

(b)

```
     (n) Numeric Notation: Scientific                                     25+/F9
     Notation options: Scientific  Decimal  Either
     --------------------NUMERIC FORMAT: fs----------------------------------
     Comment:                       scientific notation
     Numeric Notation:              Scientific
     Significant Digits:            15
     Decimal Places:                1
     Padding:                       None
     Decimal Point Symbol:          .
     Digit Grouping Symbol:
     Zero Representation:           0E0
     +/- Notation:                  - Only
     Prefix:
     Suffix:
     Justification:                 Left
     Left Margin Width:             0
     Right Margin Width:            0
```

(c)

FIGURE 8.20 *(continued)*.

the **Comment field.** Once a format is defined, it can be associated with one or more variables.

Selecting Formats The variable values in Figure 8.19 fall into at least three major classes: Very small values (y_ul, ymax_ul), very large (E), and in between (Sb_max). Therefore, it would be appropriate to display each of these values with different number of digits after the decimal place. For instance, y_ul and ymax_ul with four digits, I, S with two digits, and E in scientific notation would be an appropriate set of choices. This means three different numeric formats should cover the values in the figure.

Let's define three formats, one for displaying values to two decimal places (f2), another to four decimal places (f4), and a third for displaying values in scientific notation (fs). We have chosen f2, f4, and fs as the format names to better reflect the roles of the formats. We enter f2 in the Name field of the Numeric Format Sheet (Figure 8.20a) and the corresponding description in the Comment field. We bring up the Numeric Format Subsheet with the Dive command (>).

The Numeric Format Subsheet in Figure 8.20b provides the framework for setting the various parameters to control the display format of a variable. The **Numeric Notation field** denotes the notation in which values are displayed, viz., decimal, scientific, or either (depending on the magnitude, the value is displayed in either decimal or scientific notation). The **Significant Digits field** denotes the number of significant digits to be displayed. The **Decimal Places field** is used for entering the number of digits to be displayed after the decimal point; the **Decimal Point Symbol field** is used for entering the symbol to denote the decimal point. The usual symbol is a period (.), although in Europe it is not uncommon to see the comma (,) used for the same purpose.

Since we want to display the values in the decimal format, we select the Decimal option and specify 2 as the value for number of Decimal places (Figure 8.20b). Likewise, we define formats f4 and fs. The Numeric Format Subsheet for fs is shown in Figure 8.20c.

Associating Numeric Formats with Variables Having defined the three numeric formats, the next step is to associate them with variables. This is done by entering the format name in the corresponding Variable Subsheet. To associate the f4 format with y_ul, we display the Variable Subsheet for y_ul as shown in Figure 8.21a. Note the number of digits after the decimal place (15). When we enter f4 in the Numeric Format field of the subsheet, the value changes as shown in Figure 8.21b. Since the value is small and the Either option has been selected for Numeric Notation, TK displays the value in scientific notation to 4 decimal places. Although the number of digits displayed after the decimal place is only 4, TK still maintains all the 16 digits internally. In a similar manner, we associate f2 with Sb_max, Ssh_max, c, I, S and Area, and fs with E. The resulting formatted Variable Sheet is shown in Figure 8.22.

Thus TK's Numeric Formatting feature can be utilized for improving the display format of variable values in a TK model. The Numeric Format Sheet can be saved using the Storage command (/SN) and loaded into other models for subsequent use.

```
             (o) Output Value: .1778650746803948                              25+/F9

    ---------------------VARIABLE: y_ul------------------------------------------
    Status:
    First Guess:
    Associated List:
    Input Value:
    Output Value:             .177865074680395
    Numeric Format:
    Display Unit:             in
    Calculation Unit:         in
    Comment:                  deflection at desired point (x)
```

(a)

```
    ---------------------VARIABLE: y_ul------------------------------------------
    Status:
    First Guess:
    Associated List:
    Input Value:
    Output Value:             1.7787E-1
    Numeric Format:           f4
    Display Unit:             in
    Calculation Unit:         in
    Comment:                  deflection at desired point (x)
```

(b)

FIGURE 8.21 Associating numeric format with variable. (a) Variable Subsheet for y_ul; unformatted variable value. (b) Associating the numeric format f4 with y_ul.

8.5 REPRESENTING DESIGN DATA: THE USER-DEFINED FUNCTION FEATURE

In design problems, we often refer to data relationships that cannot be stated in the form of formulas or equations. Such relationships are usually determined experimentally or represent some given norm. Examples of such relationships include the Periodic Table of Elements, table of moduli of various woods (experimentally determined), and stress-strain curve of a polymeric material (experimentally determined). Such data usually exist in design or reference handbooks. What we need is a mechanism for storing and referencing such tables of information or databases[1] during problem solving. TK provides the user-defined feature for this purpose.

[1] A database is a collection of related information. See Chapter 10 for more.

```
---------------------VARIABLE SHEET----------------------------------------
St-Input-----Name----Output----Unit------Comment---------------------------
    5.5      b                 in        beam width
   11.5      d                 in        beam depth (height)
   12        L                 ft        beam length
   70        w                 lb/in     uniform load on beam
    1.6E6    E                 lb/in^2   elastic modulus of beam material
    2        x                 ft        beam end to desired point
             y_ul     1.7787E-1 in       deflection at desired point (x)
             ymax_ul  3.5139E-1 in       maximum deflection
             Mx       100,800  in·lb     bending moment at x
             Mmax     181,440  in·lb     maximum bending moment
             Sbx      831.48   lb/in^2   bending stress at x
             Sb_max   1,496.67 lb/in^2   maximum bending stress
             Vx       3,360    lb        shearing force at x
             Vmax     5,040    lb        maximum shearing force
             Ssh_max  119.53   lb/in^2   maximum shearing stress
             c        5.75     in        maximum distance to neutral axis
             I        697.07   in^4      moment of inertia of beam section
             S        121.23   in^3      section modulus
             Area     63.25    in^2      area of cross-section
```

FIGURE 8.22 Variable Sheet after numeric formats are associated with variables.

Representing Wood Properties: Creating an Interactive Table The properties of wood vary with the type of wood. For example, hickory wood has a higher modulus of elasticity than maple wood. Consequently, beams made from different types of wood will respond differently to similar loads. This means that in the engineering design of beams, we can select a particular type of wood depending on the operating conditions (load or stress levels) to meet performance and cost requirements. Some typical properties of different types of wood are shown in Table 8.2.

In our model, until now, we have not considered the effect of wood type on the performance of the beam. Let's say we now want to expand the scope of the model to include the properties of wood. We can then use the model to find the stress or deflection values for a beam made from a specific type of wood selected from the

TABLE 8.2 PROPERTIES OF WOOD

Material	Modulus, lb/in^2	Bending stress, lb/in^2	Shear stress, lb/in^2	Mass density, lb/ft^3
Southern pine	1.8E6	2500	150	40
Douglas fir	1.8E6	2100	125	32
Western hemlock	1.5E6	1500	210	28
Hickory	2.1E6	3500	190	48
Maple	1.6E6	3200	140	40
Oak	1.5E6	2600	160	45

database, or to select a particular type of wood to meet certain performance requirements (stress and deflection).

The data in Table 8.2 can be represented in TK, with each column in the table corresponding to a TK list. The lists can then be displayed as a table. We can use TK's Interactive Table feature for this purpose—directly entering and storing the information in a table. As we will see in Section 8.5.3, we can access these values in the table and use them in calculations.

Symbolic Values in TK Note that the values in the Material column are nonnumeric, i.e., they are **symbolic**. Symbolic values can be represented in TK by prefixing them with an apostrophe ('). This distinguishes them from variable names. Variables in TK, unlike in many programming languages, are "type-less." This means that a variable x can be assigned a value of 23.45 (a numeric value) or 'Aluminum (a symbolic value). Inconsistencies in value types are checked by TK during model resolution. For example, multiplying 44.56 by 'Aluminum will cause an error leading to the termination of the solution.

Creating an Interactive Table The various steps to be followed in creating an interactive table of values are as follows:

- Choose a name for the table and enter it in the Table Sheet.
- Choose names for the various lists and enter them in the corresponding Table Subsheet.
- Use the Dive command and display the Interactive Table.
- Enter the values in the corresponding cells.

Figures 8.23a through 8.23c show the various stages in creating an interactive table from the values in Table 8.2. The name of the table is Wood_Prop to reflect the contents of the table. The list names are seen in Figure 8.23b. Numeric formats have been associated with the lists. The wood types in the Material column are prefixed by an apostrophe denoting that they are symbolic values. Note the matrixlike notation for the addresses of the various "cells" or elements in the table in Figure 8.23c. For example, (6,5) at the top of the figure, corresponds to the address of the element (value = 45) at the intersection of the sixth row and fifth column in the table, while (4,1) refers to 'Hickory.

The Interactive Table feature can thus be utilized to create databases in TK. The next step is to utilize the created database in the model by defining relationships between wood type and specific properties, e.g., wood type—modulus of elasticity or wood type—mass density and so on. As we will see in 8.5.1, TK provides the facility for defining such relationships between sets of values (or lists) in the form of **user-defined functions.**

8.5.1 Functions in TK

Functions in TK can be classified into two major groups based on how they are defined:

- Built-in functions
- User-defined functions

```
(2n) Name: Wood_Prop                                                25+/OK


--------------------TABLE SHEET------------------------------------------
Name-----------Title-----------------------------------------------------
Beam1           Deflection and Stresses in Beams
Wood_Prop       Properties of Wood
```

(a)

```
(s) Screen or Printer: Screen                                        25+/OK
Display options: Screen   Printer
---------------------TABLE: Wood_Prop------------------------------------
Screen or Printer:        Screen
Title:                    Properties of Wood
Vertical or Horizontal:   Vertical
Row Separator:
Column Separator:
First Element:            1
Last Element:
List------Numeric Format---Width---Heading-------------------------------
Material                    15
Modulus     fs              10
B_Stress    f2              10
Sh_Stress   f2              10
Mass_Dens   f4              10
```

(b)

```
(6,5) Mass_Density: 45                                               25+/OK

---------------------TABLE: Wood_Prop------------------------------------
Title:        Properties of Wood
Element-Material--------Modulus----B_Stress---Sh_Stress--Mass_Densi------
1            'South_pine   1.8E6    2,500       150         40
2            'Douglas_fir  1.8E6    2,100       125         32
3            'Wes_hemlock  1.5E6    1,500       210         28
4            'Hickory      2.1E6    3,500       190         48
5            'Maple        1.6E6    3,200       140         40
6            'Oak          1.5E6    2,600       160         45
```

(c)

FIGURE 8.23 Creating an interactive table of values. (*a*) Specifying the table name in the Table Sheet (Wood_Prop). (*b*) Entering names for lists in corresponding Table Subsheet. (*c*) Entering values in corresponding "cells."

Built-in functions are predefined functions that can be directly accessed in TK. Just as a programming language (e.g., FORTRAN or BASIC) has a set of library functions, TK has a host of built-in mathematical, trigonometric, hyperbolic, and financial functions that can be used in rules. In addition, TK provides a set of functions that can be used for carrying out specific actions or commands. These include blanking the elements in a list (Blank), copying values from one list to another (Listcopy), displaying customized error messages (Errmsg), or checking the status of a variable (Given). Table 8.3 summarizes the major built-in functions in TK. You will recall from Section 7.4.1 that built-in functions are not case-sensitive, i.e., Errmsg is the same as ERRMSG, ErRmSg, etc.

Built-in functions can be used as expressions in rules or can be invoked using a **Call** statement (some functions, however, can be invoked only through a Call statement). For example, y = sin(theta) is valid and so is Call sin(theta). In this example, however, the function reference is preferable to using the Call statement.

The problem-solving power of TK is further enhanced by the **User-defined function feature.** As the name implies, you can define specific relationships between lists of values, or sequences of computations, in the form of user-defined functions. Once defined, these functions can be invoked in rules with the appropriate arguments. For example, the relationship between wood type and modulus of elasticity can be expressed in the form of a user-defined function in TK. Likewise, other relationships such as between the material and resistivity of electrical conductors, temperature and viscosity of solutions, or individual earnings and tax rate, can all be expressed in the form of user-defined functions.

Types of User-Defined Functions Three types of user-functions can be defined in TK: **List, Rule,** and **Procedure functions.** The relationship between two lists of values is defined in the form of a List function. A Rule function is generally used to combine a set of one or more rules that can be collectively invoked by referring to the function name (see Section 9.1 for more). Likewise, a Procedure function is used to define a sequence of computations in a programming-language-like syntax (see Section 9.3 for more). We will limit our discussion only to List functions here.

Anatomy of a List Function In general, a function has an argument and a corresponding function value. For example, in:

$$y = \sin(x)$$

x is the function argument, y corresponds to the function value, and sin is the name of the built-in sine function in TK. Thus, the value of y would depend on the value of the argument x.

Similarly, a List function is used for defining the relationship between two lists of values, known as the **Domain** and **Range lists.** The Domain list contains the function arguments, while the Range list contains the corresponding function values. For example, in a List function defining the wood-type to modulus relationship (lists Material and Modulus in Figure 8.23), Material would correspond to the Domain list,

TABLE 8.3 TK'S BUILT-IN FUNCTIONS*

Invertible Mathematical Functions

Gamma(n)	$(n-1)$ factorial
Root(x,n)	nth root of x
Sqrt(x)	Positive square root of x

Noninvertible Mathematical Functions

Abs(x)	Absolute value of x
Ceiling(x)	Smallest integer value greater than or equal to x
Floor(x)	Largest integer value less than or equal to x
Int(x)	Integer part of x
Mod(x,y)	Remainder of x/y
Round(x)	Even rounding to nearest integer to x
Sgn(x)	Signum of x: $(-1, +1,$ or $0)$
Step(x,y)	If $x \geq y$ returns 1, else 0

Complex Functions

Power((x,y),n)	Calculates the complex pair equal to the complex pair (x, y) raised to the nth power
Ptor(r,φ)	Polar to rectangular form: Calculates a complex number in rectangular form corresponding to the complex number in polar form where r is the radius and ϕ is the angle in radians
Rtop(x,y)	Rectangular to Polar—inverse of Ptor

For arguments in degrees, add suffix d to Ptor and Rtop.

Boolean Functions

Not(x)	Boolean NOT
And(x1,x2,x3..)	Boolean AND
Or(x1,x2,x3..)	Boolean OR
Eqv(x,y)	If $x = y$ (both 0 or both 1) returns 1, else 0
Imply(x,y)	If $x = 1$ and $y = 0$ returns 0, else 1

List-Oriented Functions

Blank(listname)	Blanks specified list
Blank(listname,n)	Blanks the nth element in the list
Count(listname)	Number of nonblank elements in a list
Delete(listname)	Deletes specified list from List Sheet
Listcopy(x,y)	Copies elements of list x into list y
Length(listname)	Length of list (includes blank elements)
Dot(listname, x1,x2,..)	Dot product of list elements with $x_1, x_2, \ldots$
Max(x1,x2,x3..)	Maximum value of arguments
Min(x1,x2,x3..)	Minimum value of arguments
Npv(r, x1,x2,x3,..)	Net present value of a series of cash flow values $x_1, x_2, x_3, \ldots,$ at interest rate r
Sum('x1)	Sum of elements in list

TK-Specific Functions

Apply(fun_name,x1,x2,x3..)	Returns the value of the specified function with given arguments
Element(listname,n,expr)	Returns the value of the nth element of list; if element is blank returns expr

Errmsg(message)	Displays message in argument when an error is encountered
Evltd(var_name)	Returns 1 if specified variable is evaluated during current run of Direct Solver; else 0
Given(var_name)	Returns 1 if specified variable is assigned an input value
Known(var_name)	Returns 1 if specified variable is assigned an input value or is evaluated during current run of Direct Solver
Place(listname,n)	Places a value into the *n*th element of specified list
Solved( )	Returns true if the model has been solved, else false
Value(var_name)	Returns the value of the specified variable

*The standard trigonometric and hyperbolic functions and their inverses are not included in the table.

while Modulus would correspond to the Range list. Of course, the roles of the lists Material and Modulus could be reversed. However, in practice we are more likely to know the type of wood and would want to find out the value of the modulus. Therefore, we could select Material as the Domain list and Modulus as the Range list for our function.

List Functions in Rules The usage of a List function in a TK rule is similar to that of a built-in function. If Fun_Name, Fun_Val, and Fun_Arg correspond to the function name, function value, and function argument, respectively, a List function is used in a TK rule as follows:

$$\text{Fun\_Val} = \text{Fun\_Name}(\text{Fun\_Arg})$$

When the rule is solved for a given value of Fun_Arg, TK returns a value for Fun_Val.

Types of Relationships or Mapping The relationship between the Domain and Range lists can be one of four types: **Table look-up, Step mapping, Linear interpolation,** and **Cubic interpolation.**

Table Mapping The Table Mapping option is used for representing one-to-one relationships between the values in the Domain and Range lists. In Table 8.2, the underlying relationship between wood types and the various properties is discrete. This means that every type of wood has a specific value for each of the properties. It is meaningless to "infer" or interpolate the properties of one type of wood from those of two other wood types. For example, we cannot calculate the modulus of elasticity of southern pine wood from those of hickory and maple. Therefore, the relationship between wood type and properties is one of direct mapping or one-to-one look-up, i.e., for a given wood type, we can scan (look up) the table and find out the corresponding modulus of elasticity, bending stress, etc. TK provides the **Table Mapping option** for representing such relationships.

We can also scan the table to find out the inverse relationship, i.e., find the type of wood that matches a specific value of the modulus of elasticity, mass density, etc. For this reason, Table mapping is invertible. However, there is a caveat. Let's say we specify the modulus as 1.8E6 and let TK find the type of wood. TK will return southern pine, which is correct (Table 8.2). However, we see in the table that Douglas fir also has the same modulus, but TK will always return southern pine. This is because TK scans the Domain (or Range) list top to bottom, and once it finds a match, it returns the corresponding Range (or Domain) value. Since southern pine precedes Douglas fir in the table, TK returns southern pine, the first range value corresponding to the specified domain value (1.8E6).

Step Mapping In the Step Mapping option, for a given argument, TK finds the starting value of the interval in the Domain list within which the argument lies and returns the corresponding value from the Range list. For example, consider the relationship between points obtained and course grades shown in Table 8.4. Whether the score is 91, 94, or 99, TK identifies the interval 91 through 100 in the points-obtained list (the Domain list) and returns the grade, A, from the grade list (the Range list) corresponding to the starting value. Likewise, whether the score is 72, 76, or 80, TK returns grade C. Thus, the assigned grade depends on the *step* or interval within which the student's points fall. Hence the name **Step mapping.**

Note that the Step Mapping option is noninvertible. For example, there is no single (unique) score corresponding to a particular grade, i.e., if a student's grade is B, we (and therefore, TK) cannot determine the exact number of points scored — it could lie anywhere from 81 through 90. Consequently, TK does not return a value, and the function remains noninvertible.

Linear and Cubic Interpolations The Linear Interpolation option approximates the relationship within each pair of consecutive points in the two lists by a straight line, whereas the Cubic Interpolation option uses a cubic polynomial approximation instead of the linear approximation. The choice of either of these options depends on the nature of the relationship (either existing or assumed) between the values in the Domain and Range lists. Both options are invertible. These options could be used to reference experimental data in the model. For example, if the stress-strain curve for a polymeric material is experimentally obtained, the values could be stored as lists of discrete *x-y* pairs. Depending on the material and the observed nature of relationship between stress and strain values, Linear or Cubic interpolation could be chosen.

8.5.2 Using Wood Properties Database in the Model

We will utilize the wood properties database (Figure 8.23) in our model by creating List functions. These functions can then be referenced in rules. The various steps involved in creating a List function are as follows:

• Select the **Domain** and **Range lists** for the function. The Domain list contains the function arguments, and the Range list, the corresponding function values.

TABLE 8.4 POINTS AND GRADES: ILLUSTRATION OF TK'S STEP MAPPING

Points obtained	Grade
91–100	A
81–90	B
71–80	C
61–70	D
60 and below	F

8.5 REPRESENTING DESIGN DATA: THE USER-DEFINED FUNCTION FEATURE

- Decide the type of mapping (Table, Step, Linear, and Cubic) between the Domain and Range lists.
  - Choose a name for the function and enter it in the Function Sheet.
  - Select the List Function option for the function type.
  - Use the Dive command and display the List Function Subsheet.
  - Enter the list names in the Domain and Range List fields.
  - Enter the desired type of mapping in the Mapping field.
  - If the values have not been entered in the lists (using the Interactive Table feature), use the Dive command and enter the values in the corresponding List Subsheets.

Creating a List Function: Wood-Type to Modulus Relationship We will illustrate the various steps by creating a List function to express the relationship between wood type and modulus of elasticity.

Domain and Range Lists In Figure 8.23, the values of wood type are stored in list Material, while the corresponding modulus values are entered in list Modulus. Therefore, we select Material as the Domain list and Modulus as the Range list for our function.

Type of Mapping As mentioned in Section 8.5.1, the relationship between wood type and modulus is a discrete (one-to-one) relationship. Therefore, we can select the Table Mapping option.

Selecting a Function Name We can select any name for the function, provided it does not correspond to one of the built-in function names in TK. A name that best characterizes the purpose or role of the function should be selected. Likewise, using the same name for the function and its argument will lead to confusion. For the present example, we choose Elastic_Mod because the name describes the purpose or nature of the relationship being defined.

Declaring the Function: The Function Sheet We invoke the Select command and display the Function Sheet. The Function Sheet is similar to the Table Sheet. Function names are declared on the sheet and are defined in corresponding subsheets. Once the Function Subsheet is set up, the function definition is complete and the functions can be referred to in rules.

The **Name field** on the Function Sheet is used for entering the name of the function and the **Type field** is used for declaring the function type: List, Rule, or Procedure (Figure 8.24a). The **Comment field** is used optionally to enter a brief description of the purpose of the function. Once the function is defined, TK automatically fills the value in the **Arguments field.**

We enter the name of the function Elastic_Mod in the Name field; since we want the function to be a List function, we enter L (for List) in the Type field. TK immediately displays 1;1 in the Arguments field denoting the Domain and Range lists, respectively (Figure 8.24a). We also enter an appropriate comment in the Comment field.

The List Function Subsheet To define the function, we invoke the Dive command and display the List Function Subsheet. We enter Material in the Domain List field (Figure 8.24b). Immediately, the values defined earlier in the list (while creating the interactive table) appear in the Domain column. Since the relationship is a table

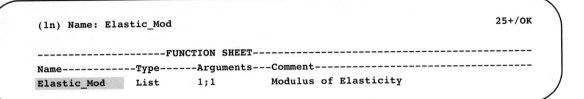

```
(1n) Name: Elastic_Mod                                            25+/OK

--------------------FUNCTION SHEET------------------------------------
Name-----------Type------Arguments---Comment-------------------------
Elastic_Mod    List      1;1         Modulus of Elasticity
```

(a)

```
(m) Mapping: Table
Mapping options: Table  Step  Linear  Cubic
--------------------LIST FUNCTION: Elastic_Mod-------------
Comment:             Modulus of Elasticity
Domain List:         Material
Mapping:             Table
Range List:          Modulus
Element---Domain--------------Range-----------------------
   1         'South_pine       1.8E6
   2         'Douglas_fir      1.8E6
   3         'Wes_hemlock      1.5E6
   4         'Hickory          2.1E6
   5         'Maple            1.6E6
   6         'Oak              1.5E6
```

(b)

```
--------------------FUNCTION SHEET------------------------------------
Name-----------Type------Arguments---Comment-------------------------
Elastic_Mod    List      1;1         Modulus of Elasticity
Bending        List      1;1         Allow. Bending Stress Parallel to Grain
Shear          List      1;1         Allowable Shear Stress Parallel to Grain
Density        List      1;1         Mass Density of Wood
```

(c)

FIGURE 8.24 User-defined functions. (a) Function Sheet with name and type of function. (b) Declaring the function arguments in Function Subsheet. (c) Function Sheet with all function names.

look-up, we type T (for Table look-up) in the Mapping field to select Table mapping. Note that the four possible options, viz., Table, Step, Linear, and Cubic are displayed at the top of the screen. In the Range List field, we enter Modulus for the name of the list containing the modulus of elasticity values. The resulting screen with the complete function definition is shown in Figure 8.24b.

Creating Other List Functions In a similar manner, we declare and define functions for relating wood type to allowable bending stress (Bending) and shear stress (Shear), respectively. Figure 8.24c shows the complete list of functions in the model. Thus it is easy to define relationships between lists of values in the form of user functions in TK. The defined functions can be saved using the Save Function option (/SF) of the Storage command.

8.5.3 Incorporating User-Defined List Functions in Rules

Having defined the various functions, the next step is to use them in the rules in the model. The Elastic_Mod function can be entered in the Rule Sheet in the following form:

$$E = \text{Elastic\_Mod(Material)}$$

where E corresponds to the modulus of elasticity (as defined earlier) and Material is a new variable. Note that in the function definition, the names of the Range and Domain lists are Modulus and Material, respectively. It is not necessary to use the list names for the corresponding variables in the rules. For illustration, we have chosen to have one common name between the list and variable (Material), and one that is different (E and Modulus). However, it is good practice to have different names for the two to avoid mixing up variable names and list names.

Figure 8.25a shows the additional rules in the model with the user-defined functions. An equation is included for calculating the allowable deflection in the beam. The variables y_all, Sb_all, and Ssh_all correspond to the allowable limits, i.e., upper limits of the variables based on wood properties and theoretical considerations. In engineering practice, the maximum values for the variables (ymax_ul, Sb_max, and Ssh_max) should be below these allowable values, because designs should incorporate a margin-of-safety factor.

Example Consider the following problem: A 5.5- by 11.5-in simply supported rectangular beam, 12 ft long, is subjected to a uniformly distributed load of 100 lb/in. If the beam is made from hickory wood, estimate the deflection, bending moment, and shearing force at a distance of 2 ft from the end of the beam. Also, calculate the maximum values for these three variables along with the allowable limits.

We specify the known variable values. Since the value of Material (Hickory) is a symbolic value, we prefix it with an apostrophe (') when entering it in the Input field. Figure 8.25b shows the solution of the problem. TK uses the value of Material ('Hickory) and returns the corresponding value of E (2.1E6), which is then used in other rules to evaluate additional variables.

In Figure 8.25b, the computed maximum values for deflection, shearing, and bending stresses are below the allowable limits estimated by the new rules in the model. Therefore, the beam is expected to be safe under the load, and the design is acceptable. Note, however, that we have not included a design safety factor in the model.

User-Function Error While entering the inputs, if we specify the value of material incorrectly, e.g., 'Hickor (missing y), TK will not find the value in the user-defined

```
--------------------RULE SHEET--------------------------------------------------
S-Rule--------------------------------------------------------------------------
    "               Using Functions in Rules

    * E      = Elastic_Mod(Material)      " Modulus of Elasticity
    * y_all  = L/360                       " Allowable Deflection
    * Sb_all = Bending(Material)           " Allowable Bending Stress
    * Ssh_all= Shear(Material)             " Allowable Shear Stress
```

(a)

```
--------------------VARIABLE SHEET----------------------------------------------
St-Input-----Name----Output----Unit------Comment--------------------------------
    5.5      b                 in         beam width
    11.5     d                 in         beam depth (height)
    12       L                 ft         beam length
    100      w                 lb/in      uniform load on beam
    'Hickory Materia                      beam material
             E        2.1E6    lb/in^2    elastic modulus of beam material
    2        x                 ft         beam end to desired point
             y_ul     1.9359E-1 in        deflection at desired point (x)
             ymax_ul  3.8247E-1 in        maximum deflection
             Mx       144,000  in·lb      bending moment at x
             Mmax     259,200  in·lb      maximum bending moment
             Sbx      1,187.83 lb/in^2    bending stress at x
             Sb_max   2,138.1  lb/in^2    maximum bending stress
             Vx       4,800    lb         shearing force at x
             Vmax     7,200    lb         maximum shearing force
             Ssh_max  170.75   lb/in^2    maximum shearing stress
             c        5.75     in         maximum distance to neutral axis
             I        697.07   in^4       moment of inertia of beam section
             S        121.23   in^3       section modulus
             Area     63.25    in^2       area of cross-section
             y_all    .4       in         allowable deflection
             Sb_all   3500     lb/in^2    allowable bending stress
             Ssh_all  190      lb/in^2    allowable shearing stress
```

(b)

FIGURE 8.25 User-defined functions in a model. (*a*) Referencing user-defined functions in rules. (*b*) Variable Sheet: solution for 'Hickory wood.

function (Figure 8.24*b*) and will return an error message. Likewise, if a material not included in the function (e.g., rosewood) is specified, TK will encounter an error and display an appropriate message.

Thus, we can use the user-defined List Function feature to store and reference databases during the course of problem solving. We will now look at ways of further enhancing the model.

8.6 CONTROLLING THE FLOW OF COMPUTATION: CONDITIONAL RULES

In the current example, we specified the type of wood that is listed in the database (Figure 8.24b), and TK returned the corresponding values for E, Sb_all, and Ssh_all. However, there are situations when the value of E is known, while the type of wood is unknown. Likewise, we may want to specify the values for allowable limits for shearing and/or bending stresses in the beam, instead of letting TK compute them. Under these circumstances, if the specified values do not correspond to those listed in the user-defined List functions, TK will encounter an error and terminate the solution.

In Figure 8.26a, E is assigned an input value of 2.2E6 and the model is solved. Since this value of E does not exist in the function (Figure 8.24b), TK signals a user function error. The rule is also marked as shown in Figure 8.26b. However, this response of TK, while technically correct, may be incomprehensible to the model user and is not always desirable; moreover, the scope of the model will be restricted to those materials listed in the database. A more desirable solution would be to use the value of E if it were specified as input; otherwise, the rule with the user-defined List function should be executed and the value obtained from the database. TK provides a built-in function for checking the status of the variable — the Given function. This function can be used for implementing the desirable solution.

8.6.1 The Given Function

The Given function, that is, Given($v_1, v_2, \ldots, v_n, x_1, x_2$) checks the status of the variables $v_1, \ldots, v_n$ and, if all are specified as input, returns the value of expression or variable x_1; otherwise, even if one of the variable values is not specified as input, the value of expression or variable x_2 is returned. To check the status of just one variable, the syntax reduces to Given(v_1, x_1, x_2).

For example, Given('E,E,Elastic_Mod(Material)) returns the value of E if E is specified as input and the value of Elastic_Mod(Material) if E is not specified as input.

Consider the following rule:

$$E = \text{Given}('E, E, \text{Elastic\_Mod}(\text{Material}))$$

If an input value is specified for E, the Given function returns the value of E, which is the same as the left-hand side. That is, the rule reduces to a simple equality, E = E. However, if E is not specified as input, the Given function returns the value of the second expression, Elastic_Mod(Material), and this value is assigned to the unknown E. In other words, the rule becomes E = Elastic_Mod(Material). Thus, the Given function can be used to control the execution of rules in a model.

Modifying the Rules in the Model We use the Edit command (Function key F5) and edit the three rules with user-defined List functions to include the Given function. The resulting rules are shown in Figure 8.27a, and the corresponding solution is shown in Figure 8.27b. Since neither the allowable limits for Sb_all and Ssh_all, nor material is specified, the corresponding two rules are not resolved.

```
(5s) Status:   > Error                                                 25+/F9
User function: Argument error
---------------------VARIABLE SHEET------------------------------------------
St-Input-----Name----Output----Unit------Comment----------------------------
    5.5      b                 in        beam width
    11.5     d                 in        beam depth (height)
    12       L                 ft        beam length
    100      w                 lb/in     uniform load on beam
>            Materia                     beam material
    2.2E6    E                 lb/in^2   elastic modulus of beam material
    2        x                 ft        beam end to desired point
             y_ul     1.8479E-1 in       deflection at desired point (x)
             ymax_ul  3.6508E-1 in       maximum deflection
             Mx       144,000  in·lb     bending moment at x
             Mmax     259,200  in·lb     maximum bending moment
             Sbx      1,187.83 lb/in^2   bending stress at x
             Sb_max   2,138.1  lb/in^2   maximum bending stress
             Vx       4,800    lb        shearing force at x
             Vmax     7,200    lb        maximum shearing force
             Ssh_max  170.75   lb/in^2   maximum shearing stress
             c        5.75     in        maximum distance to neutral axis
             I        697.07   in^4      moment of inertia of beam section
             S        121.23   in^3      section modulus
             Area     63.25    in^2      area of cross-section
             y_all             in        allowable deflection
             Sb_all            lb/in^2   allowable bending stress
             Ssh_all           lb/in^2   allowable shearing stress
```

(a)

```
(17s) Status: > User function: Argument error                          25+/F9
User function: Argument error
---------------------RULE SHEET---------------------------------------------
S-Rule----------------------------------------------------------------------
    "                         Allowable Limits

>   E = Elastic_Mod(Material)          " Modulus of Elasticity
*   y_all = L/360                      " Allowable Deflection
*   Sb_all = Bending(Material)         " Allowable Bending Stress
*   Ssh_all = Shear(Material)          " Allowable Shear Stress
```

(b)

FIGURE 8.26 User function error. (a) Input value of E (2.2E6) does not exist in the Material list. (b) The Elastic_Mod rule is also marked with >.

```
---------------------RULE SHEET------------------------------------------------
S-Rule-------------------------------------------------------------------------
      "                    Allowable Limits

  * E = Given('E,E,Elastic_Mod(Material))       " Modulus of Elasticity
  * y_all = L/360                               " Allowable Deflection
  * Sb_all = Given('Sb_all, Sb_all, Bending(Material)) " Allowable Bending Stress
  * Ssh_all = Given('Ssh_all, Ssh_all, Shear(Material)) " Allowable Shear Stress
```

(a)

```
---------------------VARIABLE SHEET--------------------------------------------
St-Input-----Name----Output----Unit------Comment-------------------------------
    5.5      b                  in        beam width
    11.5     d                  in        beam depth (height)
    12       L                  ft        beam length
    100      w                  lb/in     uniform load on beam
             Materia                      beam material
    2.2E6    E                  lb/in^2   elastic modulus of beam material
    2        x                  ft        beam end to desired point
             y_ul    1.8479E-1  in        deflection at desired point (x)
             ymax_ul 3.6508E-1  in        maximum deflection
             Mx      144,000    in·lb     bending moment at x
             Mmax    259,200    in·lb     maximum bending moment
             Sbx     1,187.83   lb/in^2   bending stress at x
             Sb_max  2,138.1    lb/in^2   maximum bending stress
             Vx      4,800      lb        shearing force at x
             Vmax    7,200      lb        maximum shearing force
             Ssh_max 170.75     lb/in^2   maximum shearing stress
             c       5.75       in        maximum distance to neutral axis
             I       697.07     in^4      moment of inertia of beam section
             S       121.23     in^3      section modulus
             Area    63.25      in^2      area of cross-section
             y_all   .4         in        allowable deflection
             Sb_all             lb/in^2   allowable bending stress
             Ssh_all            lb/in^2   allowable shearing stress
```

(b)

FIGURE 8.27 Checking the status of variables: the Given function. (a) Rules modified with Given function. (b) Variable Sheet with solution.

8.6.2 Selecting Sound Designs: Decision-Making Statements

The addition of rules in Figure 8.27a provides the mechanism for estimating the allowable limits for the values of deflection, shearing, and bending stresses. In engineering design practice, beams are designed on the basis of one or more of these conditions or criteria: maximum deflection, maximum bending, and shearing stresses. This means that if the computed maximum value of one of the specified (or basis) parameters exceeds its allowable limit, the beam is deemed unsuitable even though the other parameters may be within the allowable limits. Thus, for a design based on deflection, the beam design is rejected if the computed maximum deflection exceeds the allowable limit.

For example, in our present model, both the maximum and allowable limits are computed for deflection, bending, and shearing stresses. However, if the computed maximum values exceed the allowable limits, the model does not automatically point it out. We need to examine the values ourselves and come to a conclusion. Instead, we will now shift this burden of checking to TK so that it makes the comparisons and alerts us when one or more of the limits is exceeded. This checking should be performed only *after* all the equations have been processed.

Shifting the Burden: The Algorithm The additional tasks to be performed for checking the limits of each of the three parameters, deflection, bending stress, and shearing stress, are:

1 Compare computed maximum value (e.g., ymax_ul) with allowable limit (e.g., y_all).
2 IF ymax_ul ≥ y_all, THEN display an error message (e.g., deflection too high).

Several key constructs or features are necessary for implementing this simple algorithm in TK; they are:

1 A conditional statement analogous to the IF-THEN statement in a conventional programming language (FORTRAN or BASIC).
2 A mechanism or function that will delay the execution of the conditional statements until all the equations have been processed. Recall that in TK, all the rules are cycled through one after another by the solvers (Section 7.4.1).
3 A mechanism or function for displaying a customized error message.

TK provides all the three features for implementing the algorithm. See Capsule 8.3 for specifying conditional rules in TK.

Expressing the Algorithm in Conditional Rules The first step in the algorithm is to compare the maximum and allowable values. As an initial attempt, we can express the comparison in TK's conditional rule syntax (see Capsule 8.3) as follows:

IF ymax_ul => y_all, THEN <display "deflection too high">

However, the action in the THEN clause (display "deflection too high") needs to be expressed in valid TK syntax. We turn to TK's built-in functions in Table 8.3 and find the **Errmsg function.** This function (Errmsg<message>) displays the message specified in the argument. The conditional rule then becomes:

IF ymax_ul => y_all, THEN Errmsg('Deflection_Too_High)

> **CAPSULE 8.3**
>
> **CONDITIONAL RULES IN TK**
>
> TK's conditional rule is very similar to the IF-THEN-ELSE statement in BASIC (Section 3.2.4). It takes the form:
>
>   IF <condition> THEN <action1> [ELSE <action2>]
>
> where <condition> is a logical expression.
>   <action1> is a valid TK rule.
>   <action2> is a valid TK rule.
>
> The ELSE <action2> clause is optional, and this is denoted by [ ].
>
>   Relational operators such as less than (<), greater than (>), less than or equal to (≤), greater than or equal to (≥), etc., are used for checking the condition in the IF clause. If <condition> evaluates to true, <action1> following the THEN clause is executed. Otherwise, if an ELSE clause is present, <action2> is executed. If there is no ELSE clause, TK proceeds with the next rule, if any. The keywords IF, THEN, and ELSE are not case-sensitive.
>
> **Example**
>
> The formula for computing the cross-sectional area of a bar depends on the shape of the cross-section. If it is rectangular, the area is given by length × width; otherwise, the cross-section is assumed to be circular, and the area is given by $\pi d^2/4$. This condition can be expressed in a TK conditional rule as follows:
>
>   IF shape = 'rectangular THEN area
>   = length * width ELSE area = pi() * d^2/4
>
> When TK processes this rule, depending on the value assigned to the variable shape, the THEN or ELSE clause is executed, and the area computed.
>
> **Another example**
>
> The two roots of a quadratic equation $ax^2 + bx + c = 0$ are given by
>
> $$\frac{-b \pm \sqrt{b^2 - 4ac}}{2a}$$
>
> However, if we try to take the square root of a negative argument, TK will return an error message (see also Section 3.2.4). To avoid the error, we can check to make sure that the argument is not negative and only then evaluate the root. A rule of the following form can be used:
>
>   If b^2 => 4 * a * c, then root1
>   = (-b + sqrt(b^2 - 4 * a * c))/(2 * a)
>
> If the condition (b^2 ≥ 4 * a * c) is true, the THEN clause is processed and root1 is evaluated; if the condition fails, the THEN clause is not executed and TK continues with the solution process.
>
>   The IF-THEN-ELSE construct in TK can thus be used to control the flow of computation in the model and to execute select rules depending on conditions.

To avoid the underscores (_) between the words in the message, the rule can be expressed as:

   IF ymax_ul => y_all, THEN Errmsg('Deflection,'Too,'High)

Problem with the Rule   If this rule is appended to the existing Rule Sheet, it will be executed every time the solver makes a pass through the set of equations (see Section 7.4.1). However, this rule should be executed only *after* all the other equations have been processed. In other words, we need to delay the execution of the rule until the model is solved. Stated in pidgin English, the conditional rule would be:

IF (model is solved AND ymax_ul ≥ y_all), THEN Errmsg('Deflection,'Too,'High).

We need two mechanisms to implement this conditional rule: one mechanism to check if all the equations have been solved and a second mechanism which would be a boolean operator (AND) so that the THEN clause will be executed only if both conditions are true. TK provides both mechanisms in the form of the Solved( ) function and the boolean operator AND.

The Solved and Boolean Functions TK's built-in function **Solved( )** returns true if the model has been solved and false if the equations have not been solved (Table 8.3). So we can use this function to check if all the other equations in the model have been solved. If they have been solved, i.e., Solved( ) is true, then the second part of the IF clause can be executed, viz., the comparison of variable values. Depending on the result of the comparison, the THEN clause will be executed.

TK provides the AND, OR, NOT, EQV, and IMPLY **boolean** operators (see Table 8.3). We make use of the AND construct to come up with the following rule:

If and(Solved( ), ymax_ul => y_all), Then Call Errmsg('Deflection,'Too,'High)

This rule meets our requirements because the comparison is made only after the model is solved and the error message is displayed only if the logical expression is true. We enter the above rule in the Rule Sheet along with the following ones for shear and bending stresses:

If and(Solved( ), Sb_max => Sb_all), Then Call errmsg('Bending, 'Stress, 'Too, 'High)

If and(Solved( ), Ssh_max => Ssh_all),
      Then Call errmsg('Horizontal, 'Shear, 'Stress, 'Too, 'High)

Increasing the Load on the Beam To ensure the correct working of the model after the addition of the new rules, we solve the 'Hickory wood problem. The solution obtained is the same as the one in Figure 8.25b. Since the computed maximum values are below the allowable limits, no error messages are displayed.

Example Let's now increase the load on the beam from 100 lb/in (Figure 8.25b) to 110 lb/in and see what happens. We assign 110 to w in the Variable Sheet and solve. The resulting solution is shown in Figure 8.28a. The computed maximum deflection of 0.4207 in is greater than the allowable limit of 0.4 in. Since all the rules have been solved and ymax_ul is greater than y_all, the deflection error message rule is triggered and the message is displayed (with a beep) on the prompt/error line at the top of the window. The rule triggering the error is also marked as shown in Figure 8.28b. Thus the model is now capable of flagging designs that violate the design criteria.

Note that the maximum bending and shearing stresses are below the corresponding allowable limits. However, the beam will be unsuitable due to excessive deflection, the design basis for the current problem. We save the model under the name <u>beamul3</u>. We can play with the model and try out other limiting conditions of bending and shearing stresses.

Modular Approach to Problem Solving The engineering beam design and analysis example illustrates the modular or step-by-step approach to problem solving with TK: starting small and expanding the scope, power and usefulness of the model

8.6 CONTROLLING THE FLOW OF COMPUTATION: CONDITIONAL RULES

```
(4i) Input: 110                                                      25+/F9
Deflection Too High
---------------------VARIABLE SHEET---------------------------------------
St-Input-----Name----Output----Unit------Comment--------------------------
    5.5      b                 in        beam width
   11.5      d                 in        beam depth (height)
   12        L                 ft        beam length
   110       w                 lb/in     uniform load on beam
   'Hickory  Materia                     beam material
             E       2.1E6     lb/in^2   elastic modulus of beam material
    2        x                 ft        beam end to desired point
             y_ul    2.1295E-1 in        deflection at desired point (x)
             ymax_ul 4.2071E-1 in        maximum deflection
             Mx      158,400   in·lb     bending moment at x
             Mmax    285,120   in·lb     maximum bending moment
             Sbx     1,306.62  lb/in^2   bending stress at x
             Sb_max  2,351.91  lb/in^2   maximum bending stress
             Vx      5,280     lb        shearing force at x
             Vmax    7,920     lb        maximum shearing force
             Ssh_max 187.83    lb/in^2   maximum shearing stress
             c       5.75      in        maximum distance to neutral axis
             I       697.07    in^4      moment of inertia of beam section
             S       121.23    in^3      section modulus
             Area    63.25     in^2      area of cross-section
             y_all   .4        in        allowable deflection
             Sb_all  3500      lb/in^2   allowable bending stress
             Ssh_all 190       lb/in^2   allowable shearing stress
```

(a)

```
(24s) Status: > User-defined error                                   25+/F9

---------------------RULE SHEET-------------------------------------------
S-Rule--------------------------------------------------------------------
      "              Error Messages

> If and(solved(),ymax_ul=>y_all) Then Call errmsg('Deflection,'Too,'High)
* If and(solved(),Sb_max=>Sb_all) Then Call errmsg('Bending, 'Stress, 'Too, 'Hi
* If and(solved(),Ssh_max=>Ssh_all) Then Call errmsg('Horizontal, 'Shear, 'Stre
```

(b)

FIGURE 8.28 User-defined error message: increasing the load (w) on the beam. (a) Variable Sheet: maximum deflection exceeds allowable limit. (b) Rule Sheet with marked error.

in a step-by-step manner. At any logical point along the way, we could have terminated the expansion process and would still have been left with a useful model, though narrower in scope. The advantages associated with modular programming are true for TK as well. By focusing on one issue at a time (units, plotting, functions, etc.), we can minimize the risk of committing errors and not being able to debug the model.

This chapter also illustrates another important point: Just as you wouldn't need to use all the features of a spreadsheet when solving a problem, you might not need to use all of TK's features for solving every problem. For instance, if we were not interested in including the properties of wood in the model, or did not have the table of properties, we would not have used the user-defined Function feature. The model would still have been useful for solving problems involving simply supported beams. The features you would use in a model would depend on the scope and complexity of the problem.

SUMMARY

We created a basic model for analyzing the response of a simply supported beam subjected to a uniformly distributed load and gradually expanded its scope to tackle a wide range of problems. We introduced the concept of Calculation and Display units of variables and looked at the Unit Conversion feature in TK. We discussed the repetitive solution of a model (List Solving) and ways of presenting the results: tables and graphs.

We created a database of wood properties using the Interactive Table feature in TK and defined functions for accessing these properties during the course of problem solving. We further expanded the scope of the model by including conditional (IF-THEN) rules to check if the computed maximum values of some of the design parameters exceeded the allowable limits and to return appropriate messages, if necessary.

We also introduced some TK-specific built-in functions that contribute to enhanced flexibility during problem solving. Through the beam design example, you should have learned about the power of an equation solver and the ease with which it can be used in science and engineering for tackling a wide spectrum of problems of varying complexities and to varying depths.

CHAPTER 9

EQUATION-SOLVING APPLICATION EXAMPLES

In this chapter, we present several application examples to illustrate the power of equation-solving tools in science and engineering. We reinforce the methodology introduced in Chapter 7 for using equation solvers. While developing a model for electric circuit analysis, we explore the ability to define and use sets of rules as functions — Rule functions — in TK. We create another model dealing with the elements of a triangle using Rule functions.

TK's powerful equation-solving capability, i.e., its backsolving feature, is further enhanced by the facility to define **procedures** (or sequence of computations) similar to subroutines in conventional programming languages. We illustrate this procedural capability by creating a Procedure function for evaluating factorials. We also create a model for computing the greatest common divisor of two integers using the Procedure Function feature.

We then discuss the seamless integration of declarative and procedural capabilities of TK by developing a model for dealing with problems in stoichiometry. Finally, we take a brief look at TK's library of models that spans a wide variety of fields from engineering to optimization.

9.1 ELECTRIC CIRCUIT ANALYSIS AND RULE FUNCTIONS

The four major elements of an electric circuit are resistors, inductors, capacitors, and a voltage source. The engineer is typically involved in selecting the individual circuit components for meeting certain performance requirements or analyzing a given circuit to determine the various circuit parameters such as the current through the various components or the voltage drop across them. The elements in a circuit can be connected in several different ways. When the same current flows through the various circuit elements, they are said to be connected in **series** (see Figure 9.1a). If the potential

314 EQUATION-SOLVING APPLICATION EXAMPLES

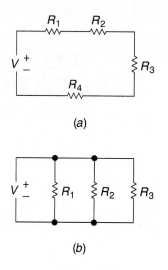

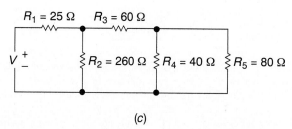

FIGURE 9.1 Electric circuits. (a) Resistors in series. (b) Resistors in parallel. (c) Series-parallel circuit.

difference across the various elements is the same, they are said to be connected in **parallel** (see Figure 9.1b). Finally, series-parallel circuits (see Figure 9.1c) are also very common.

We will create a TK model to determine the total resistance in the circuit shown in Figure 9.1c.

9.1.1 Problem Analysis and Model Design

Following the problem-solving methodology discussed in Chapter 2, the first step is to develop the mathematical model. The total resistance of a set of resistors $R_1, R_2, R_3, \ldots, R_n$, connected in series is given by the sum of the individual resistances[1] i.e.,

$$R_{\text{series}} = R_1 + R_2 + R_3 + \cdots + R_n \tag{1}$$

---

[1] See any electrical engineering textbook, e.g., *Introduction to Electrical Engineering* by C. R. Paul, S. A. Nasar, and L. E. Unnewehr, McGraw-Hill, New York, 1986.

The total resistance of a set of resistors $R_1, R_2, R_3, \ldots, R_n$, connected in parallel is computed as follows:

$$\frac{1}{R_{\text{parallel}}} = \frac{1}{R_1} + \frac{1}{R_2} + \frac{1}{R_3} + \cdots + \frac{1}{R_n} \tag{2}$$

Using these two basic relationships, we analyze the circuit in Figure 9.1c and the result is shown in Figure 9.2. Given five of the six variables in the last equation in the figure, we should be able to evaluate the unknown using the Direct Solver in TK.

Identifying a Pattern There are two distinct sets of computations, one each for series and parallel elements, as shown by the last equation in Figure 9.2. If *ser* is used to denote the computation for two resistors in series, and *par* denotes the elements in parallel, the last equation can be expressed as:

$$R12345 = \text{ser}(R1,\text{par}(R2,\text{ser}(R3,\text{par}(R4,R5)))) \tag{3}$$

For example, in Equation (3), par(R4,R5) implies that R_4 and R_5 are in parallel and the computation is a sum of the reciprocals of the individual resistances, as per Equation (2). The other parts of the equation can be interpreted in a similar manner. The sequence of computations in Equation (3) beginning with the innermost set of parentheses follows the sequence listed in Figure 9.2.

If we can represent the series and parallel computation relationships [given by Equations 1 and 2] in TK as individual functions, they can be invoked as many times as necessary (with appropriate arguments) in the model. TK provides the **Rule Function** feature for this very purpose.

9.1.2 Creating the Model: The Rule Function Feature

We will now utilize the Rule Function feature in TK (see Capsule 9.1) to create two functions, one for computing the equivalent resistance of elements in series and another for elements in parallel.

Creating Rule Functions Figure 9.3 lists the sequence of steps for creating a rule function. We have already defined the purpose of the two functions, viz., compute equivalent resistances for elements in series and parallel, respectively. The next step is to identify the types and numbers of variables. From the mathematical relationships in Figure 9.2, we see that two resistors are combined at each step to determine the equivalent resistance. Likewise, the Rule function should be defined to accept two values and compute the third. The two values of resistances can be passed to the function through Argument variables and the result returned through the Result variable.

Selecting and Entering Function Names The next step in Figure 9.3 is to select an appropriate name for the function. The chosen name should reflect the purpose or role of the function. For example, *series* or *ser* would be appropriate for circuit elements in series, while *coslaw* would be appropriate for a Rule function dealing with

RELATIONSHIPS

R_4 and R_5 are in parallel resulting in $R_{4,5}$:

$$\frac{1}{R_{4,5}} = \frac{1}{R_4} + \frac{1}{R_5}$$

R_3 and $R_{4,5}$, are in series resulting in $R_{3,4,5}$:

$$R_{3,4,5} = R_3 + R_{4,5}$$

R_2 and $R_{3,4,5}$ are in parallel resulting in $R_{2,3,4,5}$:

$$\frac{1}{R_{2,3,4,5}} = \frac{1}{R_2} + \frac{1}{R_{3,4,5}}$$

R_1 and $R_{2,3,4,5}$ are in series resulting in $R_{1,2,3,4,5}$:

$$R_{1,2,3,4,5} = R_1 + R_{2,3,4,5}$$

or

$$R_{1,2,3,4,5} = R_1 + \cfrac{1}{\cfrac{1}{R_2} + \cfrac{1}{R_3 + \cfrac{1}{\cfrac{1}{R_4} + \cfrac{1}{R_5}}}}$$

CALCULATIONS

$$\frac{1}{R_{4,5}} = \frac{1}{40} + \frac{1}{80}$$

$$R_{3,4,5} = 60 + \cfrac{1}{\cfrac{1}{40} + \cfrac{1}{80}}$$

$$\frac{1}{R_{2,3,4,5}} = \frac{1}{260} + \cfrac{1}{60 + \cfrac{1}{\cfrac{1}{40} + \cfrac{1}{80}}}$$

$$R_{1,2,3,4,5} = 25 + \cfrac{1}{\cfrac{1}{260} + \cfrac{1}{60 + \cfrac{1}{\cfrac{1}{40} + \cfrac{1}{80}}}}$$

$$= 90 \; \Omega$$

FIGURE 9.2 Mathematical relationships for the series-parallel circuit in Figure 9.1c.

CAPSULE 9.1

RULE FUNCTIONS IN TK

A **Rule function** consists of a set of one or more rules whose syntax is the same as the syntax of rules in the Rule Sheet. The Rule function is backsolvable—the rules are handled in the same manner as rules on the Rule Sheet so that known and unknown variables can appear on either side of the equals sign; moreover, the unknown variables will be evaluated regardless of the order in which the rules appear in the subsheet. Therefore, the Rule function may be viewed as an extension of the Rule Sheet. A Rule function can be saved and used in other models. The Rule function is analogous to a function subprogram in FORTRAN (see Section 3.5.3) and has the same advantages associated with subprograms, viz., repeated use of a set of rules (each time with a different argument), compact model, etc.

Defining a Rule function

The function name is declared on the Function Sheet; the function arguments and rules are defined on the Rule Function Subsheet, which is accessed from the Function Sheet with the Dive command. Up to three types of variables, **Parameter, Argument,** and **Result,** can be associated with a Rule function. These variables serve for carrying variable values to and from the Rule function.

A **Parameter variable** is common to the Rule function and other parts of the model and provides a direct link between the Variable and Rule Sheets and the Rule Function Subsheet. The value of a Parameter variable is automatically available to the Rule function when the function is processed by the solvers. A Rule function, however, **cannot** assign a value to an unknown Parameter variable, i.e., it cannot return a value. Thus, the Parameter link can be thought of as a one-way street. Up to 20 Parameter variables can be associated with each Rule function.

An **Argument variable name** is local to the Rule function, i.e., the Argument variable has no meaning outside the function and has no connection to an identically named variable in the Variable Sheet and in other Function Subsheets. For example, a variable *time* on the Variable Sheet can be assigned a value different from the variable *time* on the Rule Function Subsheet. The Argument variable can be thought of as an input variable to the function because it generally carries a value from the Variable Sheet to the function. Up to 20 Argument variables can be associated with each Rule function.

The **Result variable name** is also local to the Rule function and generally carries the result of the function computations back to the Variable Sheet. In keeping with TK's philosophy of backsolvability, a Result variable can also carry a value *to* the function; likewise, an Argument variable can be evaluated in the function and carry a value out of the function. Thus, the Argument and Result variables can be regarded as two-way **communication variables** between the Variable Sheet and Rule Function Subsheet. Depending on how the function is invoked, up to 20 Result variables can be associated with each Rule function.

In addition to these three types of declared variables, local or auxiliary variables can be used in the rules in the Rule function. Unlike the declared variables, these variables do not appear on the Variable Sheet and have no meaning outside the function.

Invoking a Rule function

The Rule function can be invoked in one of two ways from the Rule Sheet. First, it can be referred to in a rule with appropriate arguments, just as a TK built-in or user-defined List function, viz.,

$$\text{function\_name\{arguments\}}$$

The sequence of arguments in the function reference should correspond to the sequence of Argument and Result variables specified on the Rule Function Subsheet.

Second, a Rule function can be invoked with a **Call statement,** viz.,

Call fun_name(arg_exp1, arg_exp2, ..., arg_expn;
                res_exp1, res_exp2, ..., res_exp3)

where arg_exp1, arg_exp2, etc., correspond to Argument variables and res_exp1, res_exp2, etc., correspond to Result variables in the Rule function. In this type of usage, TK allows up to 20 Result variables.

A Rule function can also be invoked from within a Procedure function on the Procedure Function Subsheet, either as a function reference or through a Call statement. A Rule function cannot call or refer to itself either directly, i.e., recursively, or indirectly through intermediary rule and/or procedure functions. A Rule function can be invoked even if the Parameter variable is unknown; however, the Parameter variable cannot be assigned a value by the Rule function.

Passing values to and from Rule functions

Because the names of Argument and Result variables are local to the function and have no meaning outside the Rule Function Subsheet, values for these variables are communicated to and from the Variable Sheet through the arguments in the function reference or Call statement. Arguments preceding the optional semicolon correspond to Argument variables on the Rule Function Subsheet, while those following the semicolon correspond to Result variables. Therefore, while invoking Rule functions, care should be exercised to ensure that variables are listed in the proper sequence in which they are defined in the function.

When to use Rule functions

Just as function subprograms in FORTRAN are a way of improving the elegance and modularity of the program code, Rule functions enhance the elegance of the model. Rule functions are appropriate in situations where a set of rules is repetitively used throughout the model. In big models with large numbers of equations, Rule functions provide the means to partition the model into manageable parts which can be separately tested and debugged. Finally, Rule functions are useful for creating a library of commonly used functions that can be subsequently brought into other models.

the law of cosines in a triangle, and so on. There is no limit on the length of the function name. However, the Rule function name should not duplicate the name of a built-in function. We select *ser* for series and *par* for the parallel elements Rule functions, respectively.

We enter the two names on consecutive rows of the Function Sheet and enter $\underline{R}^1$ in the corresponding Type field to denote Rule functions. We enter the function description in the Comment field.

Defining the Functions With the cursor highlighting *ser* on the Function Sheet, we invoke the Dive command to display the Rule Function Subsheet. The description entered in the Function Sheet is automatically displayed in the Comment field of the Rule Function Subsheet. We enter the names $\underline{a}$ and $\underline{b}$ in the Argument Variables field and $\underline{z}$ in the Result Variables field. In the Rule field, we enter the rule as shown in Figure 9.4a. TK parses the rule, and since there are no syntax errors, they are accepted by TK. On the other hand, if there were syntax errors, they would be flagged, and we would be placed in the edit mode.

In a similar manner, we define the *par* function as shown in Figure 9.4b. Note that we have used the same variable names for the Argument and Result variables in the two functions. Since the variables are local to the respective functions, a value assigned to the variables in one function will not affect the value assignment in the other function.

Using the Rule Functions We enter the variable names for the resistors (R1, R2,..., R5) and the equivalent resistance (R12345) in the Variable Sheet. We also associate $\underline{ohms}$ as the unit with each of the variables. Since the chosen Rule function names correspond to the notation used in Equation (3), we enter Equation (3) as a rule in the Rule Sheet. We assign the known values to the variables and press F9 to solve the model. The resulting solution is shown in Figure 9.5. The equivalent resistance in the circuit is 90 Ω. This value is reasonable from a practical standpoint, and the model can be considered to be working correctly. Moreover, the same result was obtained in

[1] All entries that we type at the computer are shown underlined in text.

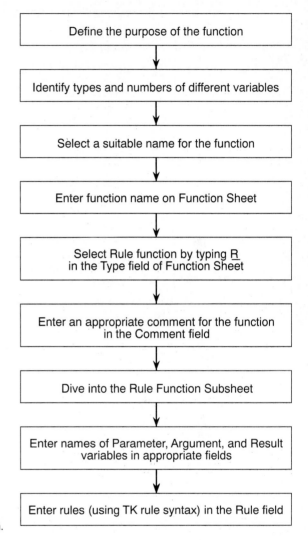

FIGURE 9.3 Steps in defining a Rule function.

the paper-pencil solution shown in Figure 9.2. We save the model under the name *circuit1*.

Modification of the Problem Let's consider a modification of the current example: If the circuit should have an equivalent resistance of 100 Ω, how should the value of one of the resistors, say, R4, change, if others are to remain the same?

We blank the value of R4 from the Input field and assign 100 to R12345 and press F9. From the resulting solution in Figure 9.6, we can see that R4 should increase to 105 Ω (from 40 Ω in Figure 9.5). Thus, the Rule functions are backsolvable.

How Does TK Tick? In Section 7.4.1 we saw how TK solves the equations one after another (Direct Solver is invoked). Even when Rule functions are present in

```
---------------------RULE FUNCTION: ser------------------
Comment:                    Two components in series
Parameter Variables:
Argument Variables:   a,b
Result Variables:     z
S-Rule---------------------------------------------------
    z = a + b
```

(a)

```
---------------------RULE FUNCTION: par------------------
Comment:                    Two components in parallel
Parameter Variables:
Argument Variables:   a,b
Result Variables:     z
S-Rule---------------------------------------------------
    1/z = 1/a + 1/b
```

(b)

FIGURE 9.4 Defining Rule functions. (*a*) Resistors in series. (*b*) Resistors in parallel.

a model, TK works in the same manner. As soon as a Rule function is encountered, the solver transfers control to the function. TK treats the Rule function as a model and the solver (Direct or Iterative) repeatedly cycles through the equations in the function until one of the conditions for the termination of the solver is reached. After this,

FIGURE 9.5 Variable and Rule Sheets for series-parallel circuit.

```
---------------------VARIABLE SHEET----------------------------------
St-Input-----Name----Output----Unit------Comment---------------------
   25        R1                ohms      resistor
   260       R2                ohms      resistor
   60        R3                ohms      resistor
   40        R4                ohms      resistor
   80        R5                ohms      resistor
             R12345   90       ohms      equivalent resistance


---------------------RULE SHEET--------------------------------------
S-Rule---------------------------------------------------------------
    "    * * *    Series-Parallel Circuits    * * *

    R12345 = ser(R1,par(R2,ser(R3,par(R4,R5))))
```

9.1 ELECTRIC CIRCUIT ANALYSIS AND RULE FUNCTIONS 321

```
--------------------VARIABLE SHEET-----------------------------------------
St-Input-----Name----Output----Unit------Comment--------------------------
    25       R1                 ohms      resistor
   260       R2                 ohms      resistor
    60       R3                 ohms      resistor
             R4       105       ohms      resistor
    80       R5                 ohms      resistor
   100       R12345             ohms      equivalent resistance
```

FIGURE 9.6 Solution to problem with known R12345 and unknown R4.

control returns to the rule from which the Rule function was invoked, and TK proceeds with the solution process making use of values evaluated in the function, if any.

Thus, we have illustrated the concept and use of Rule functions by creating two simple functions for solving problems dealing with series-parallel electric circuits. These functions can be saved with the Save Function option of the Storage command (/SF) and used in other models.

9.1.3 The Wheatstone Bridge

The basic techniques of series-parallel circuit reduction used in the previous example for computing the equivalent resistance of a circuit are not always applicable. For example, in the Wheatstone Bridge circuit shown in Figure 9.7a, the paths from nodes B and C are neither of the series nor parallel kind. The resistor across BC, viz., Rd_3 is the "problem" resistor. Consequently, it is not possible to use the basic techniques of circuit reduction. In such situations, the Δ-Y transformation technique is used. Circuit ABC in the figure is known as a Δ network. The idea behind the Δ-Y transformation is to convert the Δ network to a wye (Y) network and in the process remove the problem

FIGURE 9.7 Wheatstone Bridge circuit. (a) Original circuit: Δ network: ABC. (b) Wye network. (c) Transformed circuit.

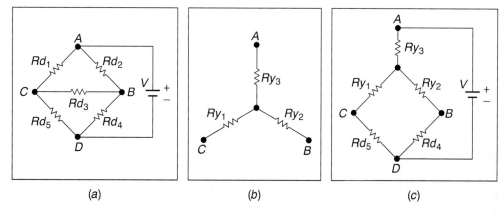

(a)   (b)   (c)

resistor (Figure 9.7*b*). The resulting network in Figure 9.7*c* can then be analyzed using series-parallel reduction techniques.

We will expand the scope of the model *circuit1* to perform the Δ-Y transformation and analyze the circuit in Figure 9.7*a* to solve the following problem: If $V = 220$ V, $Rd_1 = 10$ Ω, $Rd_2 = 25$ Ω, $Rd_3 = 15$ Ω, $Rd_4 = 9$ Ω, and $Rd_5 = 30$ Ω, calculate the corresponding Y resistances, the total equivalent resistance, and the current through the circuit.

Mathematical Relationships The equations for computing the corresponding resistances in the wye network in Figure 9.7*b* are as follows:

$$Ry_1 = \frac{Rd_1 Rd_3}{Rd_1 + Rd_2 + Rd_3} \qquad (4)$$

$$Ry_2 = \frac{Rd_2 Rd_3}{Rd_1 + Rd_2 + Rd_3} \qquad (5)$$

$$Ry_3 = \frac{Rd_1 Rd_2}{Rd_1 + Rd_2 + Rd_3} \qquad (6)$$

These equations can be derived from first principles or obtained from any standard electrical engineering textbook.

On analyzing Figure 9.7*c*, we see that resistors Ry_1 and Rd_5 are in series and so are Ry_2 and Rd_4; the equivalent resistances of these two sets of resistors are in parallel. Using *ser* and *par* as before, the equivalent resistance between nodes *AD* is given by:

$$Rad = Ry_3 + \text{par}(\text{ser}(Ry_2, Rd_4), \text{ser}(Ry_1, Rd_5)) \qquad (7)$$

and the current (*I*) from the source (*V*) to the network is given by:

$$I = \frac{V}{Rad} \qquad (8)$$

The set of Equations (4 through 8) can thus be used to perform Δ-Y transformations and compute the equivalent resistance of the circuit in Figure 9.7*c*. Let's expand the model *circuit1* to perform these computations.

Expanding the Model Circuit1 An examination of Equations (4 through 6) shows that the same arithmetic operations are being performed in all equations, but on different variables. This type of situation is ideal for the creation and use of a Rule function (see Capsule 9.1). This function can then be invoked with appropriate arguments to compute the equivalent wye resistor values.

Figure 9.8 shows the definition of the Rule function, *deltay*. There are three Argument variables and one Result variable in the function. Note the similarity between

```
--------------------RULE FUNCTION: deltay------------------------------------
Comment:               Computes equivalent wye resistance
Parameter Variables:
Argument Variables:    d1,d2,d3
Result Variables:      y
S-Rule-----------------------------------------------------------------------
  y = d1*d2 / (d1+d2+d3)
```

FIGURE 9.8 Rule function for Δ-Y transformation.

this rule and Equations (4 through 6). The Argument variables $d1$ and $d2$ correspond to the variables in the numerators of the respective equations while $d3$ corresponds to the remaining variable in the denominator. In the Rule Sheet (Figure 9.9), we invoke this function to perform the three Δ-Y transformations; we also enter Equations (7 and 8) in the Rule Sheet.

Solving the Model We assign known values to the variables and press F9 to solve the model. The solution is shown in Figure 9.10. The equivalent resistance in the circuit is 16 Ω, while the current from the source is 13.75 A. Note that we have assigned amp as the unit for the current variable, I.

Y-Δ Transformation and Iterative Solving Let's say for the circuit in Figure 9.7a, we are given resistances, $Ry_1 = 10\ \Omega$, $Ry_2 = 8\ \Omega$, $Ry_3 = 40\ \Omega$, $Rd_4 = 4\ \Omega$ and $Rd_5 = 25\ \Omega$, and $V = 220$ V, and we need to compute the other resistances, Rd_1, Rd_2 and Rd_3 respectively.

Since the Δ resistances are unknown, the solution calls for Y-Δ transformations. Let's try to solve the problem with the existing model. When we assign the known

FIGURE 9.9 Rule Sheet with additional equations for circuit in Figure 9.7a.

```
--------------------RULE SHEET-----------------------------------------------
S-Rule-----------------------------------------------------------------------
  "    * * *   Series-Parallel Circuits   * * *

  R12345 = ser(R1,par(R2,ser(R3,par(R4,R5))))


  " * * *   Delta to Wye Transformation   * * *
  Ry1 = deltay(Rd1,Rd3,Rd2)
  Ry2 = deltay(Rd2,Rd3,Rd1)
  Ry3 = deltay(Rd1,Rd2,Rd3)

  Rad = Ry3 + par(ser(Ry2,Rd4),ser(Ry1,Rd5))     "equivalent resistance
  I = V/Rad                                      "current
```

```
---------------------VARIABLE SHEET----------------------------------------
St-Input-----Name----Output----Unit------Comment----------------------------
    10       Rd1               ohm       \
    25       Rd2               ohm        Delta resistances
    15       Rd3               ohm       /

     9       Rd4               ohm
    30       Rd5               ohm
   220       V                 volt       voltage source

             Ry1      3        ohm       \
             Ry2      7.5      ohm        Wye resistances
             Ry3      5        ohm       /

             Rad      16       ohm        equivalent resistance (D-Y)
             I        13.75    amp        current
```

FIGURE 9.10 Variable Sheet with solution of circuit in Figure 9.7a.

values and solve, only the last two rules in Figure 9.9 are solved, and *Rad* and *I* are evaluated. The Δ resistances Rd_1, Rd_2, and Rd_3 are not evaluated. **Reason:** When the Direct Solver processes the first Δ-Y transformation equation (the second rule in Figure 9.9), control is transferred to the Rule function *deltay*. Since the value of only one of the four variables has been passed to the function, the rule in *deltay* remains unresolved, control is returned to the Rule Sheet and the rule is marked Unsatisfied. Similarly the next two rules on the Rule Sheet remain unresolved and the values of Rd_1, Rd_2 and Rd_3 remain unknown.

Iterative Solving   There are eleven variables (Figure 9.10) and five equations in the model for dealing with the circuit in Figure 9.7a. Though we assigned input values to six of these variables, all unknowns have not been evaluated. This means we need to use the Iterative Solver (see Section 7.5.1). The candidates for guess variables are the unknowns Rd_1, Rd_2, and Rd_3 (see Capsule 7.1). If we select just one guess variable, the equations will remain unresolved because there will still be two unknowns in each of the equations. Therefore, we select two guess variables, say Rd_1 and Rd_2, so that the solution will propagate through the set of equations.

We type G in the Status fields of Rd_1 and Rd_2; the values from the previous problem (10 and 25) appear in the respective Input fields and will serve as reasonable initial guess values for iteration (Figure 9.11a). When we trigger the solution process, the Iterative Solver is invoked and the results shown in Figure 9.11b are obtained after 6 iterations.

Thus we carry out Y-Δ transformations without adding rules or Rule functions to the existing Δ-Y model. This example also illustrates how the Iterative Solver can be used in conjunction with Rule functions in TK.

We save the complete model under the name *circuit2*. Thus, after creating a basic model with Rule functions for analyzing series-parallel circuits, we have expanded its

```
--------------------VARIABLE SHEET-------------------------------------
St-Input-----Name----Output----Unit------Comment------------------------
G   10       Rd1               ohm       \
G   25       Rd2               ohm         Delta resistances
             Rd3               ohm       /

    4        Rd4               ohm
    25       Rd5               ohm
    220      V                 volt        voltage source

    10       Ry1               ohm       \
    8        Ry2               ohm         Wye resistances
    40       Ry3               ohm       /

             Rad   48.93617    ohm         equivalent resistance (D-Y)
             I     4.4956522   amp         current
```

(a)

```
--------------------VARIABLE SHEET-------------------------------------
St-Input-----Name----Output----Unit------Comment------------------------
             Rd1     100       ohm       \
             Rd2     80        ohm         Delta resistances
             Rd3     20        ohm       /

    4        Rd4               ohm
    25       Rd5               ohm
    220      V                 volt        voltage source

    10       Ry1               ohm       \
    8        Ry2               ohm         Wye resistances
    40       Ry3               ohm       /

             Rad   48.93617    ohm         equivalent resistance (D-Y)
             I     4.4956522   amp         current
```

(b)

FIGURE 9.11 Y-Δ transformations: Iterative Solving. (*a*) Variable Sheet with initial guess values. (*b*) Equivalent Δ resistances computed after Iterative Solving.

scope to carry out extensive network analysis with an additional Rule function. As you can see, Rule functions are easy to create and use and greatly reduce the size of the model while enhancing its elegance. The Rule functions created in this model, especially *ser* and *par,* can be saved and used in other models.

9.2 ELEMENTS OF A TRIANGLE AND RULE FUNCTIONS

Problems involving the evaluation of the elements of a triangle are frequently solved by scientists and engineers. The four common combinations of known elements from which a triangle is solved are as follows:

1. Two angles and one side
2. Two sides and the included angle
3. Two sides and the angle opposite one of them
4. Three sides

Among the elements evaluated are the unknown sides and angles, area, perimeter, and radii of escribed and inscribed circles. We will follow the methodology discussed in Chapter 7 and create a TK model to solve for the elements of a triangle for various combinations of known elements.

9.2.1 Mathematical Relationships

Figure 9.12 shows the basic mathematical relationships relating the different elements of a triangle. The three equations in Figure 9.12b express the basic relationships between the six elements — three sides and three angles — of the triangle. From Section 7.4.2, we know that given any three of the values, the Direct Solver should be able to solve the model. However, in the equation for the law of cosines, b and c appear twice, meaning that they cannot be evaluated from that equation by the Direct Solver. Moreover, knowing the three angles would not constitute a set of independent input values representing a unique triangle.

To avoid the use of the Iterative Solver for solving the triangle for any combination of inputs, we can add the other versions of the laws of cosines and sines (Figure 9.12c). These additional equations that obviate the need for the Iterative Solver are known as **redundant equations** in TK terminology. Figure 9.12d shows the equations for computing the other parameters of the triangle.

Identifying Patterns An examination of the equations in Figure 9.12 shows that the use of Rule functions for representing the cosine and sine theorems would be appropriate: we have the same set of computations with different arguments each time. Likewise, we can create Rule functions to include other versions of the equations for computing the area and escribed circle radius. Note that there can be only one version of the equations for perimeter and inscribed circle radius.

9.2.2 Creating the Model

We follow the sequence of steps outlined in Figure 9.3 and create the Rule functions shown in Figure 9.13. In defining the functions, we use the degree version of the trigonometric built-in functions, sin$d$ and cos$d$. This means the arguments can be specified in degrees. Recall from Section 7.4.1 that the trigonometric functions sin, cos, etc. assume the arguments to be in radians. By using the degree version of the functions, we obviate the need to define unit conversion between degrees and radians.

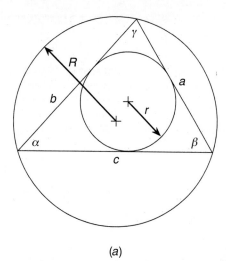

(a)

Sum of angles: $\alpha + \beta + \gamma = \pi$

Law of cosines: $a^2 = b^2 + c^2 - 2bc \cos \alpha$

Law of sines: $\dfrac{a}{\sin \alpha} = \dfrac{b}{\sin \beta}$

(b)

Law of cosines: $b^2 = c^2 + a^2 - 2ca \cos \beta$

Law of cosines: $c^2 = a^2 + b^2 - 2ab \cos \gamma$

Law of sines: $\dfrac{b}{\sin \beta} = \dfrac{c}{\sin \gamma}$

(c)

Area: $A = \dfrac{1}{2} ab \sin \gamma$

Perimeter: $P = a + b + c$

Inscribed circle radius: $r = \dfrac{2A}{P}$

Escribed circle radius: $R = \dfrac{a}{2 \sin \alpha}$

(d)

FIGURE 9.12 Elements of a triangle. (a) Triangle. (b) Basic relationships. (c) Other versions of sine and cosine theorems. (d) Additional relationships.

328 EQUATION-SOLVING APPLICATION EXAMPLES

```
-------------------RULE FUNCTION: coslaw---------------
Comment:                The Law of Cosines
Parameter Variables:
Argument Variables:     p,q,theta
Result Variables:       s
S-Rule--------------------------------------------------
  s^2 = p^2 + q^2 - 2*p*q*cosd(theta)
```

(a)

```
-------------------RULE FUNCTION: sinlaw---------------
Comment:                The Law of Sines
Parameter Variables:
Argument Variables:     side1,side2,ang1,ang2
Result Variables:
S-Rule--------------------------------------------------
  side1/sind(ang1) = side2/sind(ang2)
```

(b)

```
-------------------RULE FUNCTION: area-----------------
Comment:                Area of triangle
Parameter Variables:
Argument Variables:     p,q,theta
Result Variables:       A
S-Rule--------------------------------------------------
  A = 0.5*p*q*sind(theta)
```

(c)

```
-------------------RULE FUNCTION: rsine----------------
Comment:                Radius of escribed circle
Parameter Variables:
Argument Variables:     a,alpha
Result Variables:       R
S-Rule--------------------------------------------------
  R = 0.5*a/sind(alpha)
```

(d)

FIGURE 9.13 Rule functions for the triangle model. (a) The law of cosines. (b) The law of sines. (c) Area computation. (d) Rule of escribed circle. *Note:* The degree versions of trigonometric functions are used, i.e., the function arguments can be specified in degrees, instead of the default radians.

```
---------------------RULE SHEET------------------------------------------------
S-Rule-------------------------------------------------------------------------
    alpha + beta + gamma = 180      " sum of angles is 180 degrees
    perim = a + b + c               " perimeter of triangle

    a = coslaw(b,c,alpha)           " law of cosines
    b = coslaw(c,a,beta)
    c = coslaw(a,b,gamma)

    call sinlaw(a,b,alpha,beta)     " law of sines
    call sinlaw(b,c,beta,gamma)

    A = area(a,b,gamma)             " compute area of triangle
    A = area(b,c,alpha)
    A = area(c,a,beta)

    call rsine(a,alpha;R)           " compute radius of escribed circle
    call rsine(b,beta;R)
    call rsine(c,gamma;R)

    r = 2*A/perim                   " compute radius of inscribed circle
```

FIGURE 9.14 Rule Sheet for the triangle model.

The Rule Sheet Figure 9.14 shows the Rule Sheet for the model. We have used the **Call statement** to invoke the *sinlaw* and *rsine* Rule functions. Since there is no Result variable associated with *sinlaw* (Figure 9.13b), the function needs to be invoked with a Call statement. If we attempt to invoke the function through a reference (as the rule for the cosine law), TK will display an error message and place us in the edit mode. The sequence of arguments in the function Call statement should correspond to the sequence of Argument variables defined in the Rule function. The unknown argument will be evaluated by the function.

In Figure 9.13d, a Result variable is associated with Rule function *rsine*; therefore, the function call includes a variable name after the semicolon (see Capsule 9.1). Note that it is not necessary to use a Call statement to invoke *rsine* — a function reference would suffice, as in the case of Rule function *area* in the Rule Sheet.

To illustrate the concept of local variables, we have used the same variable name A for area on the Rule Sheet and for the Result variable in Rule function *area* (Figure 9.13c). Since the Result variable is local to the function, the value of A on the Variable Sheet will not conflict with the value of A in the Rule function. The same argument applies to the variable R.

9.2.3 Utilizing the Model

We will now use the model to solve some problems dealing with the triangle shown in Figure 9.12a. We have associated the unit centimeters (cm) with the sides, radii, and perimeter of the triangle and square centimeters (cm^2) with area. We have not shown any unit conversion definitions in the present model, though they can be easily added so that different units can be associated with the variables. In the following examples, all values of length (distance) are assumed to be in centimeters (cm).

Example 1 Given $AB = 15$, $BC = 14$, and $\alpha = 55°$. Compute the unknown parameters.

We assign the values of a, b, and α in the Variable Sheet and press F9, the Solve key. TK displays the solution shown in Figure 9.15a.

```
---------------------VARIABLE SHEET----------------------------------------
St-Input-----Name----Output----Unit------Comment-----------------------------
    15       a                    cm         side1 (AB)
    14       b                    cm         side2 (BC)
             c        17.69868    cm         side3 (CA)
    55       alpha                deg        angle apposite to side1
             beta     49.866255   deg        angle apposite to side2
             gamma    75.133745   deg        angle apposite to side3
             perim    46.69868    cm         perimeter of triangle
             A        101.48537   cm^2       area of triangle
             R        9.1558094   cm         radius of escribed circle
             r        4.3463914   cm         radius of inscribed circle
```

(a)

FIGURE 9.15 Variable Sheets with solutions for different combinations. (a) Given two sides and an angle. (b) Given one side and two angles. (c) Given three sides. (d) Given a side, an angle, and the radius of the escribed circle. (e) Given an angle, radii of inscribed and escribed circles. (f) Given two sides and an angle: TK spots out-of-range arcsin argument.

Example 2 Given $CA = 8$, $\beta = 40°$, and $\gamma = 55°$.

We blank out the values of a, b, and α in the Variable Sheet, assign the new values and solve. The solution shown in Figure 9.15b is obtained.

```
---------------------VARIABLE SHEET----------------------------------------
St-Input-----Name----Output----Unit------Comment-----------------------------
             a        9.7290334   cm         side1 (AB)
             b        6.2775902   cm         side2 (BC)
     8       c                    cm         side3 (CA)
             alpha    85          deg        angle apposite to side1
    40       beta                 deg        angle apposite to side2
    55       gamma                deg        angle apposite to side3
             perim    24.006624   cm         perimeter of triangle
             A        25.014808   cm^2       area of triangle
             R        4.8830984   cm         radius of escribed circle
             r        2.0839922   cm         radius of inscribed circle
```

(b)

FIGURE 9.15 (continued).

Example 3 Given $AB = 3$, $BC = 4$, and $CA = 5$.
The resulting solution in Figure 9.15c shows that it is a right-angled triangle.

```
---------------------VARIABLE SHEET-----------------------------------
St-Input-----Name----Output----Unit------Comment-----------------------
    3         a                   cm        side1 (AB)
    4         b                   cm        side2 (BC)
    5         c                   cm        side3 (CA)
              alpha    36.869898  deg       angle apposite to side1
              beta     53.130102  deg       angle apposite to side2
              gamma    90         deg       angle apposite to side3
              perim    12         cm        perimeter of triangle
              A        6          cm^2      area of triangle
              R        2.5        cm        radius of escribed circle
              r        1          cm        radius of inscribed circle
```

(c)

FIGURE 9.15 *(continued).*

Example 4 Given $CA = 19$, $\beta = 60°$, and $R = 10$.

```
---------------------VARIABLE SHEET-----------------------------------
St-Input-----Name----Output----Unit------Comment-----------------------
              a        14.908327  cm        side1 (AB)
              b        17.320508  cm        side2 (BC)
    19        c                   cm        side3 (CA)
              alpha    48.194872  deg       angle apposite to side1
    60        beta                deg       angle apposite to side2
              gamma    71.805128  deg       angle apposite to side3
              perim    51.228835  cm        perimeter of triangle
              A        122.6544   cm^2      area of triangle
    10        R                   cm        radius of escribed circle
              r        4.7884908  cm        radius of inscribed circle
```

(d)

FIGURE 9.15 *(continued).*

The resulting solution is shown in Figure 9.15d. The radius of the inscribed circle (r) is 4.7885 cm. However, let's say we want the radius to be 4.5 cm instead of 4.7885 cm, but wouldn't mind relaxing the value of c. We assign 4.5 to r and blank the value of c. When we solve the model, only the value of b is computed from the rule

call rsine(b,beta;R)

and the other rules remain unresolved. Since three variables have been specified and the model is not solved by the Direct Solver, we need to help the solution process by selecting a guess variable to trigger the Iterative Solver.

Using the guidelines discussed in Section 7.5.2, we select c as the guess variable and type a G in its Status field. The old value of 19 (from the previous problem) appears in the Input field; this is a logical starting or initial value for the Iterative Solver since the desired value of r (viz., 4.5) is not far from the value obtained in the previous example when c was 19. We trigger the solution process and after six iterations, the solution shown in Figure 9.15e is obtained. The value of c is 19.58 cm.

```
--------------------VARIABLE SHEET--------------------------------------------
St-Input-----Name----Output----Unit------Comment------------------------------
             a       13.331984 cm        side1 (AB)
             b       17.320508 cm        side2 (BC)
             c       19.576982 cm        side3 (CA)
             alpha   41.805128 deg       angle apposite to side1
   60        beta              deg       angle apposite to side2
             gamma   78.194872 deg       angle apposite to side3
             perim   50.229473 cm        perimeter of triangle
             A       113.01632 cm^2      area of triangle
   10        R                 cm        radius of escribed circle
   4.5       r                 cm        radius of inscribed circle
```

(e)

FIGURE 9.15 (continued).

Example 5 Given $AB = 50$, $BC = 20$, and $\beta = 30°$.

```
(4s) Status: >                                                       25+/F9
ASIN: Argument Error
--------------------VARIABLE SHEET--------------------------------------------
St-Input-----Name----Output----Unit------Comment------------------------------
   50        a                 cm        side1 (AB)
   20        b                 cm        side2 (BC)
             c                 cm        side3 (CA)
   >         alpha             deg       angle apposite to side1
   30        beta              deg       angle apposite to side2
             gamma             deg       angle apposite to side3
             perim             cm        perimeter of triangle
             A                 cm^2      area of triangle
             R                 cm        radius of escribed circle
             r                 cm        radius of inscribed circle
```

(f)

FIGURE 9.15 (continued).

When the input values are assigned and the model solved, TK beeps and displays the error message "ASIN: Argument Error" on the prompt/error line (see Figure 9.15*f*). None of the variables is evaluated. On the Rule Sheet, the rule

call sinlaw(a,b,alpha,beta)

is marked with the error symbol (>).

Reason When the Direct Solver processes the *sinlaw* Rule function, *alpha* is the only unknown. While attempting to evaluate *alpha* (Argument variable *ang1* in the Rule function), TK tries to take the inverse of the sind function and finds the argument to be greater than 1—an invalid value. Therefore, it marks the rule and terminates the solution process. We can correct the problem by increasing the value of *c*, side *BC*.

Thus, we have utilized TK's Rule Function feature for creating a general model dealing with the elements of a triangle, and these five examples illustrate the flexibility and ease with which a wide variety of problems can be solved, in quick succession.

Two-Stage Problem Solving A plot of land (*ABCD*) is shown in Figure 9.16. One of the sides is known (125 m), and some angles are known as marked. The problem is to find the value of the unknown sides in the figure and its area.

Problem Analysis We can modify the triangle model to add a rule or two to solve the problem in Figure 9.16. However, a brief analysis of the problem yields an alternate approach: The triangle model can be used in two stages to compute the unknowns. First, in triangle *ABD*, two angles and a side are known; therefore, the unknowns including *BD* and *AD* can be evaluated, along with the area of the triangle. In the second stage dealing with triangle *BCD*, we can use the computed value of *BD* along with known angles *BCD* and *CBD*. Solving the second triangle will result in values for *BC*, *CD*, and area; all the desired unknowns (*BC*, *CD*, and *DA*) would be evaluated, and the values of the two areas can be summed up to give the total area.

FIGURE 9.16 A plot of land.

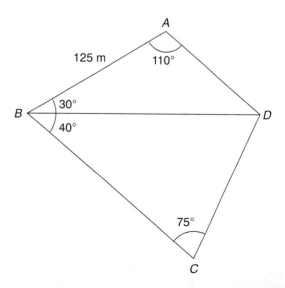

334 EQUATION-SOLVING APPLICATION EXAMPLES

```
------------------------VARIABLE SHEET----------------------------------------
St-Input-----Name----Output----Unit------Comment---------------------------
   125        a                   m        side1 (AB)
              b       182.73778   m        side2 (BC)
              c        97.232739  m        side3 (CA)
              alpha    40         deg      angle apposite to side1
   110        beta                deg      angle apposite to side2
    30        gamma               deg      angle apposite to side3
              perim   404.97051   m        perimeter of triangle
              A      5710.5555    m^2      area of triangle
              R        97.232739  m        radius of escribed circle
              r        28.202327  m        radius of inscribed circle
```

(a)

```
------------------------VARIABLE SHEET----------------------------------------
St-Input-----Name----Output----Unit------Comment---------------------------
              a       171.45899   m        side1 (AB)
              b       121.60517   m        side2 (BC)
  182.73778   c                   m        side3 (CA)
              alpha    65         deg      angle apposite to side1
    40        beta                deg      angle apposite to side2
    75        gamma               deg      angle apposite to side3
              perim   475.80194   m        perimeter of triangle
              A     10069.922     m^2      area of triangle
              R        94.592033  m        radius of escribed circle
              r        42.328208  m        radius of inscribed circle
```

(b)

FIGURE 9.17 Finding the area of Figure 9.16: two-stage problem solving. (a) Computing BD, AD, and area. (b) Computing BC, CD, and area.

Solution The Variable Sheet in Figure 9.17a shows the result of the first step in the solution process. *AD* is found to be 97.232 m, and *BD*, the diagonal common to both triangles, is equal to 182.738 m, while area (*A*) is equal to 5,710.5555 m^2.

We use this value of *BD* as input in the second stage and solve the model with values of angles *BCD* and *CBD*. The resulting Variable Sheet in Figure 9.17b shows that sides *BC* and *CD* are evaluated along with the area. Thus, all the unknown sides *BC*, *CD*, and *DA* are evaluated. Adding the values of *A* in the two figures gives us the total area viz., 15,780.48 m^2.

This example illustrates how the triangle model can be used for solving problems in other areas. We have also shown how a problem can be solved in stages using the output from one stage as the input in the next, all without modifying the model.

9.3 FACTORIALS AND PROCEDURE FUNCTIONS

As engineers and scientists, we deal with problems in probability and statistics in which we are frequently required to compute the factorials of numbers. Factorials are used for computing the number of combinations of objects taken several at a time or the number of permutations of objects taken in order, several at a time.

We will follow the methodology discussed in Chapter 7 and create a TK model for computing the factorial of a number and the number of combinations. We will introduce the concept of **Procedure functions** in TK and use it in the model.

9.3.1 Mathematical Relationships and Problem Analysis

The factorial of a number n is defined as follows:

$$n! = 1 \qquad\qquad n = 0$$
$$n! = 1 * 2 * 3 * \cdots * (n-1) * n \qquad n > 0 \qquad (9)$$

e.g., $6! = 1 * 2 * 3 * 4 * 5 * 6 = 720$.

The number of combinations of n objects, taken r at a time is given by:

$$C(n, r) = \frac{n!}{r!(n-r)!} \qquad r < n \qquad (10)$$

$$C(6, 2) = \frac{6!}{2!(6-2)!} = \frac{720}{2 * 24} = 15$$

Problem Analysis From Equation (9) it is clear that computing the factorial requires a sequence of statements or a procedure to carry out the multiplications and accumulate the product at each stage. In other words, a looplike construct, as found in conventional programming languages, is required. Once a procedure or sequence of steps is defined, the number of combinations can be computed by invoking the procedure with different arguments. TK's **Procedure Function feature** is well-suited for this type of computation.

9.3.2 Creating the Model: The Procedure Function Feature

We will utilize the Procedure Function feature (see Capsule 9.2) to create a function for computing the factorial of a number. We will then use this function to compute the number of combinations based on Equation (10).

Creating a Procedure Function The sequence of steps for creating a Procedure function closely matches the sequence listed for creating a Rule function in Figure 9.3. Since creating a Procedure function is analogous to programming in a conventional language, there is an additional step of developing an algorithm (see also Section 3.1). We have identified the purpose of the function, viz., compute the factorial of a number.

CAPSULE 9.2

PROCEDURE FUNCTIONS IN TK

A **Procedure function** is essentially a sequential program written in TK Programming Language (TKPL); **statements** in TKPL define the sequence of actions to be executed by a Procedure function. The syntax of TKPL closely parallels that of the programming language BASIC. TKPL includes the four major programming constructs: (1) assignment statement, (2) loop, (3) flow control (conditional and unconditional), and (4) Call. In addition, the various built-in functions and user-defined List, Rule, and Procedure functions can be used in the procedure. TK's **Procedure interpreter** is similar to the BASIC interpreter and ensures the integrity of the procedure syntax. A Procedure function in TK can also be thought of as a subroutine in conventional programming languages. It can be saved and used in other models. It can be called from the Rule Sheet with the appropriate number of input and output arguments.

Defining a Procedure function

The function name is declared on the Function Sheet; the function arguments and statements are defined on the Procedure Function Subsheet, which is accessed from the Function Subsheet with the Dive command. Up to three types of variables, **Parameter, Input,** and **Output,** can be associated with a Procedure function. These variables serve for carrying variable values to and from the function.

A **Parameter variable** is common to the Procedure function and other parts of the model and provides a direct link between the Variable and Rule Sheets and the Procedure Function Subsheet. The value of a Parameter variable is automatically available to the Procedure function when the function is processed by the solvers. A Procedure function, however, **cannot** assign a value to an unknown Parameter variable. Likewise, unless the values of all the declared Parameter variables are known, the solver does not enter the procedure. Thus, the Parameter link can be thought of as a one-way street. Up to 20 Parameter variables can be associated with each Procedure function.

The **Input variable** carries values from the Variable Sheet to the function. The Input variable is local to the Procedure function, i.e., the variable has no meaning outside the function and has no connection to an identically named variable in the Variable Sheet and in other Function Subsheets. For example, a variable *volume* on the Variable Sheet can be assigned a value different from the variable *volume* on the Procedure Function Subsheet. Unless all the Input variables associated with a function are known, the procedure is not entered by the solvers.

Note that this differs from the working of the Rule function, which is entered by the solvers irrespective of whether the Argument variables are known or not. Up to 20 Input variables can be associated with each Procedure function.

The **Output variable** carries the result of the function computations back to the Variable Sheet—similar to the Result variable in a Rule function. The Output variable name is also local to the Procedure function. Unlike the Rule function, however, an Output variable **cannot** be evaluated outside the Procedure function and the value passed to the function when it is called. In other words, there is no backsolvability in procedure functions. Depending on how the function is invoked, up to 20 Output variables can be associated with each Procedure function.

In addition to these three types of declared variables, local or auxiliary variables can be used in the statements in the Procedure function. Unlike the declared variables, these variables do not appear on the Variable Sheet and have no meaning outside the function.

Invoking a Procedure function

The Procedure function can be invoked in one of two ways from the Rule Sheet. First, it can be referred to in a rule with appropriate arguments, just as a TK built-in or user-defined List function, viz.,

Function_name(input_arguments; {output_arguments})

The sequence of arguments in the function reference should correspond to the sequence of Input and Output variables specified on the Procedure Function Subsheet. Note that Output variables are optional as denoted by the { } around the output_arguments.

Second, a Procedure function can be invoked with a **Call statement,** viz.,

Call fun_name(in_exp1, in_exp2, . . . , in_expn;

out_exp1, out_exp2, . . . , out_expn),

where in_exp1, in_exp2, etc., correspond to Input variables and out_exp1, out_exp2, etc., correspond to Output variables in the Procedure function. In this type of usage, TK allows up to 20 Input and 20 Output variables.

A Procedure function can also be invoked from within a Procedure function on the Procedure Function Subsheet, either as a function reference or through a Call statement. A Procedure function can call or refer to itself either directly, i.e., **recursively,** or indirectly through intermediary Rule and/or Procedure functions. The chain of references, or calls from one Procedure or Rule function to another, can extend down to 24 levels. A Procedure function, however, **cannot** be invoked if a Parameter variable is unknown.

The Procedure function syntax

As in conventional programming languages, the basic building block of a procedure is the **assignment statement** whose syntax is as follows:

            variable := valid_TK_expression

e.g.,   simp:= simp + r^4 - exp(s)

When the assignment statement is executed, the value of *simp* is replaced by the result of the expression on the right-hand side. So the value of a variable on the left-hand side of a procedure statement can be overwritten during the course of execution of the procedure. Recall that in contrast, the value assigned to a variable in a rule (on the Rule Sheet or Rule Function Subsheet) **cannot** be overwritten. If the right-hand-side expression cannot be evaluated (say, *r* is unknown or an illegal operation is attempted), the solution process is terminated with the error message "Unevaluated statement in procedure." To reiterate, the assignment statement differs from a rule on the Rule Sheet or Rule Function subsheet in two major ways: It is **not** backsolvable and the assignment operator is not a true equals sign.

Labeled statement

A **label** can be associated with a statement in the Procedure function. Unlike FORTRAN or BASIC where the label should be a number, the label in a TKPL statement can be a number or a symbolic name. The syntax of a labeled statement is as follows:

            label: {statement}

e.g.,   begin: i:=1

The curly or optional braces { } in the statement syntax denote that there is no need for a specific statement following the label. The label is very useful for transferring control within the procedure.

Control or GoTo statement

The flow of computation in the procedure is controlled with the use of **GoTo statements.** Control is transferred to a specific labeled statement either unconditionally or conditionally, i.e., based on testing for certain conditions. The syntax of an **unconditional GoTo statement** is as follows:

            Goto <label>

e.g.,   Goto begin

where *begin* is a label in the procedure.

The **conditional GoTo statement** makes use of the IF-THEN-ELSE construct of TK (see Capsule 8.3), and the syntax is as follows:

If <condition is true>, Then GoTo <label 1>
                                {Else GoTo <label 2>}

e.g., If x > y^q, Then GoTo <cond1> Else GoTo <cond2>

When this statement is executed in the procedure, depending on the result of the comparison test, control is transferred to either of the labeled statements <cond1> or <cond2>.

Loop construct

The **loop construct** in TKPL is used for the repetitive execution of a set of statements in a procedure, and its working is similar to that of loop constructs in conventional programming languages. The syntax of the TKPL loop construct closely matches the FOR-NEXT loop in BASIC and is as follows:

For variable_name {:}= <start> To <final>
                                {Step <increment>}
   statement[s]
Next variable_name

where <start>, <final>, and <increment> are expressions resulting in numeric values for the initial, final, and step values for the loop, respectively. The : following the variable name and the step are optional.

e.g.,   For i:= 1, To 10 Step 2
             k := i^2 - 1
             sum := sum + k * exp(i)/q* sqrt(p/k)
     Next i

where exp and sqrt are TK's built-in exponential and square root functions, respectively. This loop will be executed five times with the value of *i* going from 1 to 9, in increments of 2.

Return statement

The **Return statement** is used to transfer control back from a Procedure function to the calling rule or statement. The Return statement is optional at the end of the procedure, and, if omitted, control is automatically transferred to the calling rule or procedure. The syntax of the Return statement is as follows:

            RETURN.

Passing values to and from Procedure functions:

Because the names of Input and Output variables are local and have no meaning outside the Procedure Function Subsheet, values for these variables are communicated to and from the arguments in the Function reference or Call statement. Arguments preceding the semicolon correspond to Input variables on the Procedure Function Subsheet, while those following the semicolon correspond to Output variables. Therefore, while invoking Procedure functions, care should be exercised to ensure that variables are listed in the proper sequence in which they are defined in the function.

Purpose of Procedure functions

Procedure functions complement the powerful declarative or constraint resolution nature of TK. Together, the seamlessly integrated declarative and procedural problem-solving capabilities make TK a unique programming language. The major use of Procedure functions is to encode sequences of computations or procedures that would not fit the declarative structure available on the Rule Sheet.

For example, implementing a loop or a conditional statement to execute a block of statements is not easy on the Rule Sheet, while it is extremely simple to achieve on the Procedure Function Subsheet.

Euclid's algorithm for computing the greatest common denominator of a set of numbers is procedural in nature and cannot be implemented in a declarative framework. Likewise, evaluating a definite integral using Simpson's rule, solving a differential equation using the Runge-Kutta procedure, or optimizing a nonlinear objective function using the Nelder-Mead simplex algorithm are three of the broad range of problems that can be tackled in TK using Procedure functions.

Besides these unique enhancements arising from the use of Procedure functions, other advantages associated with Rule functions (see Capsule 9.1) are applicable for Procedure functions as well.

From Equation (9), we know that the function should accept a value and return its factorial. In other words, one Input and one Output variable should be associated with the function.

The Algorithm Based on the definition of factorial, Equation (9), we come up with the following pseudocode algorithm:

Variables: n (for number) and fact (for factorial)

Obtain n

Set fact = 1

If n equals 0 or 1, return fact (i.e., 1) & stop.

For i = 2 to n

    Compute fact = i * fact   ←Loop

Next i

Naming the Function The function name should reflect the purpose or nature of the function and should not clash with the name of an existing built-in or user-defined function. We select factoriali for the name and enter it in the Name field of the Function Sheet. We type P (for Procedure) in the Type field and enter a suitable description in the Comment field.

Defining the Function We dive into the Procedure Function Subsheet and enter the Input and Output variable names. Based on the algorithm, we enter the procedure statements; TK parses the statements and checks for any syntax errors. The resulting Procedure Function Subsheet is shown in Figure 9.18. In this procedure, the factorial is being computed in steps till the final value is reached. Since the final value is obtained after several steps (or iterations), this method of computing the factorial is known as an **iterative procedure.** Note that the term *iterative* as used here does not refer to TK's Iterative Solver.

```
    (4t) Statement: next i                                              25+/OK

    -------------------PROCEDURE FUNCTION: Factoriali----------------------
    Comment:                  Factorial, iterative definition
    Parameter Variables:
    Input Variables:      n
    Output Variables:     z
    S-Statement---------------------------------------------------------------
      z:= 1
      for i = 2 to n
        z:= i*z
      next i

    " Definition intervals:
    "     0 ≤ n ≤ 21    . . .  procedure returns a correct value
    "    22 ≤ n ≤ 170   . . .  returns an approximate value; rounding-off error
    "                          due to 16 significant digits in TK
    "         n > 170   . . .  procedure fails because of TK's limit on floating
    "                          point numbers
```

FIGURE 9.18 Computing factorials: Procedure Function Subsheet. *Note:* The definition intervals are included as comments in the sheet.

Using the Procedure Function We enter the variable names, n for number and facti for factorial in the Variable Sheet. On the Rule Sheet, we enter the rule to invoke the Procedure function. In this instance, the procedure is invoked through a function reference. To compute the factorial of 9, we assign 9 to n and press F9. The resulting solution is shown in Figure 9.19. Since 9! is indeed equal to 362880, we know that the procedure is error-free. If $n = 0$ or 1, the for-next loop is not executed and the function returns the value of z, viz., 1. To compute the factorial of a different number, all we need to do is assign the value to n and press F9—TK comes up with the answer. We save the model under the name factor1.

9.3.3 Computing Factorials: A Recursive Procedure Function

Recursion is defined as a technique of describing something partly in terms of itself. For example, the iterative definition of factorial in Equation (9) can be expressed recursively as follows:

$$n! = 1 \qquad n = 0$$
$$n! = n * (n - 1)! \qquad n > 0 \tag{11}$$

e.g.,
$$5! = 5 * (5 - 1)! = 5 * 4!$$
$$5! = 5 * 4 * (4 - 1)! = 5 * 4 * 3!$$
$$5! = 5 * 4 * 3 * (3 - 1)! = 5 * 4 * 3 * 2!$$
$$5! = 5 * 4 * 3 * 2 * (2 - 1)! = 5 * 4 * 3 * 1!$$
$$5! = 5 * 4 * 3 * 2 * 1 * (1 - 1)! = 5 * 4 * 3 * 2 * 1 * 0!$$

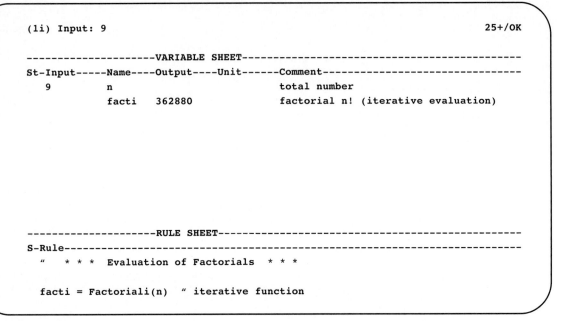

FIGURE 9.19 Variable and Rule Sheets for computing factorials.

But 0! = 1, which is the terminal condition.

So $\quad 5! = 5 * 4 * 3 * 2 * 1 * 1 = 120.$

Thus, in the recursive definition of factorial, the function calls itself repeatedly, each time the argument decreases by 1, until the terminal condition of $n = 0$ is reached, at which time the factorial is evaluated completely. If *factorialr* is the name of the function, then the recursive definition of the factorial of n is given by:

$$\text{factorialr}(n) = n * \text{factorialr}(n - 1) \tag{12}$$

Recursion in TK The major hallmark of structured programming languages such as Pascal and C is recursion. As we mentioned in Capsule 9.2, recursion is allowed in TK, i.e., a Procedure function can call or reference itself directly or indirectly. The limit of recursive nesting is 24 levels, however. We will use the definition in Equation (11) and create a recursive procedure function, *Factorialr*.

Creating a Recursive Function The sequence of steps for creating a recursive function is the same as for creating any other Procedure function. The Input variable will be the number and the Output of the function will be the computed factorial. Figure 9.20 shows the Procedure Function Subsheet for *Factorialr*, the recursive definition of factorial. We have used TK's IF-THEN-ELSE construct for defining the function. If n is greater than 1, the function calls itself repeatedly, otherwise it returns 1. Note that the function definition matches Equation (11) and the representation in

```
(c) Comment: Factorial, recursive definition                              25+/F9

--------------------PROCEDURE FUNCTION: Factorialr----------------------------
Comment:                  Factorial, recursive definition
Parameter Variables:
Input Variables:     n
Output Variables:    z
S-Statement------------------------------------------------------------------
  if n > 1 then z:= n*Factorialr(n-1) else z:= 1

  " Definition intervals:
  "    0 ≤ n ≤ 21   . . .  procedure returns a correct value
  "   22 ≤ n ≤ 25   . . .  procedure returns an approximate value; rounding-off
  "                        error due to 16 significant digits in TK
  "        n > 25   . . .  procedure fails because of TK's limit on recursive
  "                        calls nesting
```

FIGURE 9.20 Computing factorials: recursive procedure. *Note:* The definition intervals are included as comments in the sheet.

Equation (12). The recursive procedure is shorter and more elegant than the iterative procedure. It is also more "natural" or intuitive.

The Rule Sheet and Number of Combinations On the Rule Sheet, we enter the equation to invoke the newly created Procedure function, *Factorialr*. From Equation (10), the rule for computing the number of combinations (numcomb) of n objects, taken k at a time is as follows:

$$\text{numcomb} = \text{Factorialr}(n)/(\text{Factorialr}(n - k)*\text{Factorialr}(k))$$

Since Equation (10) is valid only if n > k, we can use the IF-THEN-ELSE construct to test if n is greater than k and proceed with the computation or return an appropriate error message using the built-in function Errmsg. We enter the following rule in the Rule Sheet:

If n > k then numcomb = Factorialr(n)/(Factorialr(n - k)*Factorialr(k))
                              else call errmsg('n, 'should, 'be, 'greater, 'than, 'k).

Of course, in the above rule we could use the iterative definition (Factorial*i*) for computing the number of combinations.

Example Figure 9.21 shows the Variable and Rule Sheets with the solution for n = 9 and k = 6. Note that the value of the factorial is the same (as it should be) with iterative and recursive computations. Therefore, the recursive procedure is also correct. There are 84 different combinations of selecting six objects at a time from nine. We save the model under the same name factor1.

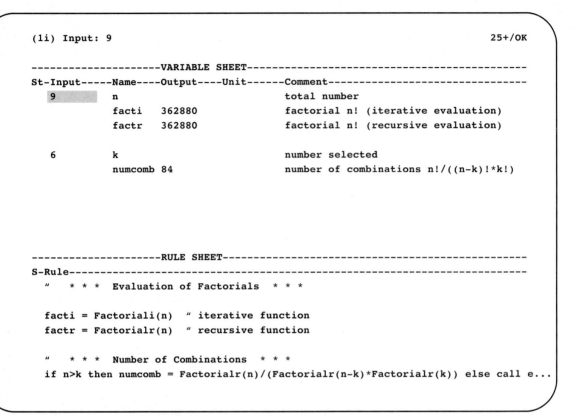

FIGURE 9.21 Factorials and number of combinations: Variable and Rule Sheets. Note: the complete rule for numcomb is not shown here.

Generating a User-Defined Error Figure 9.22 shows the Variable and Rule Sheets when we try to solve the model with n = 5 and k = 6. The factorial is computed (120); when the Direct Solver processes the rule for computing the number of combinations, the test n > k fails and the ELSE clause of the rule is activated. TK marks the rule and displays the error message in the prompt/error line with a beep.

Levels of Nesting and Context-Sensitive Help Figure 9.23 shows the Variable and Rule Sheets when the model is solved with n = 26 and k = 6. The Iterative function yields a value for *facti* (4.0329E26). When the recursive function in the second rule is invoked with n = 26, the maximum number or levels of nesting in TK (24) is exceeded. TK terminates the solution process with an error message on the prompt/error line. Since the Iterative function (*factoriali*) does not make any calls to itself, it computes the value of *facti*.

To obtain more information on the error message, we press the Help key (F1). TK displays the Help screen shown in Figure 9.24. The message is context-sensitive, i.e., the probable cause and plausible remedies for the current error are displayed. Thus, TK's context-sensitive Help facility comes in very handy in "problem" situations.

Number Representation and Computational Errors In TK a numeric constant or value is represented by a real number in standard or exponential format with up to

```
      (7s) Status: > User-defined error                                      25+/F9
      n should be greater than k
      --------------------VARIABLE SHEET------------------------------------
      St-Input-----Name----Output----Unit------Comment-----------------------
           5         n                                total number
                     facti    120                    factorial n! (iterative evaluation)
                     factr    120                    factorial n! (recursive evaluation)

           6         k                                number selected
                     numcomb                          number of combinations n!/((n-k)!*k!)



      --------------------RULE SHEET----------------------------------------
      S-Rule-----------------------------------------------------------------
           "   * * *  Evaluation of Factorials  * * *

        facti = Factoriali(n)   " iterative function
        factr = Factorialr(n)   " recursive function

           "   * * *  Number of Combinations  * * *
      > if n>k then numcomb = Factorialr(n)/(Factorialr(n-k)*Factorialr(k)) else call e...
```

FIGURE 9.22 User-defined error message displayed when n < k.

16 significant digits; the nonzero absolute value ranges from 1 E–307 to 1 E+308. Attempts to use a greater value will result in a numeric overflow with an appropriate message, while smaller numbers will result in a value of 0. In most computations with TK, we wouldn't have to worry about these limits. However, in some examples such as those dealing with factorials, the returned value becomes approximate due to rounding-off errors caused by the number of significant digits available for representing the number.

For instance, the **definition intervals** for the Iterative Factorial function (*factoriali*) for various values of n are as follows:

$0 \leq n \leq 21$	Procedure returns a correct value.
$22 \leq n \leq 170$	Returns an approximate value; rounding-off error due to 16 significant digits in TK.
$n > 170$	Procedure fails because of TK's limit on floating-point numbers.

Likewise, the **definition intervals** for the Recursive Factorial function (*factorialr*) computation are as follows:

$0 \leq n \leq 21$	Procedure returns a correct value.
$22 \leq n \leq 25$	Returns an approximate value; rounding-off error due to 16 significant digits in TK.

```
(1i) Input: 26                                                          25+/F9
Functions nested too deep
---------------------VARIABLE SHEET----------------------------------------
St-Input-----Name----Output----Unit------Comment--------------------------
    26        n                          total number
              facti   4.0329E26          factorial n! (iterative evaluation)
>             factr                      factorial n! (recursive evaluation)

    6         k                          number selected
              numcomb                    number of combinations n!/((n-k)!*k!)




---------------------RULE SHEET-------------------------------------------
S-Rule--------------------------------------------------------------------
   "   * * *  Evaluation of Factorials  * * *

   facti = Factoriali(n)   " iterative function, see the Function Subsheet for no
>  factr = Factorialr(n)   " recursive function, see the Function Subsheet for no

   "   * * *  Number of Combinations  * * *
*  if n>k then numcomb = Factorialr(n)/(Factorialr(n-k)*Factorialr(k)) else call e...
```

FIGURE 9.23 Error message: TK's limit on number of recursive calls (24) is exceeded. TK stops without evaluating numcomb.

n > 25   Procedure fails because of TK's limit on recursive calls nesting (24)—see Figure 9.23.

We can add these definition intervals as comments in the corresponding Procedure Function Subsheets (see Figures 9.18 and 9.20).

The Built-in Gamma Function   TK provides the built-in **Gamma function** which returns the (n - 1)! of its argument n. For example, to compute the factorial of 9, the TK rule would be:

$$\text{Fact} = \text{gamma}(9 + 1)$$

The advantage of the Gamma function is that it is invertible. In other words, for a given value of the factorial, we can find the corresponding number. Note, however, that the Procedure functions *factoriali* and *factorialr* are noninvertible.

Procedure or Rule Functions   On occasion, you may find that the set of equations can be expressed equally well in the form of a Procedure function or a Rule function. The option chosen will depend on the specific circumstances and/or requirements:

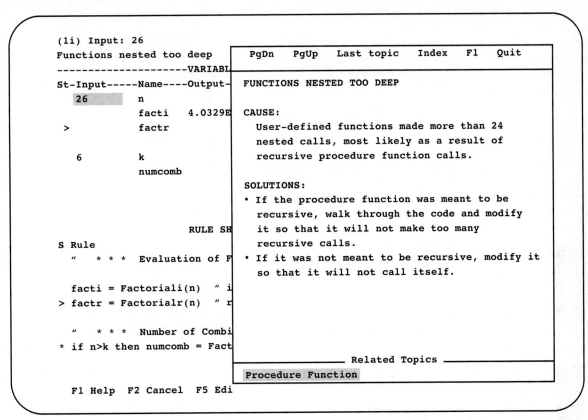

FIGURE 9.24 Context-sensitive help invoked by pressing F1. TK provides a probable cause and plausible remedies for overcoming the problem.

Rule functions are backsolvable and can be used for solving for various sets of known and unknown variables with ease. On the other hand, certain types of situations will require full control over the sequence of calculations which may speed up the problem-solving process. In such situations, a Procedure function will be the appropriate choice.

Thus, we have introduced the basic concepts of Procedure functions in TK and created a model for computing the factorial of a number and the number of combinations. We have deliberately chosen a simple example (and also refrained from using the built-in **Gamma function**) to illustrate the concepts. We can create fairly involved and lengthy procedures (similar to conventional programs) using this feature for solving complex problems.

9.4 GREATEST COMMON DIVISOR AND PROCEDURE FUNCTION

The Euclidean algorithm, named after Euclid, presents an elegant way of computing the greatest common divisor (GCD) of two positive integers. We will create a TK model using the Euclidean algorithm to compute the GCD of two positive numbers.

9.4.1 Mathematical Relationships

Figure 9.25 shows the Euclidean algorithm to compute the GCD of two integers a and b along with an example computation. Since the computation is repetitive, or loop-oriented, we need to create a Procedure function.

Creating the Procedure Function GCD The function should accept two values (a and b) and return the result. Therefore, there will be two Input variables and one Output variable associated with the function. No Parameter variables will be required for the function. Figure 9.26 shows the Procedure Function Subsheet for the procedure GCD. The statements are annotated with comments and closely match the algorithm in Figure 9.25.

The first statement in the procedure checks to see if the values are integers; if either of them is a noninteger, the **Errmsg function** returns an error message. The next two statements make use of the built-in function **abs** to ensure that the integers are positive. A local variable x is used. The next statement (loop:) is an example of a TKPL **labeled statement** (see Capsule 9.2) and serves as the address to which control can be transferred. Step 1 of the algorithm in Figure 9.25 is implemented using the built-in function **mod** in the next statement. Note that **mod**(x,z) returns the remainder of the integer division of x by z. In the next statement, the **Return construct** of TKPL is used to terminate the procedure if the remainder is 0. After the assignment of values, the **Goto** construct is used to transfer control to the labeled statement loop.

The Model Figure 9.27 shows the Variable and Rule Sheets with the variables and rule in the model. TK's result of GCD(90, 36) matches the paper-and-pencil solution in Figure 9.25 and confirms the correctness of the procedure.

FIGURE 9.25 The Euclidean algorithm for computing the GCD of two numbers a and b.

EUCLIDEAN ALGORITHM

Step 1	Divide a by b and assign remainder to r.
Step 2	If $r = 0$, then GCD = b; END.
Step 3	Set $a = b$, and $b = r$.
Step 4	Goto Step 1.

Example $a = 90$, $b = 36$. Find GCD.

Pass 1:

Step 1	90/36, remainder $r = 18$.
Step 2	r is not equal to 0.
Step 3	$a = 36$, $b = 18$.
Step 4	Goto step 1.

Pass 2:

Step 1	36/18, remainder $r = 0$.
Step 2	$r = 0$, so $GCD = 18$, END.

```
---------------------PROCEDURE FUNCTION: GCD---------------------------------
Comment:              Greatest Common Divisor, Euclidean algorithm
Parameter Variables:
Input Variables:      a,b
Output Variables:     z
S-Statement------------------------------------------------------------------
  if or(int(a)<>a,int(b)<>b) then call errmsg('Non,'integer,'argument,'for,'GCD)
  x:= abs(a)                         " take the absolute value of a
  z:= abs(b)                         " take the absolute value of b
  loop:                              " labeled statement
    remainder:= mod(x,z)             " compute remainder (Step 1)
    if remainder=0 then return       " if remainder=0 stop (Step 2)
    x:= z                            " set x = z (Step 3)
    z:= remainder                    " set z = remainder (Step 3)
    goto loop                        " transfer control (Step 4)
```

FIGURE 9.26 The Procedure function GCD.

```
---------------------VARIABLE SHEET-----------------------------------------
St-Input-----Name----Output----Unit------Comment---------------------------
   90         a                          \ integer
   36         b                          / numbers

              gcd       18               greatest common divisor


---------------------RULE SHEET---------------------------------------------
S-Rule---------------------------------------------------------------------
 "   ***  Greatest Common Divisor ***

   gcd = GCD(a,b)
```

FIGURE 9.27 Variable and Rule Sheets with solution of GCD(90, 36).

Recursive Definition of GCD The GCD of two numbers a, b may also be evaluated recursively using the following statement in a separate Procedure function (GCDrec):

If b > 0 then z:= GCDrec(b,mod(a, b)) else z:= a

While this elegant statement can replace the iterative definition in Figure 9.26, the recursive definition has the disadvantage arising out of the limit on number of recursive calls that can be made by a TK function. Therefore, the recursive procedure will not be appropriate for values requiring a large number of steps, in excess of 25.

9.4.2 Tracing the Procedure: The Debug Function

During the course of debugging a procedure, we may want to know the values of the variables during the intermediate steps. This will help us debug the procedure step by step. Recall from Section 3.2.2 that the intermediate values are printed (using the PRINT statement) when programming in a conventional language. TK's built-in function **Debug(var1, var2,..., varn)** can be called to display the values of the variables var1, var2,..., varn during the course of execution of the procedure.

In the present example, it would be a good idea to find out the values of x and z at the beginning of each loop execution. Therefore, we insert an empty row after the labeled statement (loop:) and enter the statement Call Debug(x,z). When we press F9 to solve the model, TK displays the two values 90, 36 on the prompt/error line indicating that the values of x and z are 90 and 36, respectively. When we press any key, TK continues with processing the statements until it comes to the Debug function again. This time, the next set of values 36, 18 is displayed on the prompt/error line. Since the procedure is terminated on the next line, no more intermediate values are displayed — only the result on the Variable Sheet.

Another Example   The sequence of values displayed by the Debug function for GCD(624, 222) is 624, 222; 222, 180; 180, 42; 42, 12; 12, 6. The result, of course, is 6. This sequence can be used to gain an understanding of the algorithm, as well. If one or more of the variables listed in the argument of the Debug function is unknown, TK displays ??? on the prompt/error line to signal the unknown status. The Debug function can thus be judiciously utilized for debugging complex procedures. The function can also be called from the Rule Sheet or Rule Function Subsheet.

We have once again illustrated the usefulness of the Procedure function in TK for solving problems involving procedural computations. This function can be saved separately and used in other models. A simple variation of the GCD function is the function for computing the lowest common multiplier. We have also shown the usefulness of the built-in **Debug function** in tracing the working of procedures.

9.5 STOICHIOMETRY: INTEGRATING DECLARATIVE AND PROCEDURAL COMPUTATIONS

The analysis and synthesis of chemical compounds involves chemical reactions, and the best way to describe a chemical reaction is to write it in the form of a chemical equation using the formula for each substance involved. The chemical equation describes a reaction both qualitatively and quantitatively, viz., what substances are involved and how much of each, respectively. In a chemical reaction, one or more substances called **reactants** are changed into one or more new substances called **products**. Stoichiometry is the quantitative study of chemical reactions. Typical problems solved in stoichiometry include finding the resulting mass of products from known quantities of reactants or estimating the amount of reactants required for producing desired amounts of products and other combinations.

9.5.1 Protecting the Environment: Processing Industrial Waste

The treatment and ecologically safe disposal of industrial waste are two important functions of manufacturing facilities, especially if the waste contains toxic materials.

Often, the treatment involves one or more chemical reactions. For example, if the waste contains dangerous mercury compounds (mercuric chloride: $HgCl_2$), one technique is to let the compounds react (with aluminum, for instance) to form elemental mercury (Hg), which can be easily trapped. The other product would be aluminum chloride ($AlCl_3$).

The first step toward studying this reaction is to write a balanced chemical equation for the reaction. We can then calculate the amount of the individual products and reactants involved in the reaction. For example, if we know the amount of $HgCl_2$ being processed, we can estimate the amount of aluminum that would be required and the resulting product masses. Alternately, we can find out the mass of $HgCl_2$ that can be processed for a given quantity of aluminum and so on.

We will follow the problem-solving methodology introduced in Chapter 7 and create a TK model for carrying out these computations. We will make the model as general as possible so that problems involving other reactions can be solved without extensive modifications to the model.

Example Consider the following problem: Aluminum is made to react with $HgCl_2$ in an effort to trap the mercury. Estimate the amount of aluminum required to react with 500 g of $HgCl_2$. Also calculate the amounts of Hg and $AlCl_3$ produced as a result of the reaction.

Figure 9.28 shows a step-by-step paper-and-pencil solution of part of the problem. The amount of aluminum required for processing 500 g of $HgCl_2$ is 33.127 g.

Solution Methodology Balancing the chemical reaction requires that the number of atoms of each of the elements on both sides be equal. We start by selecting a value for the coefficient of one of the compounds and correspondingly assign values to others to maintain the equality of atoms on both sides. For example, in Figure 9.28, we start with assigning 3 to $HgCl_2$. In the TK model, we would need to have separate rules for balancing each of the elements in the reaction. These rules can be grouped into a single Rule function.

Estimating the molecular mass of the individual compounds requires access to the atomic masses of elements, i.e., information from the Periodic Table of Elements. This information from the Periodic Table can be represented in the form of a user-defined List function in the model, thus providing the necessary values for computing the molecular masses. Since the computation steps are the same for each of the compounds, a function can be defined and used as many times as required. A Procedure function would be appropriate since the computation involves summation of atomic masses of individual elements — a procedural computation.

Mole balances for an individual element on both sides of the reaction involves a simple computation of ratios that can be implemented as a Rule function.

Thus, we need to create the four functions shown in Figure 9.29 as part of the model.

9.5.2 Creating the Model

As the first step in developing the model, we create the following four functions:

List Function mass Figure 9.30 shows part of the user-defined List function, *mass*. The Domain list *symbol* contains the symbols of the elements from the Periodic Table, while the Range list *atommass* contains the corresponding atomic masses. The

Step 1 Write chemical equation and balance it:

$$HgCl_2 + Al \rightarrow Hg + AlCl_3 \quad \text{(Unbalanced)}$$
$$3HgCl_2 + 2Al \rightarrow 3Hg + 2AlCl_3 \quad \text{(Balanced)}$$

Step 2 Estimate the number of moles of the given substance ($HgCl_2$); a **mole** is defined as the ratio of the mass of the substance to the molecular or formula mass of the substance.

Molecular or formula mass of $HgCl_2$

$$= \text{atomic mass of mercury} + 2 \times \text{(atomic mass of chlorine)}$$
$$= 200.59 + 2 \times (35.453)$$
$$= 271.496 \text{ g}$$

Note: The atomic masses of mercury and chlorine are obtained from the Periodic Table.

$$\text{Given mass of } HgCl_2 = 500 \text{ g}$$

$$\text{Moles of } HgCl_2 \text{ available} = \frac{\text{mass of substance}}{\text{molecular mass}}$$

$$= \frac{500}{271.496} = 1.842 \text{ moles}$$

Step 3 Determine the moles of the required substance using the balanced equation.

3 moles $HgCl_2$ react with 2 moles of aluminum

$$\therefore \text{ moles of Al required} = 1.842 \text{ moles } HgCl_2 \times \frac{2 \text{ moles Al}}{3 \text{ moles } HgCl_2}$$

$$= 1.228 \text{ moles of Al}$$

Step 4 Express the moles of the required substance in grams:

$$= 1.228 \text{ moles Al} \times \frac{\text{atomic mass of Al}}{1 \text{ mole Al}}$$

$$= 1.228 \times \frac{26.98}{1 \text{ mole Al}}$$

$$= 33.127 \text{ g}$$

Thus 33.127 g of aluminum are required to react with 500 g of $HgCl_2$.

FIGURE 9.28 Sequence of steps in solving a stoichiometry problem.

```
---------------------FUNCTION SHEET------------------------------------
Name------------Type------Arguments---Comment------------------------------
mass            List       1;1         relates atomic masses & element symbols
mmole           Rule       4;0         balances moles on both sides
gmmass1         Procedure  2;1         computes molecular masses of compounds
reaction        Rule       4;0         Chemical reaction: x HgCl(2) + y Al --> u
```

FIGURE 9.29 Function Sheet with summary of various functions in model.

```
--------------------LIST FUNCTION: mass------------------------------
Comment:               relates atomic masses & element symbols
Domain List:           symbol
Mapping:               Table
Range List:            atommass
Element---Domain--------------Range----------------------------------
1         'H                  1.0079
2         'He                 4.0026
3         'Li                 6.94
4         'Be                 9.01218
5         'B                  10.81
6         'C                  12.011
7         'N                  14.0067
8         'O                  15.9994
9         'F                  18.998403
10        'Ne                 20.17
11        'Na                 22.98977
12        'Mg                 24.305
13        'Al                 26.98154
14        'Si                 28.0855
15        'P                  30.97376
```

FIGURE 9.30 List function mass: relates elements and atomic masses. There are 105 elements in the Domain and Range lists of the function. Only 15 are shown in the figure.

two lists are related through a Table mapping since it is a discrete relationship. All the 105 elements are included in the model, though only a few are shown in the figure.

Procedure Function gmmass1 The Procedure function *gmmass1* to compute the molecular mass of compounds is shown in Figure 9.31a. It is annotated with appropriate comments. The function accepts the names of two lists as Input variables, one containing the symbols of the elements in the compound and the other the corresponding number of atoms. These two lists are used to compute the molecular mass which is returned by the function as *gmm*, the Output variable.

The built-in function **length** is used to determine the number of elements in the list containing the element symbols. Thus, the model is not restricted to handling a fixed

```
------------------------PROCEDURE FUNCTION: gmmass1------------------------------
Comment:                computes gm molecular masses of compounds
Parameter Variables:
Input Variables:        el1,n1
Output Variables:       gmm
S-Statement---------------------------------------------------------------
  a:=length(el1)                       " find number of elements in compound
  if a=0 return                        " if no elements, return.
  gmm:= 0                              " initialize mol. mass to zero
  for i = 1 to a                       " for each element, find atomic mass
      gmm:= gmm + n1[i] * mass(el1[i]) " and multiply it by number of elements
  next i                               " to compute molecular mass.
```

(a)

```
------------------------RULE FUNCTION: mmole------------------------------
Comment:                balances moles on both sides
Parameter Variables:
Argument Variables:     p,q,r,s
Result Variables:
S-Rule---------------------------------------------------------------
  p/q=r*s
```

(b)

FIGURE 9.31 Rule and Procedure functions. (a) Procedure function to compute molecular mass of compound. (b) Rule function to balance moles.

number of elements per compound. If there are no elements in the compound, that is, a = 0, no value is returned by the function. Otherwise, the user-defined List function *mass* is used to compute the molecular mass of the compound in the FOR-NEXT loop construct. Note that the computation of the molecular mass parallels that shown in Figure 9.28. An individual element in a list is referred to by the use of square brackets [ ]. For example, n1[i] refers to the *i*th element in list referenced by n1. This is similar to accessing elements in an array in BASIC or FORTRAN.

Rule Function mmole Figure 9.31b shows the Rule function *mmole* used for balancing the moles on either side of the reaction. The variables p, q, and s correspond respectively to the mass, gram molecular mass, and coefficient of the reactant (or product) whose moles are being balanced. The variable r represents the balancing factor and is common for all reactants and products. All the variables have been specified as Argument variables so that any of the unknowns can be evaluated. The Rule function thus provides backsolvability and can be invoked with different arguments for each of the elements involved in the reaction.

Rule Function reaction We have created the Rule function *reaction* in which the unbalanced chemical reaction is specified. Each rule in the function (Figure 9.32) corresponds to an element; the number of atoms on one side is equated to the corre-

```
------------------------RULE FUNCTION: reaction------------------------------------
Comment:                Chemical reaction: x HgCl(2) + y Al --> u Hg + v AlCl(3)
Parameter Variables:
Argument Variables:     cr1,cp1,cr2,cp2
Result Variables:
S-Rule-------------------------------------------------------------------------------
    cr1   =   cp1     " Balancing the # of Mercury atoms
    cr1*2 =   cp2*3   " Balancing the # of Chlorine atoms
    cr2   =   cp2     " Balancing the # of Aluminum atoms
```

FIGURE 9.32 Rule function for entering the chemical reaction: Balance the individual elements on both sides.

sponding number on the other. For example, in the second rule, two chlorine atoms in $HgCl_2$ are equated to the three chlorine atoms in $AlCl_3$.

By creating a separate function for entering the chemical reaction (in the form of rules in *reaction*), we have isolated the portion of the model that needs to change as the chemical equation changes. This model can thus be used for solving most chemical reactions without any major modifications.

The Rule Sheet Figure 9.33 shows the Rule Sheet for the stoichiometry model. Provision has been made for handling up to three reactants and three products in the chemical reaction. If there are only two reactants (as in the present example), the third set of equations is automatically ignored without affecting the final results.

Data Entry and Interactive Table We make use of the Interactive Table feature in TK to facilitate easy data entry for specifying the elements and the number of atoms involved in the reaction. Figure 9.34*a* shows the list names in the reactants table (*gmmcalR*). For example, the list eR1 contains the symbols of elements in the first reactant, while nR1 contains the corresponding number of atoms. An interactive table for products (*gmmcalP*) is similarly created (Figure 9.34*b*).

To solve the current chemical reaction, we display the Interactive Table *gmmcalR* and enter the information as shown in Figure 9.35*a*. For example, since $HgCl_2$ is the first reactant, the elements Hg and Cl are listed in eR1, while the corresponding number of atoms 1 and 2, respectively, are entered in nR1. Since aluminum is the only other reactant, Al is entered in eR2, and so on. In a similar manner, the Interactive Table *gmmcalP* is filled in with information on the products of the reaction (see Figure 9.35*b*). Note that all the elements are prefixed with an apostrophe (') to denote symbolic values (see Section 8.5).

The Variable Sheet We create the Variable Sheet with the appropriate variable names, units, and comments in the desired order (see Figure 9.36). We use cRi and cPi to denote the coefficients of reactants and products, respectively, and mRi and mPi for the masses of the reactants and products, respectively. The coefficients are based on the following notation for the balanced chemical equation:

$$[cR1] \ [reactant1] + [cR2] \ [reactant2] + [cR3] \ [reactant3] \rightarrow$$
$$[cP1] \ [product1] + [cP2] \ [product2] + [cP3] \ [product3] \qquad (13)$$

```
------------------------RULE SHEET-----------------------------------------
S-Rule---------------------------------------------------------------------
   "  * * *  Compute molecular mass of compounds  * * *
   gmmR1 = gmmass1('eR1,'nR1)
   gmmR2 = gmmass1('eR2,'nR2)
 * gmmR3 = gmmass1('eR3,'nR3)

   gmmP1 = gmmass1('eP1,'nP1)
   gmmP2 = gmmass1('eP2,'nP2)
 * gmmP3 = gmmass1('eP3,'nP3)

   "  * * *  Compute coefficients  * * *
   call mmole(mR1,gmmR1,n,cR1)
   call mmole(mR2,gmmR2,n,cR2)
 * call mmole(mR3,gmmR3,n,cR3)

   call mmole(mP1,gmmP1,n,cP1)
   call mmole(mP2,gmmP2,n,cP2)
 * call mmole(mP3,gmmP3,n,cP3)

   "  * * *  Invoke rule function with reaction equation  * * *
   call reaction(cR1,cP1,cR2,cP2)
```

FIGURE 9.33   Rule Sheet for stoichiometry problems.

In the present example, the equation is:

$$[cR1]\ HgCl_2 + [cR2]\ Al \rightarrow [cP1]\ Hg + [cP2]\ AlCl_3 \tag{14}$$

Solving the Model   Since the mass of the first reactant $HgCl_2$ is 500 g, we assign 500 as input to mR1. From the solution methodology discussed earlier, we know that the coefficients of one of the compounds should be assigned to trigger the solution process. Looking at the unbalanced equation, we realize that assigning 3 as the coefficient to $HgCl_2$ would make an appropriate choice. Note that it would be beneficial to use our chemistry knowledge while selecting which coefficient to assign and what value. Of course, we could select a different compound and/or a different value for the coefficient.

When we press F9, the solution shown in Figure 9.36 is obtained. The unknown coefficients for the three compounds in the reaction are evaluated. Substituting the coefficients in Equation (14), the balanced chemical equation is:

$$3HgCl_2 + 2Al \rightarrow 3Hg + 2AlCl_3$$

which is the same as the one in Figure 9.28. The individual molecular masses of the reactants and products are also displayed. The amount of aluminum required to handle

```
--------------------TABLE: gmmcalR-----------------------
Screen or Printer:              Screen
Title:                          Reactants: Elements and Numbers
Vertical or Horizontal:         Vertical
Row Separator:
Column Separator:
First Element:                  1
Last Element:
List------Numeric Format---Width---Heading-----------------
eR1                             10
nR1                             10
eR2                             10
nR2                             10
eR3                             10
nR3                             10
```

(a)

```
--------------------TABLE: gmmcalP-----------------------
Screen or Printer:              Screen
Title:                          Products: Elements and Numbers
Vertical or Horizontal:         Vertical
Row Separator:
Column Separator:
First Element:                  1
Last Element:
List------Numeric Format---Width---Heading-----------------
eP1                             10
nP1                             10
eP2                             10
nP2                             10
eP3                             10
nP3                             10
```

(b)

FIGURE 9.34 Table Subsheets. (*a*) Reactants. (*b*) Products.

the mercuric chloride is 33.127 g. Since this value corresponds to the paper-and-pencil solution, the model is working correctly. We save it under the name CHEM.

Modification of the Problem In the existing example, what happens if there are only 30 g of aluminum and we want to find out how many grams of $HgCl_2$ can be safely processed? We blank the existing value of $HgCl_2$, assign 30 to mR2 and press F9. The resulting solution is shown in Figure 9.37—only 452.80 g of $HgCl_2$ can be processed.

```
--------------------TABLE: gmmcalR----------------------------------------
Title:        Reactants: Elements and Numbers
Element-eR1--------nR1--------eR2--------nR2--------eR3--------nR3-------------
1        'Hg        1         'Al        1
2        'Cl        2
```

(a)

```
--------------------TABLE: gmmcalP----------------------------------------
Title:        Products: Elements and Numbers
Element-eP1--------nP1--------eP2--------nP2--------eP3--------nP3-------------
1        'Hg        1         'Al        1
2                             'Cl        3
```

(b)

FIGURE 9.35 Interactive tables. (a) Elements in the reactants. (b) Elements in the products.

9.5.3 Another Chemical Reaction

Electrolysis is a common commercial process used to bring about the reaction between common salt and water to produce sodium hydroxide, chlorine, and hydrogen. Calculate the amount of salt required to produce 200 g of sodium hydroxide by this process. Also estimate the amount of hydrogen and chlorine produced in the process.

Solution We follow the sequence of steps given below for solving the problem using the CHEM model:

- Write the chemical equation with the unknown coefficients.
- Enter the elements in the table of reactants (*gmmcalR*).
- Enter the elements in the table of products (*gmmcalP*).
- Modify the rule function (*reaction*) to balance the atoms on both sides of the reaction; change the number of Argument variables, if necessary.
- On the Rule Sheet, modify the arguments in the rule invoking the Rule function *reaction*.
- Assign a value to one of the reactant coefficients.
- Assign known values of variables.
- Solve the model.

The Chemical Equation The chemical equation for the electrolysis reaction is as follows:

$$[cR1]\ NaCl + [cR2]\ H_2O \rightarrow [cP1]\ NaOH + [cP2]\ Cl_2 + [cP3]\ H_2 \quad (15)$$

where [cR1], [cR2], [cP1], [cP2], and [cP3] are the unknown coefficients.

```
---------------------VARIABLE SHEET-----------------------------------------
St-Input-----Name----Output----Unit------Comment---------------------------------
                                          *  *  *   Stoichiometry   *  *  *

                                          Reaction:

                                          x HgCl(2) + y Al --> u Hg + v AlCl(3)

                                          Coefficients of compounds:
      3        cR1                mole    coefficient of 1st reactant, HgCl(2)
               cR2       2        mole    coefficient of 2nd reactant, Al
               cR3                mole    coefficient of 3rd reactant, (None)

               cP1       3        mole    coefficient of 1st product, Hg
               cP2       2        mole    coefficient of 2nd product, AlCl(3)
               cP3                mole    coefficient of 3rd product, (None)

                                          Molecular masses of compounds:
               gmmR1   271.496    g/mole  molecular mass of 1st reactant
               gmmR2   26.98154   g/mole  molecular mass of 2nd reactant
               gmmR3              g/mole  molecular mass of 3rd reactant

               gmmP1   200.59     g/mole  molecular mass of 1st product
               gmmP2   133.34054  g/mole  molecular mass of 2nd product
               gmmP3              g/mole  molecular mass of 3rd product

                                          Masses involved in the reaction:
    500        mR1                g       mass of 1st reactant
               mR2     33.126995  g       mass of 2nd reactant
               mR3                g       mass of 3rd reactant

               mP1     369.41612  g       mass of 1st product
               mP2     163.71087  g       mass of 2nd product
               mP3                g       mass of 3rd product

               n       .61388259          balancing factor by which chem. eqn.
                                          is multiplied.
```

FIGURE 9.36 Variable Sheet for reaction: x HgCl$_2$ + y Al → u Hg + AlCl$_3$. Given: mass of HgCl$_2$, computing Al.

Interactive Tables The new Interactive tables for the reactants and products in the chemical equation given by (15) are shown in Figures 9.38*a* and *b*. Note that three products are produced in the reaction, which is reflected in *gmmcalP*.

Rule Function reaction Figure 9.38*c* shows the new rules in the Rule function *reaction*. The elements on both sides of the reaction are balanced. Since there are three products, we need to add another Argument variable cP3, the coefficient for the third product, to the function.

```
-------------------------VARIABLE SHEET----------------------------------
St-Input-----Name----Output----Unit------Comment-----------------------
                                          * * *  Stoichiometry  * * *

                                          Reaction:

                                          x HgCl(2) + y Al --> u Hg + v AlCl(3)

                                          Coefficients of compounds:
     3       cR1                 mole     coefficient of 1st reactant, HgCl(2)
             cR2       2         mole     coefficient of 2nd reactant, Al
             cR3                 mole     coefficient of 3rd reactant, (None)

             cP1       3         mole     coefficient of 1st product, Hg
             cP2       2         mole     coefficient of 2nd product, AlCl(3)
             cP3                 mole     coefficient of 3rd product, (None)

                                          Molecular masses of compounds:
             gmmR1   271.496     g/mole   molecular mass of 1st reactant
             gmmR2    26.98154   g/mole   molecular mass of 2nd reactant
             gmmR3               g/mole   molecular mass of 3rd reactant

             gmmP1   200.59      g/mole   molecular mass of 1st product
             gmmP2   133.34054   g/mole   molecular mass of 2nd product
             gmmP3               g/mole   molecular mass of 3rd product

                                          Masses involved in the reaction:
             mR1     452.80292   g        mass of 1st reactant
     30      mR2                 g        mass of 2nd reactant
             mR3                 g        mass of 3rd reactant

             mP1     334.5454    g        mass of 1st product
             mP2     148.25752   g        mass of 2nd product
             mP3                 g        mass of 3rd product

             n        .55593565           balancing factor by which chem. eqn.
                                          is multiplied.
```

FIGURE 9.37 Variable Sheet for reaction: x HgCl$_2$ + y Al → u Hg + v AlCl$_3$. Given: mass of Al, computing HgCl$_2$.

Rule Sheet The call to *reaction* on the Rule Sheet is modified as follows to account for the additional Argument variable:

Call reaction(cR1, cP1, cR2, cP2, cP3)

Variable Sheet We assign a value of 2 to cR1, the coefficient of NaCl, with the idea that the chlorine atoms on both sides would be balanced. Since 200 g of NaOH

```
--------------------TABLE: gmmcalR--------------------------------------
Title:          Reactants: Elements and Numbers
Element-eR1--------nR1--------eR2--------nR2--------eR3--------nR3-----------
1          'Na          1         'H         2
2          'Cl          1         'O         1
```

(a)

```
--------------------TABLE: gmmcalP--------------------------------------
Element-eP1--------nP1--------eP2--------nP2--------eP3--------nP3-----------
1          'Na          1         'Cl        2         'H         2
2          'O           1
3          'H           1
```

(b)

```
--------------------RULE FUNCTION: reaction--------------------------
Comment:               Chemical Reaction: x NaCl+y H2O --> u NaOH + v Cl2 + w H2
Parameter Variables:
Argument Variables:    cr1,cp1,cr2,cp2,cp3
Result Variables:
S-Rule---------------------------------------------------------------
  cr1   =   cp1          " Balancing the # of Sodium atoms
  cr1   =   cp2*2        " Balancing the # of Chlorine atoms
  cr2*2 =   cp1 + cp3*2  " Balancing the # of Hydrogen atoms
  cr2   =   cp1          " Balancing the # of Oxygen atoms
```

(c)

FIGURE 9.38 Setting up another chemical equation problem. (a) Interactive table with reactants. (b) Interactive table with products. (c) Rule function with balanced elements.

are desired, we assign 200 as input to mP1, the mass of the first product. When we press F9, the solution shown in Figure 9.39 is obtained. 292.23 g of common salt are required for the reaction in which 177.28 g of chlorine and 5.03 g of hydrogen are also produced. The mass of water required (90.08 g) is also computed in the model.

Thus, CHEM is a general model that can be used easily for solving a wide variety of stoichiometry problems.

This application example clearly shows how the backsolvability of rules in TK can be effectively combined with its procedural capabilities for solving problems involving declarative and procedural computations. The model is also an excellent example of how databases can be accessed during the course of problem solving in TK.

```
----------------------VARIABLE SHEET----------------------------------------
St-Input-----Name----Output----Unit------Comment-------------------------------
                                         * * *   Stoichiometry   * * *

                                         Reaction:

                                         x NaCl+y H2O --> u NaOH + v Cl2 + w H2

                                         Coefficients of compounds:
    2        cR1                mole     coefficient of 1st reactant, NaCl
             cR2      2         mole     coefficient of 2nd reactant, H(2)O
             cR3                mole     coefficient of 3rd reactant, (None)

             cP1      2         mole     coefficient of 1st product, NaOH
             cP2      1         mole     coefficient of 2nd product, Cl(2)
             cP3      1         mole     coefficient of 3rd product, H(2)

                                         Molecular masses of compounds:
             gmmR1    58.44277  g/mole   molecular mass of 1st reactant
             gmmR2    18.0152   g/mole   molecular mass of 2nd reactant
             gmmR3              g/mole   molecular mass of 3rd reactant

             gmmP1    39.99707  g/mole   molecular mass of 1st product
             gmmP2    70.906    g/mole   molecular mass of 2nd product
             gmmP3    2.0158    g/mole   molecular mass of 3rd product

                                         Masses involved in the reaction:
             mR1      292.23526 g        mass of 1st reactant
             mR2      90.082599 g        mass of 2nd reactant
             mR3                g        mass of 3rd reactant

    200      mP1                g        mass of 1st product
             mP2      177.27799 g        mass of 2nd product
             mP3      5.0398692 g        mass of 3rd product

             n        2.5001831          balancing factor by which chem. eqn.
                                         is multiplied.
```

FIGURE 9.39 Variable Sheet for reaction: x NaCl + y H_2O → u NaOH + v Cl_2 + w H_2.

9.6 TK'S LIBRARY OF FUNCTIONS

In addition to the rich repertoire of functions built into TK and its powerful procedure and rule function capabilities, TK comes with a library of predefined models for handling problems ranging from matrix multiplication to the solution of differential equations using the Runge-Kutta procedure.

TK's model library is organized as shown in Figure 9.40. Each model is documented illustrating the usage of the model. To solve a problem using one of these

Library name	Coverage
ROOTS	Roots of equations
DIFFINT	Differentiation and integration
DIFFEQ	Differential equations
SPECIAL	Special functions
COMPLEX	Complex variables
OPTIM	Optimization
MATRIX	Matrix manipulation
STAT	Statistics
ENG	Engineering and science
FINANCE	Financial
UTIL	Utilities
GRAPH	Special graphs and plots
TABLES	Interactive tables/Data exchange

FIGURE 9.40 Organization of TK's library of models.

models, all we need to do is to load the model (/SL command), set up the problem, and press F9.

This library of functions can be thought of as a mini-knowledge-based system involving subject knowledge in each of the areas listed.

Codifying Bodies of Knowledge TK's problem-solving and knowledge management abilities are fully captured in the various supplements to classic textbooks in several areas. For example, the body of knowledge — equations and tables — in Roark and Young's text on stress-strain formulas has been crafted into a set of TK models which can be used "as is" for solving a wide variety of problems in civil engineering without having to access text or reference books. Other bodies of knowledge codified in TK include heat transfer and physics.

SUMMARY

We followed the methodology introduced in Chapter 7 for creating models to solve a wide range of problems in science and engineering. First, we used TK's Rule function feature to create a model for electric circuit analysis. We expanded the scope of the model to carry out Δ-Y and Y-Δ transformations. To further reinforce the concept of Rule functions, we created a general model for dealing with problems involving the elements of a triangle.

We introduced TK's programming capabilities by developing a model for computing the factorial of a number. In doing so, we also looked at the recursion feature in TK. We looked at the other constructs of TKPL while developing a model for computing the greatest common divisor of two numbers.

TK's ability to seamlessly integrate declarative and programming computations was used in developing a general model for solving problems in stoichiometry. This model shows how a database can be represented in the form of a List function and combined easily with rules, Rule and Procedure functions. Finally, we briefly covered the library of models accompanying TK for handling problems in a wide range of applications.

CHAPTER 10

DATABASE MANAGEMENT

In this chapter, we introduce the basic concepts of databases and discuss their terminology. We present a structured methodology consisting of six major steps for creating and utilizing databases. We then use this methodology for developing a database for the Periodic Table of Elements. We also discuss some major characteristics of database management software programs.

10.1 DATABASE CONCEPTS

The development of engineering and scientific databases is one of the most important information processing activities of engineers and scientists. Data is increasingly regarded as a vital resource, and like any other resource (e.g., human, machinery, or materials), it must be organized and utilized efficiently to maximize its value for the organization. Data, by itself, has no intrinsic value; only when it is transformed and used as information or knowledge is it of value.

The availability of powerful, yet easy-to-use, database management software on personal computers has helped scientists and engineers build and utilize large databases. These programs assist you in organizing, interrogating, and searching databases without the cumbersome chore of writing programs in conventional languages.

10.1.1 What Is a Database?

A **database** is a model or collection of interrelated data that are systematically organized without unnecessary redundancy to facilitate easy access, retrieval, and modification. A database can be thought of as a repository of information needed for carrying out several functions. For example, a corporate database containing information about

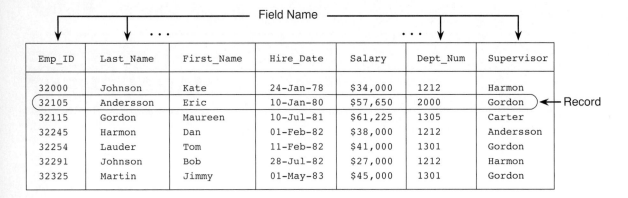

Answers to Some Queries

The *seniormost employee* is: Kate Johnson.
The employee with the *highest salary* is: Maureen Gordon.
Dan Harmon's *supervisor* is: Eric Andersson.

FIGURE 10.1 Part of a corporate database with employee information.

its employees (see Figure 10.1) can be used to find out the senior-most employee or the employee with the highest salary or the supervisor of a particular employee.

An everyday example of a database is the telephone directory with the names, addresses, and telephone numbers of individuals and businesses. Since the names are arranged or sorted by the last name, you can easily find the telephone number of an individual if it is listed in the phone book. However, to find the name of an individual in the phone book corresponding to a certain telephone number is another matter. The phone book is a good example of a database organized in a manner consistent with its usage, i.e., finding numbers for known names.

A database containing the Periodic Table of Elements can be used by a student for performing stoichiometry calculations while an engineer can use it to determine if a particular element is radioactive or not (see Figure 10.2). Likewise, a design engineer consults an engineering parts database in order to select components for an assembled product, while an inventory control manager uses the same database to order parts. Therefore, the database should be set up in such a way that it can be utilized to suit different end uses—without having to maintain a separate copy of the data for each intended use.

10.1.2 Database Terminology

By definition, a database is an organized collection of interrelated data. Data are typically stored in individual **files**. For example, Figure 10.1 corresponds to an employee information file in the corporate database. In general, a database is a collection of files. However, for our explanation, we will use the term *database* to refer to a single

Element	Symbol	Atomic number	Atomic weight	Room state	Radioactive
Plutonium	Pu	94	244	S	Yes
Aluminum	Al	13	26.982	S	No
Mercury	Hg	80	200.59	L	No
Antimony	Sb	51	121.75	S	No
Hydrogen	H	1	1.0079	G	No
Bromine	Br	35	79.904	L	No
Tin	Sn	50	118.69	S	No
Thorium	Th	90	232.04	S	Yes
Oxygen	O	8	15.999	G	No
Phosphorus	P	15	30.974	S	No
Radium	Ra	88	226.03	S	Yes

Room state: S, solid; L, liquid; G, gas.

FIGURE 10.2 A few elements from the Periodic Table. *Note:* The mnemonic field names in the database are as follows: Element, Symbol, Atom_Num, Atom_Wt, Room_Stat, and Rad_Act.

Files, Records, and Fields Figure 10.1 shows information pertaining to seven employees, where each row corresponds to an employee. Each of the rows in the figure is referred to as a **record**. So a record is a unit of data or information within a file. For example, the second record in the figure is made up of the employee ID number (32105), the last name (Andersson), the first name (Eric), the date of hire (10-Jan-80), and so on. Each of these items of data, viz., 32105, Andersson, etc., is stored in a **field**. Thus, a database is a collection of data files. A data file is a collection of records. And a record is a collection of fields.

Field Name The fields in a record are its basic data elements. A **name** is associated with a field in the record. The field name must begin with a letter from the alphabet; it can be several characters long (usually limited to 10), and it is typically made up of letters, numbers, and some special characters. Thus, in Figure 10.1, the field name *Emp_ID* stands for employee ID number, while *Salary* stands for the employee's salary. Each of these fields in a record is filled in with the appropriate data for a particular employee. Thus, for *Emp_ID* 32245, the entry in the *Salary* field is $38,000. It is a good practice to select field names that reflect field contents.

Field Type The **field type** determines the type of data that can be entered in a particular field. The three major types of frequently used data are **character** or **text** (string), **numeric**, and **logical**. In addition, **date** and long text, or **memo**, are two data types used in conjunction with databases. Therefore, database programs typically allow you to declare five different field types.

Table 10.1 shows examples of data or field types from Figures 10.1 and 10.2. Thus, the *Last_Name* field is a character field which accepts a string of alphanumeric characters, just letters in the present example. Note the *Emp_ID* field: its contents, though numbers, are treated as a string since it is declared as a character field. It is not possible to perform arithmetic operations using this value.

The *Salary* field is declared as a numeric field; it will accept any number. It is possible, however, to specify the range of values that can be entered in the field. For example, since an employee salary cannot be a negative number, you could specify

TABLE 10.1 FIELDS IN A DATABASE: NAMES, TYPES, LENGTHS, AND CHARACTERISTICS

Field name	Field type	Field length	Explanation
Last_Name	Character	10	String of letters.
Emp_ID	Character	5	String of numbers. No arithmetic computations are performed with employee ID number.
Hire_Date	Date	9	Calendar date.
Salary	Numeric	8	Numeric value. Limits can be assigned.
Supervisor	Character	10	String of letters.
Rad_Act	Logical	1	Binary choice: Can be true or false, e.g., material is radioactive or not.

that the field accept only positive values. The field contents can be used in arithmetic computations.

Hire_Date is an example of a date field: its contents are calendar dates. Dates can be displayed in different formats.

Finally, *Rad_Act* is an example of a logical field type. It will accept one of two values: true or false. If an element, e.g., thorium, is radioactive, the value is true (or Yes) as shown in Figure 10.2.

Field Length The **field length** determines the number of characters that can be entered in a field. Thus, if the *Last_Name* field is restricted to 10 characters, the field will accept only 10 characters. When you enter names longer than the field length, the program will either truncate the entry or refuse to accept it. If you have a long first or last name, you have probably experienced this on your utility and credit card bills. Field *width* is often interchangeably used with field *length*.

10.2 DESIGNING AND CREATING A DATABASE

Designing and creating a database are analogous to building a house — the form, function, and cost should be judiciously optimized. In a database system, the data should be organized in such a way that it meets the user's needs and can be economically stored and retrieved with ease. Figure 10.3 shows six major steps in the design, implementation, and utilization of a database.

10.2.1 Analyzing Data and User Needs

The objective of database design is to define and organize data to meet our current and projected requirements within system constraints (e.g., access time, reliability) and the user (e.g., degree of flexibility, ease of use).

The first step in the database development process is **analysis**: analysis of available data and our requirements leading to identification of basic data items in the database that will meet our objectives as users. Since this analysis is based on our logical data manipulation and retrieval needs and not on any specific computer hardware and software, this phase is also known as the **logical** or **conceptual design phase**.

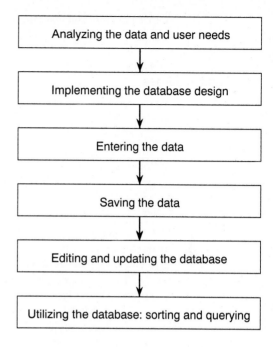

FIGURE 10.3 Six steps in the creation and utilization of databases.

At the end of this phase, the following should have been identified:

- The number of fields in the database
- The field types (character, numeric, logical, date, or memo)
- The key fields to be used for sorting and searching (more in Section 10.2.6)

10.2.2 Implementing the Database Design

The next step is to implement the conceptual database design in a specific computer hardware-software configuration. Since we are primarily concerned with personal computers (and available database management software), this step involves matching the database structure with the program's capabilities and limitations. This includes the physical organization and layout of the fields, their widths, and the user interface.

The outcome of this phase is the database framework, including the following:

- Physical layout of the fields
- The exact number and names of fields in a record
- The field lengths or widths and types
- Definition of the user interface

10.2.3 Entering Data

Once the database structure or framework is created, the next step is to enter data in appropriate fields. This process of creating database records is known as **data entry**. The time required to enter data depends on the nature and amount of data, the speed with which the data can be typed, and the number of fields in a record. The result of this phase is the first version of the database.

10.2.4 Saving the Database

Saving the created database is the next important step; since the entered data typically reside in the computer's random access memory (RAM), they will be lost if the system power is accidentally interrupted or if the system breaks down. Therefore, it is good practice, as with any data entry operation, to save entered data periodically on a hard or floppy disk. Once the database is saved, it can be subsequently retrieved and utilized.

10.2.5 Editing the Database

Data change over time. Consequently, the database should reflect current status and meet our needs. This is achieved through the database editing process. For example, new elements may be discovered that need to be added to the Periodic Table (Figure 10.2), or salaries of employees in the corporate database (Figure 10.1) may change and need to be updated. Employees may leave the company, resulting in deletion of their records from the database. Or there may be errors during data entry that must be corrected. All these activities, viz., **adding, deleting, updating,** and **correcting** records are carried out as part of the **database editing process.**

10.2.6 Utilizing the Database

The last and perhaps most useful step from the user's perspective is utilization of data entered in the database. The entire database can be displayed on screen or a hard copy generated for record keeping. The database can be viewed from different perspectives. Only a few fields in the records can be selectively displayed, e.g., the last name and salary. Or only a portion of the database, for example, the first 100 records, can be viewed. This process of displaying selective parts of the database in the desired format is known as **reporting**.

Sorting The database can be organized according to our needs. For example, the Periodic Table database can be organized to display and print the elements in ascending order of their atomic weights. This process is known as **sorting**. The field used as the criterion for sorting is known as the **key field** or simply, **key**. Here, atomic weight would be the key field.

You can specify more than one key for sorting. In that case, the main key is known as the **primary key,** and the other keys are known as **secondary keys.** For example, there are two Johnsons in the employee database in Figure 10.1. Sorting only by the last name (primary key) will not be adequate, because there is more than one employee with Johnson as the last name. We need some rules (or criteria) to determine which of these employees should be listed first. Since the specified criterion will be used for sorting after the primary key, it is known as the secondary key. More than one secondary key can be specified. An appropriate key for this example would be the first name. Recall that the names in a telephone directory are sorted first by last name (primary key) and then by first name (secondary key).

Searching The database can be **queried** or **searched** to display records meeting certain criteria. For example, the corporate database (Figure 10.1) can be queried to display all employees who joined the company after January 1, 1982, *and* who make more than $40,000 per year. From the data in Figure 10.1, only two employees meet these criteria; they are Tom Lauder and Jimmy Martin.

The ease and flexibility with which this data manipulation can be carried out greatly depends on the database design and software features and limitations.

10.2.7 Data Security and Integrity

One of the primary objectives of creating databases is to improve the accessibility and dissemination of information that will be useful to a large number of people in an organization. With this, however, come the issues of data security and integrity. These issues are especially critical since databases are increasingly accessible over computer networks.

For example, a corporate employee database containing employee performance rating and salary information is confidential and should be accessible only to authorized individuals in the organization. The legal ramifications of the improper use of information from credit bureaus and other information gathering agencies has brought a new dimension to an individual's right to privacy and freedom of information. Likewise, an engineering database with the parts specifications for the design of a highly successful commercial product should be carefully guarded against unauthorized access.

A system for controlling access to the database should be developed as part of the database development process. This is necessary to ensure the integrity and security of the database. The authority to modify the database should be carefully controlled and granted only to a few individuals. Likewise, the access to read the database should also be regulated.

10.2.8 Characteristics of a Database Management Program

Figure 10.4 shows the characteristics of database programs that let you create and use databases on personal computers. Note that these features closely match the development process shown in Figure 10.3.

Database Creation and Data Entry First and foremost, the programs provide well-designed helpful user interfaces that assist in the database development process. The program allows you to define the database structure in a straightforward manner: field names, types, lengths, and number of fields in a record. Typically, a screen- or form-oriented interface is provided for adding records to the database. The form displays all fields in a record (see Figure 10.5); after data is entered in a field, the cursor can be moved to the next field in a single keystroke, e.g., with the Tab key.

370 DATABASE MANAGEMENT

> **Defining the database structure:**
>   Field names, types, and lengths
>   Number of fields per record
> **Adding records to the database:**
>   Screen- or form-oriented Append function
> **Editing the database:**
>   Correcting errors in fields
>   Updating fields and records
>   Adding fields and records
>   Deleting fields and records
>   Retrieving a previously saved file for editing
> **Saving the database:**
>   Saving the database on disk
> **Utilizing the database:**
>   Browsing through the database
>   Looking at individual records
>   Sorting the database according to criteria
>   Searching the database according to criteria
>   Generating reports from the database (hard copy and text files)
> **Database security:**
>   Controlling Read access
>   Controlling Update access

FIGURE 10.4 The characteristics of database management programs.

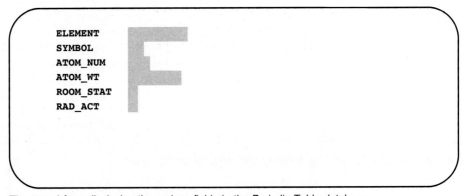

FIGURE 10.5 The record form displaying the various fields in the Periodic Table database.

Edit Facility The edit facility in the program allows you to edit and update fields and records. This also enables you to add or delete individual fields or records. This capability is essential for keeping the database current and for meeting your changing needs. You can retrieve a file for editing and save the database on a disk for subsequent use.

Sorting and Searching The usefulness of a database is largely determined by the ease with which you can manipulate the information it contains. The program lets you browse through the entire database and view it from different perspectives. The Sort feature can be used for organizing the database according to specific criteria. Specific information can be retrieved from the database using the Search or Query features. And you can print the database contents for reporting and archiving.

Data Interchange   Typically, database management programs are designed with data import-export features that let you exchange data with other programs. For example, you can move data from a spreadsheet into a database for archival purposes or for analysis. Similarly, the database (or parts thereof) can be written to a file that can be incorporated into a text document on a word processor. This feature minimizes duplication of data entry efforts and also reduces the potential for associated errors.

Database Security   Finally, the program also provides some password protection features for controlling access to the database for reading and updating. All these features assist you in effectively building and using large databases on personal computers. You should evaluate these characteristics when selecting a database program for your needs.

Major Database Management Programs   There are several database management programs for personal computers on the market. Among the popular ones are dBASE IV (and its previous versions dBASE III, dBASE III Plus) from Ashton-Tate; Paradox and Reflex from Borland; and Oracle from Oracle Corporation. As in previous chapters, we will provide the commands for two of these packages, viz., dBASE IV and Reflex.

Database Size Limitations: The size of the database — i.e., the number of files per database, the number of records per file, and the number of fields per record — that can be created and used depends on the particular database program. For example, you can have 65,000 records per file and 250 fields per record in Reflex. Moreover, each field can be 254 characters long while a record can have a maximum of 8000 characters. In other words, you can build fairly large databases on personal computers using these programs. Of course, the size of the database is also limited by the computer's storage capacity.

10.3 A DATABASE FOR THE PERIODIC TABLE OF ELEMENTS

We will illustrate the methodology in Figure 10.3 by creating a database for the Periodic Table of Elements shown in Figure 10.2.

10.3.1 Data and User Needs Analysis

Data Analysis   From Figure 10.2, we know that the database should contain six fields corresponding to each of the columns. Moreover, the values of Element, Symbol, and Room State are all strings making them Character or Text fields. The Atomic

372 DATABASE MANAGEMENT

Number and Atomic Weight fields should be numeric. Finally, the Radioactivity field should be a logical field since the values are either True or False (Yes or No).

Needs Analysis In trying to organize the database to suit our needs, let's enumerate the various purposes for which we could use the Periodic Table of Elements. These include:

- Displaying all the elements in alphabetical order
- Displaying the elements in ascending (or descending) order of the atomic number (or atomic weight)
- Finding all elements that are in a specific state at room temperature (e.g., liquids)
- Finding all elements that are not radioactive
- Finding elements that are radioactive *and* of certain atomic weight (e.g., greater than 220)

In other words, we should be able to sort the database according to certain criteria (e.g., element name and atomic number) and search for elements meeting these criteria (e.g., room state, radioactivity, atomic weight, and other combinations). Therefore, the major key fields are Element name, Atomic weight, and Atomic number. The field length is based on the longest entry in the field. Thus, a length of 12 might be appropriate for Element name, while a length of 2 would suffice for the Element symbol.

Figure 10.6*a* shows a worksheet used in this analysis for laying out the structure of the database; the result is shown in Figure 10.6*b*.

10.3.2 Implementing the Database Design

The next step is to implement the structure shown in Figure 10.6. Database management programs are typically menu-driven. Figure 10.7 shows the typical initial screen layout for a database program. It displays the program menu at the top. The menu is always displayed or can be invoked by pressing the Slash (/) key or one of the Function keys (F10 in dBASE IV and Reflex).

The Command Menu Program commands are grouped under categories shown in Figure 10.7. A particular command is selected either by moving the cursor to the command and pressing the Enter key (see Figure 10.7) or by pressing the letter corresponding to the command, e.g., C for Catalog in dBASE IV. Command options are displayed as shown in the figure, and the desired option can be selected from the menu. Some commands can be directly selected with the Function keys.

The Function key overlays for the two major programs, viz., dBASE IV and Reflex, are shown in Figure 10.8. The program's **Help facility** can also be invoked by pressing one of the Function keys.

Creating the Database File When we select the Database Create (or New) option from the program's main menu, we are prompted for a filename and we specify

10.3 A DATABASE FOR THE PERIODIC TABLE OF ELEMENTS

Field	Name	Type	Length	Major key[1]

[1]This key will be generally used for sorting the database.

(a)

Field	Name	Type	Length	Major key
Element	Element	Character	12	Yes
Symbol	Symbol	Character	2	No
Atomic Number	Atom_Num	Numeric	3	Yes
Atomic Weight	Atom_Wt	Numeric	8	Yes
Room State	Room_Stat	Character	1	No
Radioactive	Rad_Act	Logical	1	No

(b)

FIGURE 10.6 Designing the database structure. (a) Blank worksheet for analysis. (b) Results of the analysis.

PERTAB[1] for Periodic Table. The commands for creating a new database file for two database management programs are shown in Quick Reference Table 10.2.[2]

Defining Fields The program displays a table for defining database field characteristics and layout as shown in Figure 10.9a. Since we have already defined the layout (Figure 10.6), it is easy to translate it to suit the specific program restrictions. Figure 10.9a is also known as the "Field Properties window."

We enter Element under Field Name and use the **Tab key** to move the cursor to the Type field. Since this is a Character or Text field, we enter T or C (for text or character, respectively), and the program automatically fills in Text or Character. We assign a field width of 12. Likewise, we enter the information pertaining to the other fields; the corresponding screen is shown in Figure 10.9b. Depending on the program, you

---

[1]All entries that we type at the computer are shown underlined in text.
[2]While the specific command sequences and menus will vary from program to program, the underlying logic for executing the commands remains essentially the same across all programs. Likewise, when newer versions of programs are released, the capabilities are enhanced and there may be some changes in the commands. Nevertheless, to enhance the usefulness of the text, the specific command sequences for Reflex and dBASE IV are given in quick reference tables.

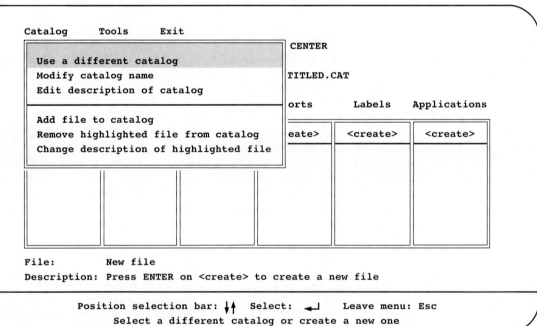

FIGURE 10.7 The program interface. (*a*) Initial screen for a database management program. Note the menu at the top of the screen. (*b*) Options under the Catalog command in the menu.

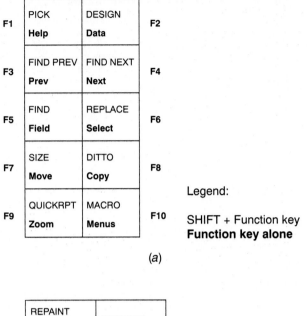

FIGURE 10.8 Important functions from the function key overlays for two programs. (a) dBASE IV. (b) Reflex.

can also specify the display format, e.g., number of decimal places to be displayed after the numeric value or alignment of text. For example, the atomic weight is formatted with four digits after the decimal place, while the atomic number is declared as an integer (see the Dec field in Figure 10.9b).

The result of this phase is the physical layout of the database with appropriate field characteristics. Depending on the program, you will be automatically placed in the data entry mode when you exit this command.

TABLE 10.2 QUICK REFERENCE TABLE FOR DATABASE CREATE COMMAND

Database	Reflex	dBASE IV
Create	ALT-Database New <name>	DATA <create> <name>
Use or retrieve	ALT-Database Open <name>	DATA <use file>

TABLE 10.3 QUICK REFERENCE TABLE FOR RECORDS COMMAND

Action	Reflex	dBASE IV
Add record	ALT-Records Add	ALT-Records Add
Delete record	ALT-Records Delete	ALT-Records Mark record for deletion
Edit record	ALT-Records Edit	From record display use Tab, Shift-Tab to select record

10.3.3 Entering Data

The next step, according to Figure 10.3, is to enter the data shown in Figure 10.2. We invoke the Command menu and select the Add or Append Records option. The commands for adding records to an existing database for two database management programs are shown in Quick Reference Table 10.3.

Figure 10.10a displays the screen for adding records. The Tab or End key is used to move the cursor from one field to the next, left to right or top to bottom. We enter first element's information from the Periodic Table in the empty row. In Figure 10.10b, the first record is shown. Likewise, we enter the information for the other elements. Corrections during data entry can be made by a combination of cursor and backspace keys. You can use the Shift-Tab or Home key to move to a preceding field in a record. Figure 10.11 shows the database with its 11 records.

Saving the Database When we finish entering the data, the program automatically saves the database. Alternatively, we can invoke the **Database Save command** from the menu and save the created database; we save it under the name PERTAB. The commands for saving the database for two programs are shown in Quick Reference Table 10.4.

10.3.4 Editing the Database

The purpose of editing the saved database is to identify and correct errors, e.g., misspelling or misplaced field entries, and to add new elements to the database or modify the database structure.

10.3 A DATABASE FOR THE PERIODIC TABLE OF ELEMENTS

```
Layout   Organize   Append   Go To   Exit
                                              Bytes remaining:   4000

Num   Field Name   Field Type   Width   Dec   Index
 1                 Character                    N










Database|C:\dbase\<NEW>        Field 1/1
           Enter the field name. Insert/Delete field:Ctrl-N/Ctrl-U
Field names begin with a letter and may contain letters, digits and underscores
```

(a)

```
Layout   Organize   Append   Go To   Exit
                                              Bytes remaining:   3974

Num   Field Name   Field Type   Width   Dec   Index
 1    ELEMENT      Character     12            N
 2    SYMBOL       Character      2            N
 3    ATOM_NUM     Numeric        3      0     N
 4    ATOM_WT      Numeric        8      4     N
 5    ROOM_STAT    Character      1            N
 6    RAD_ACT      Character      1            N







Database|C:\dbase\<NEW>        Field 6/6
              Change option to index on this field:Spacebar
```

(b)

FIGURE 10.9 Defining the database structure. (*a*) The field definition screen. (*b*) Field characteristics for the Periodic Table. *Note:* Observe the similarity between Figure 10.6 and this one; however, *Rad_Act* is declared as a Character type here.

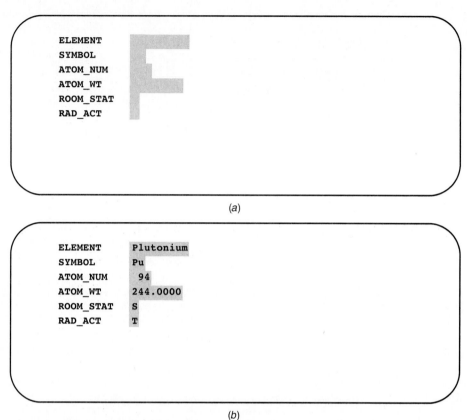

FIGURE 10.10 Adding records to the database. (*a*) The empty record. (*b*) The first record in the database; we use T (for true) to denote Yes for Radioactive.

Correcting Field Entries In Figure 10.11, we find that the element antimony has been misspelled as "antimoni." We invoke the **Edit Record command** from the main menu and select the record and the element field. We correct the misspelling and exit the edit mode. Figure 10.12 shows the corrected database. The commands for editing a database for two major programs are shown in Quick Reference Table 10.5.

Adding New Elements To add new elements to the database, we have to create additional records. This can be done using the Append or Add Records command discussed in Section 10.3.3.

Adding Fields To add new fields to a record, e.g., a field to identify metals in the database, we can use the **Add Field command** in the program. We can select the field characteristics (name, type, and length) and their placement in the record. The commands for adding fields to a database for two programs are shown in Quick Reference Table 10.5.

Modifying Field Contents This editing operation is similar to correcting errors in field entries. In this case, however, you make changes to the database to reflect new

```
| ELEMENT    | SYMBOL | ATOM_NUM | ATOM_WT  | ROOM_STAT | RAD_ACT |
|------------|--------|----------|----------|-----------|---------|
| Plutonium  | Pu     | 94       | 244.0000 | S         | T       |
| Aluminum   | Al     | 13       |  26.9820 | S         | F       |
| Mercury    | Hg     | 80       | 200.5900 | L         | F       |
| Antimoni   | Sb     | 51       | 121.7500 | S         | F       |
| Hydrogen   | H      |  1       |   1.0079 | G         | F       |
| Bromine    | Br     | 35       |  79.9040 | L         | F       |
| Tin        | Sn     | 50       | 118.6900 | S         | F       |
| Thorium    | Th     | 90       | 232.0400 | S         | T       |
| Oxygen     | O      |  8       |  15.9990 | G         | F       |
| Phosphorus | P      | 15       |  30.9740 | S         | F       |
| Radium     | Ra     | 88       | 226.0300 | S         | T       |
```

FIGURE 10.11 Database after data entry. Note the error in the spelling of antimony.

TABLE 10.4 QUICK REFERENCE TABLE FOR DATABASE FILE COMMAND

Action	Reflex	dBASE IV
Save database	ALT-Database Save <name>	ALT-E
Delete database	ALT-Database Delete <name>	ALT-C Remove

information or facts. For example, if one of the elements, say, bromine, is found to be radioactive, then we have to modify its entry in the *Rad_Act* field.

Deleting Records and Fields Let's say we want to delete the record for tin from our database. We display the records and select the record for tin. We then invoke the **Delete Record command** from the menu and delete the record. Figure 10.13*a* shows the database after deletion of the record. In many programs, a record that has been marked for deletion is not lost until a housekeeping operation known as **packing** is executed. The **Pack command** consolidates the database by removing records marked for deletion.

In a similar manner, a particular field from the record can be chosen for deletion. Figure 10.13*b* shows the database after deletion of the *Atom_Num* field from the database. The commands for deleting records and fields from a database for two programs are shown in Quick Reference Tables 10.3 and 10.5, respectively.

```
-----------------------------------------------------------------
| ELEMENT      | SYMBOL | ATOM_NUM | ATOM_WT  | ROOM_STAT | RAD_ACT |
|--------------|--------|----------|----------|-----------|---------|
| Plutonium    | Pu     |    94    | 244.0000 |    S      |    T    |
| Aluminum     | Al     |    13    |  26.9820 |    S      |    F    |
| Mercury      | Hg     |    80    | 200.5900 |    L      |    F    |
| Antimony     | Sb     |    51    | 121.7500 |    S      |    F    |
| Hydrogen     | H      |     1    |   1.0079 |    G      |    F    |
| Bromine      | Br     |    35    |  79.9040 |    L      |    F    |
| Tin          | Sn     |    50    | 118.6900 |    S      |    F    |
| Thorium      | Th     |    90    | 232.0400 |    S      |    T    |
| Oxygen       | O      |     8    |  15.9990 |    G      |    F    |
| Phosphorus   | P      |    15    |  30.9740 |    S      |    F    |
| Radium       | Ra     |    88    | 226.0300 |    S      |    T    |
-----------------------------------------------------------------
```

FIGURE 10.12   Database after error correction for antimony.

TABLE 10.5   QUICK REFERENCE TABLE FOR EDIT COMMAND

Action	Reflex	dBASE IV
Edit record	ALT-Records Edit	Select record using Tab, Shift-Tab
Select record	ALT-Edit Row or F3	ALT-G <Record#>
Edit selected record	F2	Edit
Delete a field	F3 <delete key>	Modify database structure CTRL-U CTRL-End
Add a field	ALT-Database Add <field_name> or F4 <field_name>	Modify database structure CTRL-N <field_name> CTRL-End

10.3.5 Utilizing the Database

Having created the database (see Figure 10.12), let's sort the records in alphabetical order of element name. So the primary key for sorting the database is the field *Element*. The sorted records can be written to a separate file, if necessary. We invoke the **Sort facility** in the database program (Figure 10.14*a*) and specify the field name as *Element*. The result of the sort is shown in Figure 10.14*b*. It is good practice to make a copy of the database before invoking the Sort command.

10.3 A DATABASE FOR THE PERIODIC TABLE OF ELEMENTS

Element	Symbol	Atomic number	Atomic weight	Room state	Radioactive
Plutonium	Pu	94	244	S	T
Aluminum	Al	13	26.982	S	F
Mercury	Hg	80	200.59	L	F
Antimony	Sb	51	121.75	S	F
Hydrogen	H	1	1.0079	G	F
Bromine	Br	35	79.904	L	F
Thorium	Th	90	232.04	S	T
Oxygen	O	8	15.999	G	F
Phosphorus	P	15	30.974	S	F
Radium	Ra	88	226.03	S	T

(a)

Element	Symbol	Atomic weight	Room state	Radioactive
Plutonium	Pu	244	S	T
Aluminum	Al	26.982	S	F
Mercury	Hg	200.59	L	F
Antimony	Sb	121.75	S	F
Hydrogen	H	1.0079	G	F
Bromine	Br	79.904	L	F
Thorium	Th	232.04	S	T
Oxygen	O	15.999	G	F
Phosphorus	P	30.974	S	F
Radium	Ra	226.03	S	T

(b)

FIGURE 10.13 The Delete Record and Delete Field commands. (a) The record for tin has been deleted. (b) The Atom_Num field has been deleted.

During sorting, the database records are physically reordered according to the Sort key(s). Thus, plutonium, the first record in the original database (when it was created), moves to the eighth row, or is the eighth record now, while bromine becomes the third record (moving up from the sixth position). The time taken to sort a database obviously depends on its size. It also depends on the algorithm used in the implementation of the sort facility.

Sorting by Atomic Number Figure 10.15 shows the elements sorted in descending order of their atomic number. Note that we can specify either descending or ascending order while sorting the database. Again, the records' physical positions are different from those in Figures 10.12 and 10.14b.

The commands for invoking the sort facility in the two programs are shown in Quick Reference Table 10.6.

Searching the Database One of the primary objectives of developing a database is to search and extract information in the database that meets certain search criteria.

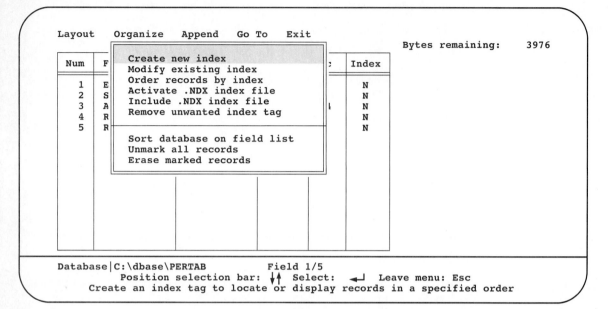

FIGURE 10.14 Sorting the database. (*a*) Specify the field for sorting: Element. (*b*) Database sorted in alphabetical order.

```
| ELEMENT    | SYMBOL | ATOM_NUM | ATOM_WT  | ROOM_STAT | RAD_ACT |
|------------|--------|----------|----------|-----------|---------|
| Plutonium  | Pu     | 94       | 244.0000 | S         | T       |
| Thorium    | Th     | 90       | 232.0400 | S         | T       |
| Radium     | Ra     | 88       | 226.0300 | S         | T       |
| Mercury    | Hg     | 80       | 200.5900 | L         | F       |
| Antimony   | Sb     | 51       | 121.7500 | S         | F       |
| Tin        | Sn     | 50       | 118.6900 | S         | F       |
| Bromine    | Br     | 35       |  79.9040 | L         | F       |
| Phosphorus | P      | 15       |  30.9740 | S         | F       |
| Aluminum   | Al     | 13       |  26.9820 | S         | F       |
| Oxygen     | O      |  8       |  15.9990 | G         | F       |
| Hydrogen   | H      |  1       |   1.0079 | G         | F       |
```

FIGURE 10.15 Database sorted according to descending order of atomic number.

TABLE 10.6 QUICK REFERENCE TABLE FOR SORT COMMAND

Action	Reflex	dBASE IV
Sort database	ALT-Records Sort <field name> . . . CTRL-<enter>	Modify database structure Organize Sort <field name> . . . <file name>

In our present example, let's find all the elements that are liquids at room temperature. We can express this search criterion or query as follows:

Display records for which *Room_Stat* = "L"

We invoke the search command from the menu and define the search condition as shown in Figure 10.16a. Note the similarity between the English-language query formulation and the figure. When we finish execution of this command, records meeting the criterion are displayed as shown in Figure 10.16b.

The search commands for two major database management programs are shown in Quick Reference Table 10.7.

Relational Operators We can use relational operators (e.g., greater than or equal to) when searching the database. For example, let's find all elements with atomic weight greater than 200. We can express this search condition as follows:

Display records for which *Atom_Wt* > 200

```
┌─────────────────────────────────────────────────────────────────────────┐
│                                                                         │
│    Layout    Fields    Condition    Update    Exit                      │
│   ┌──────────────┬──────────────┬──────────────┬──────────────┐         │
│   │ Pertab.dbf   │ ATOM_NUM     │ ROOM_STAT    │ RAD_ACT      │         │
│   ├──────────────┼──────────────┼──────────────┼──────────────┤         │
│   │              │              │   ="L"       │              │         │
│   └──────────────┴──────────────┴──────────────┴──────────────┘         │
│                                                                         │
│                                                                         │
│   ┌View──┬─────────────┬─────────────┬─────────────┬─────────────┐      │
│   │  Q5  │ Pertab->    │ Pertab->    │ Pertab->    │ Pertab->    │      │
│   │      │ ELEMENT     │ SYMBOL      │ ATOM_NUM    │ ATOM_WT     │      │
│   └──────┴─────────────┴─────────────┴─────────────┴─────────────┘      │
│   Query     |C:\dbase\Q5           Field 6/6                            │
│      Prev/Next field:Shift-Tab/Tab  Data:F2  Pick:Shift-F1  Prev/Next skel:F3/F4 │
└─────────────────────────────────────────────────────────────────────────┘
```

(a)

```
-----------------------------------------------------------------------
| ELEMENT    | SYMBOL  | ATOM_NUM | ATOM_WT   | ROOM_STAT | RAD_ACT  |
|------------|---------|----------|-----------|-----------|----------|
| Bromine    | Br      |    35    | 79.9040   | L         | F        |
| Mercury    | Hg      |    80    | 200.5900  | L         | F        |
|            |         |          |           |           |          |
|            |         |          |           |           |          |
|            |         |          |           |           |          |
|            |         |          |           |           |          |
|            |         |          |           |           |          |
|            |         |          |           |           |          |
|            |         |          |           |           |          |
-----------------------------------------------------------------------
```

(b)

FIGURE 10.16 Searching the database. (*a*) Defining the search condition; Query (at bottom of screen) can be formulated to display only desired fields in the record. (*b*) Elements that are liquids at room state.

10.3 A DATABASE FOR THE PERIODIC TABLE OF ELEMENTS

```
   Layout      Fields     Condition     Update     Exit
  ┌──────────────┬──────────────┬──────────────────┬────────────────┐
  │ Pertab.dbf   │ ELEMENT      │ SYMBOL           │ ATOM_WT        │
  ├──────────────┼──────────────┼──────────────────┼────────────────┤
  │              │              │                  │ > 200          │
  └──────────────┴──────────────┴──────────────────┴────────────────┘




   ┌View──┬───────────┬───────────┬───────────┬───────────┐
   │PERTAB│ Pertab->  │ Pertab->  │ Pertab->  │ Pertab->  │
   │      │ ELEMENT   │ SYMBOL    │ ATOM_NUM  │ ATOM_WT   │
   └──────┴───────────┴───────────┴───────────┴───────────┘
  Query    |C:\dbase\PERTAB         Field 3/6
    Prev/Next field:Shift-Tab/Tab   Data:F2   Pick:Shift-F1   Prev/Next skel:F3/F4
```

(a)

```
----------------------------------------------------------------------
| ELEMENT    | SYMBOL | ATOM_NUM | ATOM_WT  | ROOM_STAT | RAD_ACT |
|------------|--------|----------|----------|-----------|---------|
| Mercury    | Hg     |    80    | 200.5900 | L         | F       |
| Plutonium  | Pu     |    94    | 244.0000 | S         | T       |
| Radium     | Ra     |    88    | 226.0300 | S         | T       |
| Thorium    | Th     |    90    | 232.0400 | S         | T       |
|            |        |          |          |           |         |
|            |        |          |          |           |         |
|            |        |          |          |           |         |
|            |        |          |          |           |         |
|            |        |          |          |           |         |
|            |        |          |          |           |         |
----------------------------------------------------------------------
```

(b)

FIGURE 10.17 Relational operators in a search. (a) Query formulation. (b) Elements with atomic weight greater than 200.

Figure 10.17 shows the query formulation and the records meeting the criterion. The list of relational operators commonly used in database management programs is shown in Table 10.8.

Multiple Search Criteria We can search the database for records meeting more than one criterion. For example, we can find elements that are radioactive **and** have

TABLE 10.7 QUICK REFERENCE TABLE FOR SEARCH COMMAND

Action	Reflex	dBASE IV
Search database	ALT-Records Find filter conditions CTRL-<enter>	From records display GoTo Forward search <search conditions> Backward search <search conditions>

TABLE 10.8 LIST OF RELATIONAL OPERATORS

Operation	Symbol
Greater than	>
Greater than or equal to	≥
Less than	<
Less than or equal to	≤
Not equal to	<>

atomic weight greater than 230. Or we can find elements that are either liquids **or** radioactive. For searches involving multiple criteria, we make use of the boolean operators **AND, OR,** and **NOT.**

AND Operator Let's query the database to find elements that are radioactive and have atomic weight greater than 230. The English-language query can be formulated as follows:

Display records for which (Rad_Act = True) AND ($Atom\_Wt$ > 230).

This means that only records meeting both criteria (simultaneously) should be displayed. Figure 10.18 shows the query formulation and results: The records for plutonium and thorium are displayed.

OR Operator When an **OR** operator is used, only one of the search conditions needs to be true. For example, the query,

Display records for which (Rad_Act = True) OR ($Atom\_Wt$ > 230)

will result in records that meet either of the two criteria; records for plutonium, thorium, and radium are displayed in response to this query as shown in Figure 10.19. It follows that, in general, the AND search is a more restrictive search, i.e., fewer records are displayed in comparison to an OR search.

NOT Operator An example of a **NOT** query, viz.,

Display records for which NOT($Room\_Stat$ = "L")

will result in records for which the room state is not equal to "L", i.e., elements that are either solids or gases at room state will be displayed. Figure 10.20 shows the results for this query: all records except those of mercury and bromine are displayed.

10.3 A DATABASE FOR THE PERIODIC TABLE OF ELEMENTS

```
  Layout      Fields     Condition      Update      Exit

 Pertab.dbf  | ATOM_NUM      | ROOM_STAT       | RAD_ACT        |
             |               |                 | ="T"           |

              | ATOM_WT   | ATOM_NUM     | ROOM_STAT      |
              | >230      |              |                |


  View
  NEW  | Pertab->   | Pertab->   | Pertab->   | Pertab->   |
       | ELEMENT    | SYMBOL     | ATOM_NUM   | ATOM_WT    |

 Query   |C:\dbase\NEW         Field 6/6
   Prev/Next field:Shift-Tab/Tab   Data:F2   Pick:Shift-F1   Prev/Next skel:F3/F4
```

(a)

```
----------------------------------------------------------------------------
| ELEMENT    | SYMBOL  | ATOM_NUM | ATOM_WT   | ROOM_STAT | RAD_ACT |
|------------|---------|----------|-----------|-----------|---------|
| Plutonium  | Pu      |    94    | 244.0000  |    S      |    T    |
| Thorium    | Th      |    90    | 232.0400  |    S      |    T    |
|            |         |          |           |           |         |
|            |         |          |           |           |         |
|            |         |          |           |           |         |
|            |         |          |           |           |         |
|            |         |          |           |           |         |
|            |         |          |           |           |         |
|            |         |          |           |           |         |
|            |         |          |           |           |         |
----------------------------------------------------------------------------
```

(b)

FIGURE 10.18 Multiple search criteria: AND operator. (*a*) Query formulation. (*b*) Elements that are radioactive AND with atomic weight > 230.

Printing the Database We can generate a hard copy of the entire database or selected portions of it using the Print Database command shown in Quick Reference Table 10.9. The resulting printout is shown in Figure 10.21. Depending on the program, you can use the **Report feature** to add appropriate titles and headings to the output to suit your requirements.

```
┌─────────────────────────────────────────────────────────────────────────┐
│                                                                         │
│   Layout      Fields     Condition     Update      Exit                 │
│  ┌──────────┬────────────┬──────────────┬────────────────────┐          │
│  │Pertab.dbf│  ELEMENT   │   SYMBOL     │      ATOM_WT       │          │
│  ├──────────┼────────────┼──────────────┼────────────────────┤          │
│  │          │            │              │                    │          │
│  └──────────┴────────────┴──────────────┴────────────────────┘          │
│                                                                         │
│                                    ┌─ CONDITION BOX ─────────────┐      │
│                                    │ RAD_ACT ="T" .OR. ATOM_WT > 230│   │
│                                    │                             │      │
│                                    └─────────────────────────────┘      │
│                                                                         │
│   ┌View─┬────────────┬────────────┬────────────┬────────────┐           │
│   │ NEW │ Pertab->   │ Pertab->   │ Pertab->   │ Pertab->   │           │
│   │     │ ELEMENT    │ SYMBOL     │ ATOM_WT    │ ATOM_NUM   │           │
│   └─────┴────────────┴────────────┴────────────┴────────────┘           │
│                                                                         │
│   Query    |C:\dbase\NEW           Row 1/1                              │
│       Pick operators/fields:shift-F1  Data:F2  Zoom:F9  Prev/Next skeleton:F3/F4 │
└─────────────────────────────────────────────────────────────────────────┘
```

(a)

```
 --------------------------------------------------------------------------
| ELEMENT    | SYMBOL | ATOM_NUM | ATOM_WT   | ROOM_STAT | RAD_ACT |
|------------|--------|----------|-----------|-----------|---------|
| Plutonium  | Pu     |    94    | 244.0000  |    S      |    T    |
| Radium     | Ra     |    88    | 226.0300  |    S      |    T    |
| Thorium    | Th     |    90    | 232.0400  |    S      |    T    |
|            |        |          |           |           |         |
 --------------------------------------------------------------------------
```

(b)

FIGURE 10.19   Multiple search criteria: OR operator. (*a*) Query formulation. (*b*) Elements that are radioactive OR with atomic weight > 230.

Indexing the Database   Whenever we make changes to the database (add or delete records or fields), it is necessary for us to reorganize (e.g., sort) the database to maintain the desired perspective of data. For example, when we need to add the record for a new employee to an employee database sorted according to the last name,

```
  Layout      Fields     Condition     Update      Exit
 ┌──────────────┬──────────────┬────────────────────────┬──────────────┐
 │ Pertab.dbf   │ ATOM_NUM     │ ROOM_STAT              │ RAD_ACT      │
 ├──────────────┼──────────────┼────────────────────────┼──────────────┤
 │              │              │ <>"L"                  │              │
 └──────────────┴──────────────┴────────────────────────┴──────────────┘




 ┌View──────┬──────────┬──────────┬──────────┬──────────┐
 │ NEW      │ Pertab-> │ Pertab-> │ Pertab-> │ Pertab-> │
 │          │ ELEMENT  │ SYMBOL   │ ATOM_WT  │ ATOM_NUM │
 └──────────┴──────────┴──────────┴──────────┴──────────┘
  Query   |C:\dbase\NEW              Field 5/6
     Prev/Next field:Shift-Tab/Tab  Data:F2  Pick:Shift-F1  Prev/Next skel:F3/F4
```

(a)

```
 -------------------------------------------------------------------------
 | ELEMENT     | SYMBOL  | ATOM_NUM | ATOM_WT   | ROOM_STAT | RAD_ACT   |
 |-------------|---------|----------|-----------|-----------|-----------|
 | Aluminum    | Al      |     13   |   26.9820 |    S      |    F      |
 | Antimony    | Sb      |     51   |  121.7500 |    S      |    F      |
 | Hydrogen    | H       |      1   |    1.0079 |    G      |    F      |
 | Oxygen      | O       |      8   |   15.9990 |    G      |    F      |
 | Phosphorus  | P       |     15   |   30.9740 |    S      |    F      |
 | Plutonium   | Pu      |     94   |  244.0000 |    S      |    T      |
 | Radium      | Ra      |     88   |  226.0300 |    S      |    T      |
 | Thorium     | Th      |     90   |  232.0400 |    S      |    T      |
 | Tin         | Sn      |     50   |  118.6900 |    S      |    F      |
 |             |         |          |           |           |           |
 |             |         |          |           |           |           |
 |             |         |          |           |           |           |
 |-----------------------------------------------------------------------|
```

(b)

FIGURE 10.20 The NOT operator. (a) Query formulation. (b) Elements that are NOT liquids.

it is desirable to have the record in the proper position, so that the alphabetical order is maintained. This means the database should be sorted after the addition of the record. Since sorting involves physically reordering the records, the time for sorting will increase with the size of the database. Thus, with large databases, frequent sorting

TABLE 10.9 QUICK REFERENCE TABLE FOR PRINT DATABASE COMMAND

Action	Reflex	dBASE IV
Print database	ALT-U Print	Retrieve List \<printer\> \<screen\>

```
                    Periodic Table of Elements

Element         Element     Atomic      Atomic  Room    Radioactivity
Name            Symbol      Number      Weight  State
-----------     --------    -------     ------- ------  --------------

Aluminum        Al           13          26.9820   S       F
Antimony        Sb           51         121.7500   S       F
Bromine         Br           35          79.9040   L       F
Hydrogen        H             1           1.0079   G       F
Mercury         Hg           80         200.5900   L       F
Oxygen          O             8          15.9990   G       F
Phosphorus      P            15          30.9740   S       F
Plutonium       Pu           94         244.0000   S       T
Radium          Ra           88         226.0300   S       T
Thorium         Th           90         232.0400   S       T
Tin             Sn           50         118.6900   S       F
```

FIGURE 10.21 The printed output.

will be very time-consuming. In such situations, it is advisable to use the **Indexing feature** of database management programs.

Creating an index file for a database is analogous to creating an index for a book. Just as you would use the book index to look up a particular topic, the index file contains the addresses (referred to as **pointers**) of physical record positions in the database. When you add a new record to the database (with an associated index file), the program will update the index file to reflect physical placement of the record.

Figure 10.14b shows a portion of the sorted database. Let's say we want to add the element lead to the database. When we use the Add Record command in the program, it will be added to the end of the database (after tin). However, the record should come between hydrogen and mercury so that alphabetical order is maintained. If it is a small database, we could identify and "insert" the new record in the correct position, provided the program has this feature. However, this procedure would be inefficient in the case of large databases. This is where indexing comes in handy.

10.3 A DATABASE FOR THE PERIODIC TABLE OF ELEMENTS

Element	Symbol	Atomic number	Index file: Pointer to next record
Aluminum	Al	13	2
Antimony	Sb	51	3
Bromine	Br	35	4
Hydrogen	H	1	5
Mercury	Hg	80	6
Oxygen	O	8	7
Phosphorus	P	15	8
Plutonium	Pu	94	9
Radium	Ra	88	10
Thorium	Th	90	11
Tin	Sn	50	EOF

First record: 1

(a)   (b)

FIGURE 10.22 Indexing the database. (a) Part of the Periodic Table database. (b) Index file displays the location of the next record.

How Indexing Works Figure 10.22a shows a portion of the Periodic Table database with only three fields (*Element*, *Symbol*, and *Atom_Num*) sorted according to element name. The corresponding index file (indexed on the *Element* field) in Figure 10.22b contains the physical location (or number) of the next record in the pointer field. For example, the first record (aluminum) has 2 in its pointer field, which means the record that follows aluminum is physically located in the second position, viz., that of antimony, in the database. Since the records are sorted in this example, the pointers refer to succeeding records, i.e., the physical order is the same as the logical, or alphabetical, order. The last record in the database contains an end-of-file (EOF) marker, signifying that it is the last record. The position of the first record is typically shown separately as in the figure; this means ordering of the records begins with the record in the first position.

When we add the record for lead to the database, the pointer field in the index file is updated by the program as shown in Figure 10.23. The record for hydrogen now points to the 12th record (that of lead), and lead now points to the 5th record (mercury). Thus, the logical order of these three records is hydrogen, lead, and mercury — the alphabetical order is preserved. Physically, however, they are in the 4th, 12th, and 5th positions, respectively. In a similar manner, when a record is deleted, the pointer in the index file is updated to reflect the absence of the record. So the index file helps to preserve the physical order of the records in the database while arranging them in a logical order to facilitate their fast access.

Multiple Index Files The concept of indexing can be extended to several index files each with a separate index key. Thus, one index file can be based on the element name and another on atomic number. The pointers in the first index file will reflect the alphabetical order of the elements while the second file will contain pointers to the elements sorted according to atomic number.

Since the index file contains only pointers, storing several index files (each created for a different key) takes up far less space than multiple versions of the entire database

Element	Symbol	Atomic number
Aluminum	Al	13
Antimony	Sb	51
Bromine	Br	35
Hydrogen	H	1
Mercury	Hg	80
Oxygen	O	8
Phosphorus	P	15
Plutonium	Pu	94
Radium	Ra	88
Thorium	Th	90
Tin	Sn	50
Lead	Pb	82

Index file: Pointer to next record	
2	
3	
4	
5̶	12
6	
7	
8	
9	
10	
11	
EOF	
	5

First record: 1

(a)   (b)

FIGURE 10.23 Addition of a record. (a) Periodic Table database after addition of lead. (b) In the index file, hydrogen now points to the 12th record (lead), and lead points to the 5th record (mercury).

created for each key. Thus, indexing a database not only speeds up the data retrieval process but also conserves disk space. However, there is a drawback—addition or deletion of records is slower. This is because every time a record is added or deleted, the index file has to be updated, requiring some extra processing time.

When there are several changes to the database during the process of housekeeping, it is good practice to use the Sort command to sort the database and create a new index file. This will greatly enhance subsequent retrieval performance.

10.4 RELATIONAL DATABASE CONCEPTS

In a **relational database,** data is organized in a two-dimensional matrix of rows and columns. This matrix is known as a **table,** or **relation**. Thus, a database is made up of several tables, or relations, that are linked through a field common to all relations. This field is known as a **key identifier.** Each row in the table corresponds to a record, while the columns are the fields in the record. In relational database terminology, each record is referred to as a **tuple**, while columns are known as **attributes**. Thus, a database is made up of several relations, or tables, linked through a common field, the key identifier. Each relation consists of several tuples and a tuple is made up of several attributes.

Figure 10.24 shows another file (MGR_DATA) from the corporate employee database. It contains personal information about each manager: gender, date of birth, marital status, number of dependents, and type of medical insurance. Each of the records in this figure is a tuple, while the fields are the attributes. The link between this table and the one in Figure 10.1 is the *Emp_ID* field, the key identifier.

Select, Project, and Join Operations The three major operations that are frequently performed using relational database management systems are:

Mgr_ID	Gender[1]	Birth_Date	Mar_Stat[2]	Num_Dep	Med_Ins
32105	M	10-Feb-37	M	5	HMO2
32115	F	23-May-47	M	3	HMO1
32245	M	15-Jul-65	S	1	HMO2

[1] F: Female, M: Male
[2] M: Married, S: Single

FIGURE 10.24 Another file or table (MGR_DATA) from the corporate database.

1. Selecting one or more records from a table of relations (known as the **Select operation**). This is similar to the Search operation discussed earlier.
2. Selecting specific attributes or fields from tuples or rows in the table (known as **Project operation**). In contrast to the select operation where entire tuples (rows) are selected, only certain tuples or attributes can be selected with the project operation.
3. Combining records from two or more tables to create a new table (known as the **Join operation**).

Relational database management programs provide a query language (e.g., SQL) for carrying out various operations. Figure 10.25 (p. 394) shows the results of the three operations using the information in Figures 10.1 and 10.24.

Advantages The major advantage of relational databases comes from the fact that several relations are linked through the key identifier. Consequently, even when you are accessing one table (say, Figure 10.1), you can easily obtain information contained in other relations (e.g., personal information in Figure 10.24).

Moreover, the Select, Project, and Join operations provide the flexibility to view select fields from records even if these records are physically located in different files. Thus, from a user's viewpoint, the physical location of the records is not an impediment to utilization of information. Data can be better organized logically in relations. Another major advantage is that a relational database minimizes data redundancy, i.e., there is no need to maintain duplicate copies of a record to satisfy each desired information view.

SUMMARY

We introduced the basic concepts of databases and their terminology: files, records, fields, and field characteristics. We presented a structured methodology consisting of six major steps for creating and utilizing databases. Following this, we outlined the major characteristics of database management programs.

We used the structured methodology for developing a database for the Periodic Table of Elements. In doing so, we explored the various program features including adding and deleting records, sorting, searching, printing, and indexing databases. We also discussed some fundamental concepts of relational databases and their advantages.

Emp_ID	Last_Name	First_Name	Hire_Date	Salary	Dept_Num	Supervisor
32115	Gordon	Maureen	10-Jul-81	$61,225	1305	Carter
32245	Harmon	Dan	01-Feb-82	$38,000	1212	Andersson
32254	Lauder	Tom	11-Feb-82	$41,000	1301	Gordon
32325	Martin	Jimmy	01-May-83	$45,000	1301	Gordon

(a) Result of Select operation on Figure 10.1: Select records from EMP_DATA (Figure 10.1) such that (Salary > 35,000) AND (Hire_Date > 01-Jan-81).

Mgr_ID	Mar_Stat	Num_Dep
32105	M	5
32115	M	3
32245	S	1

(b) Result of Project operation on Figure 10.24: Project records from MGR_DATA (Figure 10.24) for Mgr_ID, Mar_Stat, and Num_Dep.

Emp_ID	Last_Name	First_Name	Hire_Date ...	Mgr_ID	Gender	Birth_Date ...
32105	Andersson	Eric	10-Jan-80 ...	32105	M	10-Feb-37 ...
32115	Gordon	Maureen	10-Jul-81 ...	32115	F	23-May-47 ...
32245	Harmon	Dan	01-Feb-82 ...	32245	M	15-Jul-65 ...

(c) Result of Join operation on EMP_DATA and MGR_DATA: JOIN records from EMP_DATA and MGR_DATA for Emp_ID = Mgr_ID.

FIGURE 10.25 Operations on relations EMP_DATA (Figure 10.1) and MGR_DATA (Figure 10.24).

CHAPTER **11**

WORD PROCESSING

In this chapter, we introduce the concepts of word processing and present a five-step process for creating and printing a document using word processing programs. We also cover some additional features of word processors that assist in producing better-quality documents.

11.1 WHAT IS WORD PROCESSING?

Inasmuch as scientists and engineers spend most of their time in design, development, and quantitative problem solving, the results of their activities need to be communicated to their peers, management, and the outside world. The most effective method of communication, and also the one used most often, is the written word: project proposals, competitive analysis, design evaluations, research and technical reports, and papers in professional journals.

Writing is a skill that is learned and perfected over time; however, there are tools that assist scientists and engineers in storing, packaging, and communicating their words. These tools are known as "word processors"; and the five tasks of creating, editing, formatting, saving, and printing a document are collectively referred to as "word processing." Said differently, word processing is the deft manipulation of words to produce an attractive and, hopefully, meaningful and useful document.

Word Processing Programs The typical word processor, in the past, was a stand-alone hardware-software configuration that was used exclusively for document processing by well-trained secretaries or operators. Both the hardware and the software were supplied and serviced by one vendor. Often, the different systems were incompatible, and operators had to be retrained when they switched from one system to the other. The hardware was not used for any other computations.

However, the advent of personal computers has led to the decoupling of this hardware-software system. Inexpensive word processing programs and personal computers have contributed to each other's proliferation on professionals' desks. Currently, word processing programs are just a few of the many software packages used on personal computers. In fact, word processing, along with design and equation solving, has become one of the routine tasks of scientists and engineers.

Typically, a word processing program running on a personal computer will have functions for carrying out each of the five major tasks: **document creation, editing, formatting, saving,** and **printing.** In addition, the document that is displayed on the screen will look like the final printed output. Such word processing programs are known as WYSIWYG (What you see is what you get) word processors. We use the term "word processor" to refer to a word processing program that runs on any computer. Some popular word processing programs on personal computers are WordPerfect, MS-Word, WordStar 2000, and PFS:Word.

Advantages of Word Processors Word processing is here to stay, and it has changed (forever) the writing habits of many professionals. No longer under the threat of having to retype an entire document on a typewriter due to a missed word or line or a typographical error, or cut and paste parts of a document, or worse yet, be dependent on word processing operators or secretaries, the professional can create and easily modify a document as many times and as often as needed.

Multiple copies of the document can be easily created. For example, once a college composes a letter of admission, the letter can be addressed by name to several applicants without any changes to the body of the text. Only the recipient's name and address need to be changed. Moreover, many word processing programs allow you to import data and graphics from other programs, e.g., spreadsheets. This feature is extremely useful for creating technical research reports.

11.2 EVOLUTION OF A DOCUMENT: THE FIVE-STEP PROCESS

Figure 11.1 shows the five major steps in the evolution of a document. The first step is the creation of the document which involves typing in the text, just as on a typewriter. This step is followed by the editing process where errors are corrected and changes made in the text. At this point, the contents of the document are in place; only the presentation needs to be refined. And this task is carried out in the next step, i.e., the formatting process. Changes to line spacing in the document, alignment of text at the margins, and other appearance-related improvements are made during the formatting process. The document is then saved for future use on a hard or floppy disk. And finally, the document is printed. Word processing programs provide the necessary features to carry out these tasks with relative ease.

11.2.1 Creating Text

Figure 11.2 shows parts of an article on software knowledge needed by graduating engineers; word processing is one of the four software tools mentioned in it. We will

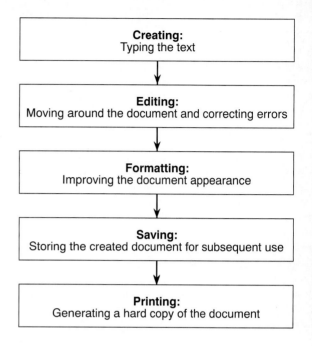

FIGURE 11.1 The five-step sequence in the evolution of a document.

explore the different features of word processing programs as we go through the five steps in making a document out of this text.

The Screen Display Figure 11.3 shows the layout of a typical word processor screen. The blinking character is the **cue** or **cursor**. When you type in text, it is entered at the cursor location. The **status line** displays the status information, i.e., the name of the document being processed and the current location of the cursor. In our example, the document name is Software, and the cursor is on page 1, line 1, and character position 1. Depending on the particular program, the status line may appear at the top or bottom of the screen.

The **ruler line** (not shown in the figure) displays the margins and tab settings for the current line. The Help menu for the commands is also displayed on the screen. However, the current trend in word processing software is to supply a keyboard overlay with the frequently used commands in the program. The remainder of the screen is for the text of the document.

Moving around the Screen When the word processor is loaded initially, the cursor is at the top left-hand corner of the display, and the program is typically in insert mode. In the **insert mode,** the characters typed are inserted at the cursor position, and the text is moved to the right. For example, in Figure 11.4a, the word "been" is missing in the second line of the text. When it is typed in the insert mode, the word is inserted and the text moves to the right as shown in Figure 11.4b. In the other mode known as the **typeover** or **replacement mode,** the typed character replaces the character at the current location of the cursor. In Figure 11.4c, the first four letters

Software Knowledge You'll Need on the Job
Nicholas Basta

Introduction

Engineers should have a proprietary air when it comes to talking about computers. After all, engineers have been among the foremost users and developers of both minicomputers and mainframes. And they are turning out to be one of the leading users of personal computers.

Today's graduating engineers understand what PCs can do, and are looked on as the de facto PC experts in an office or engineering environment simply because their education has exposed them so much to PCs. Once on the job, today's graduates can rapidly solidify their position by meeting this responsibility head-on. But to do so the new engineer needs to understand the standard types of PC programs, which nearly every PC user must have, and the special programs developed for use in the different engineering fields.

Standard Programs

We come to the point of defining the standard types of PC programs. The initial essential software is the operating system. After the operating system question is settled, you come to the four major software packages that operators use to accomplish different tasks:

i. a programming language;
ii. a word-processing program;
iii. a database program;
iv. a spreadsheet.

Conclusions

When the PC first appeared, many observers looked on it as a dressed-up calculator. But this view has not prevailed. Instead, engineers are finding that the PC is changing the very nature of their work, removing some of the drudgery (drafting, recordkeeping, repetitive calculations, etc.), while extending the range of problems addressable. By the end of this decade, market forecasters are predicting that the ready availability of extremely powerful yet inexpensive PCs will change the face of engineering even more than up to now.

FIGURE 11.2 Parts of an article on software knowledge. (From Nicholas Basta, *Graduating Engineer,* A McGraw-Hill Publication, January 1987, pp. 43–46.)

of "among" are typed over by "been." Generally, the INS key on the keyboard serves as a toggle switch between the two modes, i.e., if the program is in the insert mode, it is switched to the typeover mode and vice versa. The mode is displayed on the status line (see bottom of Figure 11.4c).

We can move the cursor around the screen with the help of the **Arrow keys** and **Function keys.**

The Arrow Keys The cursor can be moved from the current position to the adjacent character position (left, right, above, or below) using the Arrow keys on the keyboard. The Arrow keys can be used in conjunction with other keys such as the Control key to move the cursor word by word (to the left or right).

The Numeric Keypad Keys Using the Arrow keys to move the cursor character by character can be rather tedious, especially in a lengthy document. This is where the Function keys on the numeric keypad come in handy. The PgUp (page-up) and PgDn (page-down) keys enable us to scroll through the document vertically page by

```
File  Edit   Search  Layout  Mark  Tools  Font  Graphics  Help
      ┌─────────────────────────────┐
  _   │ Move (Cut)        Ctrl-Del  │
      │ Copy              Ctrl-Ins  │
      │ Paste                       │
      │ Append                      │
      │ Delete            Backspace │
      │ Undelete          F1        │
      │ Block             Alt-F4    │
      │ Select                      │
      │ Comment           Ctrl-F5   │
      │ Convert Case      Shft-F3   │
      │ Protect Block     Shft-F8   │
      │ Switch Document   Shft-F3   │
      │ Window            Ctrl-F3   │
      │ Reveal Codes      Alt-F3    │
      └─────────────────────────────┘

C:\WP\BOOK\SOFTWARE                        Doc 1 Pg 1   Ln 1"  Pos 1"
```

FIGURE 11.3 Layout of a word processor screen (WordPerfect 5.1).

page or screen by screen. The End key typically moves the cursor to the end of the current line. The Tab key moves the cursor to the next tab position. It is commonly used for indenting the first line of a paragraph.

The cursor movement commands for two major word processing programs are shown in Quick Reference Table 11.1.[1]

TABLE 11.1 QUICK REFERENCE TABLE FOR CURSOR MOVEMENT COMMANDS

Cursor movement	WordPerfect	MS-Word
Character by character	→ ← ↑ ↓	→ ← ↑ ↓
Word left	CTRL →	F7
Word right	CTRL ←	F8
End of line	End	End
Page up	PgUp	PgUp
Page down	PgDn	PgDn
Beginning of document	HomeHome ↑	CTRL-PgUp
End of document	HomeHome ↓	CTRL-PgDn
Top of screen	−	CTRL-Home
Bottom of screen	+	CTRL-End

[1] While the specific command sequences and menus will vary from program to program, the underlying logic for executing the commands remains essentially the same across all programs. Likewise, as newer versions of programs are released, the capabilities are enhanced and there may be some changes in the commands. Nevertheless, to enhance the usefulness of the text, the specific command sequences for Word-Perfect and Word are given in quick reference tables.

(a)

```
    Engineers should have a proprietary air when it comes to
talking about computers.  After all, engineers have ∧among the
foremost users and developers of both minicomputers and mainframes.
And they are turning out to be one of the leading users _
```

been

(b)

```
    Engineers should have a proprietary air when it comes to
talking about computers.  After all, engineers have been among the
foremost users and developers of both minicomputers and mainframes.
And they are turning out to be one of the leading users
```

(c)

```
    Engineers should have a proprietary air when it comes to
talking about computers.  After all, engineers have beeng the
foremost users and developers of both minicomputers and mainframes.
And they are turning out to be one of the leading users
```

Typeover   Doc 1 Pg 1 Ln 1.17" Pos 6.6"

FIGURE 11.4 The two modes in a word processor. (*a*) Text with missing "been." (*b*) In insert mode, "been" is inserted at the right place and "among" moves to the right. (*c*) In typeover or replacement mode, "been" is typed over "amon," resulting in "beeng."

Entering Text We load the word processing program. We will enter the text shown in Figure 11.2 followed by the headings. It is normal practice to signal a new paragraph by beginning the first line a few spaces (typically, five) to the right of the left margin. This is known as **indenting**. We press the Indent key, generally the Tab key, and the cursor moves five spaces to the right.

As we type the text, the cursor advances to the right, and the character position on the status line changes. Moreover, when we start typing the word "talking," it is on the first line (see Figure 11.5*a*), but when we finish, it ends up on the second line (Figure 11.5*b*). This is called **word wrap-around.** When a word is too long to fit on the current line, i.e., extends beyond the right margin, it is automatically wrapped around to the next line. This is indicated by a **soft carriage return** ([SRt]) at the end of each line that is wrapped around (see Figure 11.5*c*). This acts as a reference marker for the program but is not normally displayed on the screen.

The Word-wrap feature of word processors greatly increases the rate at which text can be typed. You can type text continuously until you reach the end of a paragraph without pressing the Enter key after each line. In contrast, on a typewriter, you have to press the Return key at the end of each line to advance the carriage to the start of the next typing line.

When using a word processing program, you press the Enter key to signal the end of the current paragraph or to enter a blank line. This inserts a **hard carriage return** at the end of the line and is denoted by a special character (e.g., <) as shown in Figure 11.5*d*.

Deleting Text There is a problem in the last line of the text in Figure 11.5*d*: The word *personal* is misspelled. We can correct this error by moving the cursor to the extra "n" and deleting it. A character is typically deleted using the Del function key. The Backspace key can also be used to delete a character. In this case, the character to the left of the cursor is deleted. In a similar manner, an entire word, line, or block of lines can be deleted. The commands for deleting text for two major word processing programs are shown in Quick Reference Table 11.2.

Help Facility and Function Key Overlay The **Help facility** in the program can be invoked for obtaining more information on the usage of a particular command or function. One of the Function keys is associated with the Help function. This feature is extremely handy when the program is used rather infrequently and for finding out about new commands. A typical Help screen is shown in Figure 11.6.

Function Key Overlay Frequently used functions and/or commands in word processing programs are bound to the Function keys on the keyboard; the specific key bindings are marked on the keyboard overlay supplied with the program. Figure 11.7 shows overlays for two major programs. For example, the Help facility in WordPerfect is invoked by pressing the F3 key.

Canceling a Command Similar to the Escape key in spreadsheets, word processing programs provide a Cancel key (one of the Function keys, F1 in WordPerfect) to exit a command. This comes in handy when you want to abandon a command in the middle of its execution, i.e., after you have initiated it.

(a)

Engineers should have a proprietary air when it comes to talk_

(b)

Engineers should have a proprietary air when it comes to
talking_

C:\WP\BOOK\SOFTWARE   Doc 1 Pg 1 Ln 1.17" Pos 1.7"

(c)

Engineers should have a proprietary air when it comes to[SRt]
talking

C:\WP\BOOK\SOFTWARE   Doc 1 Pg 1 Ln 1.17" Pos 1.7"

FIGURE 11.5 Word wrap-around and carriage returns. (*a*) The word "talking" is begun on the first line. (*b*) The word "talking" is wrapped around and ends up on the second line. Note the line number on the status line has changed from 1" to 1.17". (*c*) The soft carriage return at the end of the line signals a word wrap-around. (*d*) The hard carriage return (<) signals end of a paragraph.

```
    Engineers should have a proprietary air when it comes to
talking about computers.  After all, engineers have been among the
foremost users and developers of both minicomputers and mainframes.
And they are turning out to be one of the leading users of
personnal computers. <
<
<
```

(d)

FIGURE 11.5 (continued).

TABLE 11.2 QUICK REFERENCE TABLE FOR TEXT DELETE COMMANDS

Delete	WordPerfect	MS-Word
Character	Del key	Del key
Character to left	Backspace key	Backspace key
Word	CTRL-Backspace	
To end of line	CTRL-END	Select text
To end of page	CTRL-PgDn	Delete
Marked block	Backspace Yes	

Saving the Document Periodically Figure 11.8 shows the text as it has been entered in the word processor. This text resides in the computer's random access memory (RAM) and will be lost if the power to the system is accidentally turned off or the machine is switched off (see Section 1.3.1). Therefore, it is good practice to periodically save the text on to a hard or floppy disk, as it is being entered. Depending on the importance and length of the document, a save every 15 min should be appropriate. This way, even if something unforeseen were to happen, you would lose only 15 min of work. The File Save commands for the two major programs are shown in Quick Reference Table 11.3.

11.2.2 Editing a Document

Making changes to an existing document is known as **editing** a document. This is done by moving the cursor to the desired position and making the required changes, viz., adding (or inserting), deleting, correcting, moving, or replacing text. Figure 11.9 shows the corrections and changes necessary to make the text in Figure 11.8 look like the original document in Figure 11.2. The power and popularity of word processors lie in the ease with which these editing operations can be performed.

```
Search

    Searches forward (F2) or backward (Shft-F2) through your text for a
    specific combination of characters and/or codes.

    After entering the search text, press the Search key again to start the
    search. If the text is found, the cursor will be positioned just after
    it.

    Extended Search: Pressing Home before Search extends the search into
            headers, footers, footnotes and endnotes, graphics box captions, and
            text boxes. To continue the extended search, press Home, Search.

Note: Lowercase letters match both lowercase and UPPERCASE. UPPERCASE letters
      match only UPPERCASE.
```

FIGURE 11.6 A typical Help screen (WordPerfect).

Searching for a String   The first step in the editing process is to move the cursor to the desired position. This can be done with the cursor movement commands (see Quick Reference Table 11.1). A frequently used alternative is to "search" for the string of characters in the document. The **Search String function** in word processing programs can be used to scan through an entire document for one or more occurrences of a particular string of characters. The commands for searching for a string for two major word processing programs are shown in Quick Reference Table 11.4.

Using the Search String Function   The first error is the missing "s" in "engineer" on the first line of the second paragraph. We invoke the Search String function (F2 in WordPerfect) and specify engineer[1] as the string to be searched. The cursor is moved to the position after the string and we type s. When you specify the search string, the program does a one-for-one character match. This means it will find exact matches of the specified string, and therefore it is case-sensitive (see Note in Figure 11.6 for an exception). For instance, if the search string were engineeR, the program would not find it. Moreover, the program will search for the string from the current location of the cursor to the end of the document. The **Reverse** or **Backward Search String function** (see Figure 11.7) should be used when searching for a string in the text above the current cursor location.

Searching and Replacing a String   Another useful feature of word processors is the ability to search for a string and replace it with another string. For example, as shown in Figure 11.9, we want to replace the string PC with personal computer, but only in a few places and not all occurrences. Word processors typically provide two options for the **Search and Replace command:** (1) replacing the string only after every instance is confirmed by you and (2) replacing without your confirmation. This gives you the flexibility to selectively replace the occurrences you want to change.

[1]All entries that we type at the computer are shown underlined in text.

11.2 EVOLUTION OF A DOCUMENT: THE FIVE-STEP PROCESS

F1	SETUP / Cancel	←SEARCH / Search→	F2
F3	SWITCH / Help	→INDENT← / →Indent	F4
F5	DATE/ OUTLINE / List Files	CENTER / Bold	F6
F7	PRINT / Exit	FORMAT / Underline	F8
F9	MERGE CODES / Merge R	RETRIEVE / Save	F10
F11	Reveal Codes	Mark Block	F12

Legend:
SHIFT + Function key
Function key alone

(a)

F1	Next window / ZOOM WINDOW ON/OFF	Calculate / HEADER	F2
F3	Expand glossary name / STEP MACRO	Repeat last edit / UPDATE LIST	F4
F5	Overtype on/off / LINE DRAW	Exit selection on/off / THESAURUS	F6
F7	Previous word / LOAD	Next word / PRINT	F8
F9	Previous paragraph / REPAGINATE	Next paragraph / SAVE	F10
F11	Collapse heading	Expand heading	F12

Legend:
Function key alone
CTRL + Function key

(b)

FIGURE 11.7 Important functions from the Function key overlays for two major word processors. (a) WordPerfect 5.0 (IBM layout). (b) Microsoft Word.

Engineers should have a proprietary air when it comes to talking about computers. After all, engineers have been among the foremost users and developers of both minicomputers and mainframes. And they are turning out to be one of the leading users of personal computers.<
<
Today's graduating engineer understand what PCs can do, and are looked on as the de facto PC experts in an office or engineering environment simply because their education has exposed them so much to PCs. But to do so the new engineer needs to understand the standard types of PC programs, which nearly every PC user must have, and the special programs developed for use in the different engineering fields. Once on the job, today's graduates can rapidly solidify their position by meeting this responsibility head-on.<
<
We come to the point of defining the standard types of PC programs. The initial essential software is the operating system. After the operating system question is settled, you come to the four major software packages that operators use to accomplish different tasks:<
<
<
i.   a programming language;<
ii. a word-processing program;<
iii. a database program;<
iv. a spreadsheet.<
<
<
When the PC first appeared, many observers looked on it as a dressed-up calculator. But this view has not prevailed. Instead, engineers are finding that the PC is changing the very nature of their work, removing some of the drudgery (drafting, recordkeeping, repetitive calculations, etc.), while extending the range of problems addressable. By the end of this decade, market forecasters are predicting that the ready availability of extremely powerful yet inexpensive PCs will change the face of engineering even more than up to now.<
<

FIGURE 11.8 The typed-in text. See Figure 11.2 for original.

TABLE 11.3 QUICK REFERENCE TABLE FOR FILE COMMANDS

Action	WordPerfect	MS-Word
Save file	F10 <name>	CTRL-F10 <name>
Retrieve or load file	SHIFT-F10 <name>	CTRL-F7 <name>

With the cursor at the beginning of the text, we invoke the Search and Replace command (Table 11.4); the program displays a prompt, e.g., w/Confirm? (Y/N), to know which of the two options we wish to choose. Since we want to replace only a few of the occurrences, we respond <u>Yes</u>. We then specify the search string as <u>PC</u> and the replacement string as <u>personal computer</u>. The cursor moves to the first occurrence of <u>PC</u> with a prompt (see Figure 11.10*a*); when we confirm the replacement, the action is carried out, and the cursor moves to the next instance as shown in Figure 11.10*b*.

Software Knowledge You Will Need on the Job
Nicholas Basta

Engineers should have a proprietary air when it comes to talking about computers. After all, engineers have been among the foremost users and developers of both minicomputers and mainframes. And they are turning out to be one of the leading users of personal computers.

Today's graduating engineer understand what PCs can do, and are looked on as the de facto PC experts in an office or engineering environment simply because their education has exposed them so much to PCs. But to do so the new engineer needs to understand the standard types of PC programs, which nearly every PC user must have, and the special programs developed for use in the different engineering fields. Once on the job, today's graduates can rapidly solidify their position by meeting this responsibility head-on.

We come to the point of defining the standard types of PC programs. The initial essential software is the operating system. After the operating system question is settled, you come to the four major software packages that operators use to accomplish different tasks:

i. a programming language;
ii. a word-processing program;
iii. a database program;
iv. a spreadsheet.

When the PC first appeared, many observers looked on it as a dressed-up calculator. But this view has not prevailed. Instead, engineers are finding that the PC is changing the very nature of their work, removing some of the drudgery (drafting, recordkeeping, repetitive calculations, etc.), while extending the range of problems addressable. By the end of this decade, market forecasters are predicting that the ready availability of extremely powerful yet inexpensive PCs will change the face of engineering even more than up to now.

FIGURE 11.9 The necessary corrections on the typed-in text. See Figure 11.2 for comparison.

TABLE 11.4 QUICK REFERENCE TABLE FOR SEARCH STRING COMMANDS

Action	WordPerfect	MS-Word
Search string forward	F2	ESC-S-Up
Search string reverse	Shift-F2	ESC-S-Down
Search and replace string (with/without confirm)	ALT-F2	ESC-Replace

Since we don't want to change this instance of PC, we respond with an N; PC remains unchanged, and the cursor moves to the next occurrence (Figure 11.10c).

Block Operations One of the advantages of using a word processor for creating documents is the ease with which any part of the text, be it a word, sentence, paragraph, or even pages, can be manipulated as a block: moved, copied, deleted, saved, or retrieved. These **Block operations** are also known as **Cut-and-Paste operations;** this term (cut-and-paste) comes from pre-word processor days when you had to cut portions of a document and paste them in the desired locations.

```
      Today's graduating engineers understand what PCs can do, and
are looked on as the de facto PC experts in an office or
engineering environment simply because their education has exposed
them so much to PCs.




Confirm? (Y/N) Yes                          Doc 1 Pg 1 Ln 1" Pos 6.1"
```
(a)

```
      Today's graduating engineers understand what personal
computers can do, and are looked on as the de facto PC experts in
an office or engineering environment simply because their education
has exposed them so much to PCs.




Confirm? (Y/N) No                        Doc 1 Pg 1 Ln 1.17" Pos 6.3"
```
(b)

```
      Today's graduating engineers understand what personal
computers can do, and are looked on as the de facto PC experts in
an office or engineering environment simply because their education
has exposed them so much to PCs.




Confirm? (Y/N) No                         Doc 1 Pg 1 Ln 1.5" Pos 3.9"
```
(c)

FIGURE 11.10 The Search and Replace function. (*a*) First occurrence of PC: Program seeks confirmation to replace. (*b*) PC replaced by personal computer. Cursor moves to next occurrence of PC. Response is No replacement. (*c*) Cursor moves to the next occurrence of PC. Previous PC left unchanged.

TABLE 11.5 QUICK REFERENCE TABLE FOR BLOCK COMMANDS

Action	WordPerfect	MS-Word
Mark beginning of block	ALT-F4 or F12	Select
Mark end of block	Cursor to end	Select
Copy marked block	CTRL-F4, 1, 2	ESC-Copy
Delete marked block	Backspace, Yes	ESC-Delete

Move marked block	CTRL-F4, 1, 1	ESC-Select text
		   Delete
		     Position at desired point
		      Insert

Save marked block	F10 <name>	Same as Copy marked block
Retrieve block (file)	Shift-F10 <name>	CTRL-F7 <name>

The first step in executing a Block operation is to specify or mark the block of text. This is done by moving the cursor to the beginning of the block and marking it and repeating the action for the end of the block (see Quick Reference Table 11.5 for the commands for two major word processors). As the text is marked, it is typically displayed in inverse video or blinking image; the specific function can then be carried out.

Block Copy Operation In a Block Copy operation, the marked block of text is copied to another designated location. The original block stays in its place. This is extremely useful for copying portions of the text that can be easily modified to suit your needs. For example, equations that vary only in the suffixes can be copied and the suffixes edited with the Search and Replace command. See Quick Reference Table 11.5 for the Block Copy operations for two major word processors.

Let's say we want to make a copy of the four different tasks shown in Figure 11.11a for use in another section of the text. With the cursor highlighting the first character in the block, i, we mark the beginning of the block (ALT-F4 in WordPerfect); a blinking cursor appears. Using the Down arrow keys, we move the cursor to the end of the block on the fourth line (see Figure 11.11b). As the cursor is moved, the block is displayed in reverse video. We trigger the Block Copy function (CTRL-F4 1-2, in WordPerfect) and then indicate the destination. A copy of the marked text is displayed at that location (see Figure 11.11c).

Block Move Operation In a Block Move operation, the marked block of text is moved from the current location to a designated location. This is in contrast to the Block Copy operation where a copy of the original block stays in its place. This feature is useful for moving portions of the text around the document to improve its readability. This feature also helps to speed up the document creation process. See Quick Reference Table 11.5 for the Block Move operations for two major word processors.

Let's say we want to move the second paragraph in Figure 11.8 to the beginning and make it the first paragraph. With the cursor highlighting the first character in the block, T, we mark the beginning of the block as shown in Figure 11.12a (F-12 in WordPerfect); a blinking cursor appears. Using the Down arrow keys, we move the

```
      i.  a programming language;
     ii.  a word-processing program;
    iii.  a database program;
     iv.  a spreadsheet.

    When the PC first appeared, many observers looked on it as a
dressed-up calculator.  But this view has not prevailed.  Instead,
engineers are finding that the PC is changing the very nature of
their work, removing some of the drudgery (drafting, recordkeeping,
repetitive calculations, etc.), while extending the range of
problems addressable.  By the end of this decade, market forecasters
are predicting that the ready availability of extremely powerful
yet inexpensive PCs will change the face of engineering even more
than up to now.



Block on                                       Doc 1 Pg 1 Ln 1" Pos 1"
```

(a)

```
      i.  a programming language;
     ii.  a word-processing program;
    iii.  a database program;
     iv.  a spreadsheet.

    When the PC first appeared, many observers looked on it as a
dressed-up calculator.  But this view has not prevailed.  Instead,
engineers are finding that the PC is changing the very nature of
their work, removing some of the drudgery (drafting, recordkeeping,
repetitive calculations, etc.), while extending the range of
problems addressable.  By the end of this decade, market forecasters
are predicting that the ready availability of extremely powerful
yet inexpensive PCs will change the face of engineering even
more than up to now.



Block on                                    Doc 1 Pg 1 Ln 1.5" Pos 2.9"
```

(b)

FIGURE 11.11 The Block Copy operation. (*a*) Marking the beginning of the block. (*b*) Marking the end of the block. (*c*) Copy of block in designated location.

```
    i.   a programming language;
    ii.  a word-processing program;
    iii. a database program;
    iv.  a spreadsheet.

       When the PC first appeared, many observers looked on it as a
    dressed-up calculator.  But this view has not prevailed.  Instead,
    engineers are finding that the PC is changing the very nature of
    their work, removing some of the drudgery (drafting, recordkeeping,
    repetitive calculations, etc.), while extending the range of
    problems addressable.  By the end of this decade, market forecasters
    are predicting that the ready availability of extremely powerful
    yet inexpensive PCs will change the face of engineering even more
    than up to now.

    i.   a programming language;
    ii.  a word-processing program;
    iii. a database program;
    iv.  a spreadsheet.




    C:\WP\BOOK\SOFTWARE                                 Doc 1 Pg 1 Ln 3.5" Pos 1"
```

(c)

FIGURE 11.11   (continued).

cursor to the end of the paragraph and the paragraph is highlighted in the process (Figure 11.12*b*). We invoke the Block Move function (CTRL-F4 1-1, in WordPerfect) and then indicate the destination. The marked block is displayed at that location and it becomes the first paragraph (see Figure 11.12*c*).

Block Delete Operation   The Block Delete operation is used for deleting a block of text. As with any Block operation, the text to be deleted is first marked and then the Block Delete command is invoked (Quick Reference Table 11.5).

Block Save Operation   The Block Save operation is used for saving a block of text on to a floppy or hard disk. This feature is useful for saving portions of a document that can be used in other documents. For example, the conclusions section from a lengthy technical report can be saved as a block and used in a short memo outlining the findings of the research.

The text to be saved is first marked, and then the Block Save option is selected (Quick Reference Table 11.5).

Block Retrieve Operation   The Block Retrieve operation is used for retrieving text from a floppy or hard disk into the current document. For example, a memo "skeleton"

> Engineers should have a proprietary air when it comes to talking about computers. After all, engineers have been among the foremost users and developers of both minicomputers and mainframes. And they are turning out to be one of the leading users of personal computers.
>
> Today's graduating engineers understand what PCs can do, and are looked on as the de facto PC experts in an office or engineering environment simply because their education has exposed them so much to PCs. Once on the job, today's graduates can rapidly solidify their position by meeting this responsibility head-on. But to do so the new engineer needs to understand the standard types of PC programs, which nearly every PC user must have, and the special programs developed for use in the different engineering fields.
>
> Block on   Doc 1 Pg 1 Ln 2.17" Pos 1.5"

(a)

> Engineers should have a proprietary air when it comes to talking about computers. After all, engineers have been among the foremost users and developers of both minicomputers and mainframes. And they are turning out to be one of the leading users of personal computers.
>
> **Today's graduating engineers understand what PCs can do, and are looked on as the de facto PC experts in an office or engineering environment simply because their education has exposed them so much to PCs. Once on the job, today's graduates can rapidly solidify their position by meeting this responsibility head-on. But to do so the new engineer needs to understand the standard types of PC programs, which nearly every PC user must have, and the special programs developed for use in the different engineering fields.**
>
> Block on   Doc 1 Pg 1 Ln 3.5" Pos 2.9"

(b)

FIGURE 11.12 The Block Move command. (*a*) Marking the beginning of the block. (*b*) Marking the end of the block. (*c*) Moved block in designated location.

```
Today's graduating engineers understand what PCs can do, and
are looked on as the de facto PC experts in an office or
engineering environment simply because their education has exposed
them so much to PCs.  Once on the job, today's graduates can
rapidly solidify their position by meeting this responsibility
head-on.  But to do so the new engineer needs to understand the
standard types of PC programs, which nearly every PC user must
have, and the special programs developed for use in the different
engineering fields.

    Engineers should have a proprietary air when it comes to
talking about computers.  After all, engineers have been among the
foremost users and developers of both minicomputers and mainframes.
And they are turning out to be one of the leading users of personal
computers.




                                          Doc 1 Pg 1 Ln 1" Pos 1.5"
```

(c)

FIGURE 11.12 (continued).

(also known as a "boiler-plate") can be created once and saved for repetitive use. This boiler-plate can be retrieved and text added and sent out, greatly speeding up the process. Depending on the program, there may be an option under Block operations for retrieving a file, or it may be part of the general File Retrieve command (Quick Reference Table 11.5).

Thus, using the various Block operations, the text can be manipulated, resulting in an edited document.

11.2.3 Improving the Document Appearance: Formatting a Document

As shown in Figure 11.1, the next step is to improve the appearance of the edited document. This process is also known as **formatting** the document, and it includes:

- Designing the page layout, i.e., setting the left, right, top, and bottom margins; line spacing; and number of lines per page
- Controlling the document display, i.e., text justification; centering headings; boldface, underline, and italics; super- and subscript.

Designing the Page Layout The Page Layout or Format command of word processors can be used to design the layout of the printed page. Typically, a menu of the default settings is displayed when this command is invoked (see Figure 11.13, the WordPerfect menu). For example, the default top and bottom margins are 1 inch. To

```
Format: Page

    1 - Center Page (top to bottom)      No

    2 - Force Odd/Even Page

    3 - Headers

    4 - Footers

    5 - Margins - Top                    1"
                  Bottom                 1"

    6 - New Page Number                  1
        (example: 3 or iii)

    7 - Page Numbering                   No page numbering

    8 - Paper Size                       8.5" x 11"
                  Type                   Standard

    9 - Suppress (this page only)


Selection: 0
```

FIGURE 11.13   The Page Format (Layout) menu (WordPerfect, Shift-F8).

change the setting for an item, e.g., the left margin, you have to select the item from the menu and then set it to the desired value. The commands for invoking the Page Layout command for two major word processors are shown in Quick Reference Table 11.6.

Document Layout Options   You can change the layout of the entire document using the Page Layout command. Or you can also change the settings only in certain portions of the document. For example, in Figure 11.14, a section of the second paragraph is displayed in double spacing. Depending on the program, when you elect to change the settings only in certain portions of the document, you have to first mark

TABLE 11.6   **QUICK REFERENCE TABLE FOR PAGE FORMAT COMMANDS**

Action	WordPerfect	MS-Word
Display Page Format menu	SHIFT-F8 Line Page Document Other	ESC-F-Division Margins Page numbers Layout Line numbers

```
    Engineers should have a proprietary air when it comes to
talking about computers.  After all, engineers have been among the
foremost users and developers of both minicomputers and mainframes.
And they are turning out to be one of the leading users of personal
computers.

    Today's graduating engineers understand what PCs can do, and

are looked on as the de facto PC experts in an office or

engineering environment simply because their education has exposed

them so much to PCs.  Once on the job, today's graduates can
rapidly solidify their position by meeting this responsibility
head-on.  But to do so the new engineer needs to understand the
standard types of PC programs, which nearly every PC user must
have, and the special programs developed for use in the different
engineering fields.




                                          Doc 1 Pg 1 Ln 3.17" Pos 1"
```

FIGURE 11.14 Changing the line spacing in sections of the document.

the block of text using the Mark Block command (see Quick Reference Table 11.5) and then invoke the Page Format command to effect the changes.

Controlling the Document Display The formatting features of word processing programs can be used to control the display of the text on the screen and on the printed page. For example, headings can be centered, and selected words can be underlined or displayed in boldface or italics. Also, superscripts and subscripts can be used to display equations in the proper format.

Text Justification In textbooks and other printed documents, you will find that the text is aligned on both the left- and right-hand margins of the page. The text is said to be "justified." In this process, known as **text justification,** the empty spaces in every line are distributed, more or less evenly, between the words so that the text is "stretched" to the right margin, giving a uniform appearance. When justification is turned off, the text appears ragged at the right margin. The two examples are shown in Figure 11.15.

Many word processors support the Proportional Spacing option in which the space occupied by a particular character is proportional to its size (horizontal dimension). For

> Engineers should have a proprietary air when it comes to talking about computers. After all, engineers have been among the foremost users and developers of both minicomputers and mainframes. And they are turning out to be one of the leading users of personal computers.

(a)

> Engineers should have a proprietary air when it comes to talking about computers. After all, engineers have been among the foremost users and developers of both minicomputers and mainframes. And they are turning out to be one of the leading users of personal computers.

(b)

FIGURE 11.15 Text justification. (*a*) The text is justified: aligned at both margins. (*b*) The text is not justified: ragged at the right margin.

example, the letter m takes up more space than i. This option greatly enhances the readability of the text.

Hyphenation In certain instances when the text is justified, there will be an excessive amount of blanks between words in a line, especially when the words are long. This spoils the appearance of the document. You can use the **Hyphenation feature** in the word processing program to break such long words and improve the appearance of the document. The program typically suggests the position for the hyphenation which you can accept or override. This feature can be turned on or off as necessary.

Centering Headings Titles and headings in a document are typically centered. Once you have typed the title, you can mark it as a block and invoke the Center Text command in the word processing program. Alternatively, you can invoke the **Center Text command** and type in the heading. When you select the latter option, the cursor moves to the center of the screen, and the text is centered as you type.

Underlining Text Another commonly used feature of word processing programs is the underlining of parts of a text, e.g., titles and keywords. You can mark the block of text to be underlined and invoke the **Underline Text option** of the program. Or you can select the command and type in the text, in which case, the text is underlined as you type.

Boldfacing Text Similar to the underlining feature, this option is used for displaying parts of the text in boldface. This alerts the reader to a new or important point in the text. You can mark the block of text to be displayed in boldface and invoke the **Boldface Text option** of the program. Or you can select the command and type in the text.

Italicizing Text In technical writing, we often use Latin words, variable names (i.e., num, val) and other field-specific words that may not be readily apparent to the reader. It is good practice to make such words stand out from the text so that the reader will understand that they are different from the running text and that a dictionary may be needed. This can be done by displaying the words in *italics*.

You can mark the block of text to be italicized and invoke this display option. Alternatively, you can invoke the command and then type in the text, which is displayed in italics as it is being typed.

Superscripts and Subscripts Engineers and scientists use equations extensively, and equations typically contain subscripts and superscripts. Word processing programs have the necessary features for producing superscripts and subscripts. As with any display formatting command, the desired portion of the text should be marked and the command invoked for converting it into a superscript or subscript.

Figure 11.16 shows a portion of the text resulting from these commands. The commands for the various display options are shown in Quick Reference Table 11.7.

FIGURE 11.16 Formatting the display. (*a*) Text with centered title, underline, boldface, and italics. (*b*) Superscripts and subscripts in an equation. (From *Graduating Engineer,* A McGraw-Hill Publication, January 1987.)

Software Knowledge You Will Need on the Job[1]
Nicholos Basta

Introduction

Engineers should have a proprietary air when it comes to talking about computers. After all, engineers have been among the <u>foremost</u> users and developers of both minicomputers and mainframes. And they are turning out to be one of the leading users of personal computers.

Today's graduating engineers understand what **PCs** can do, and are looked on as the *de facto* PC experts in an office or engineering environment simply because their education has exposed them so much to PCs. Once on the job, today's graduates can **rapidly solidify** their position by meeting this responsibility head-on. But to do so the new engineer needs to <u>understand the standard types of PC programs</u>, which nearly every PC user must have, and the special programs developed for use in the different engineering fields.

[1]From *Graduating Engineer,* January 1987, A McGraw-Hill Publication.

(a)

The equation is $T_1 = T_2 * (V_2/V_1)^{gamma-1}$.

(b)

TABLE 11.7 QUICK REFERENCE TABLE FOR DISPLAY CONTROL COMMANDS

Action	WordPerfect	MS-Word
Center marked block	SHIFT-F6	ALT-C
Underline marked block	F8	ALT-U
Boldface marked block	F6	ALT-B
Italicize marked block	CTRL-F8, 2, 4	ALT-I
Superscript marked block	CTRL-F8, 1, 1	ALT+
Subscript marked block	CTRL-F8, 1, 2	ALT–
Create footnote	CTRL-F7, 1	ESC-F-Footnote

11.2.4 Saving the Document

Once the document has been formatted to achieve the desired appearance, the finished version should be saved onto a hard or floppy disk. Even during the earlier steps, the document should be saved periodically in case of unexpected loss of power or system failure. The File Save commands for the two major programs are shown in Quick Reference Table 11.3. When the command is invoked, the program prompts you for the filename; it is a good idea to specify a name that matches the document contents, e.g., memo1, sales, or paper. If a file with the same name already exists, the program will seek your confirmation before overwriting it. It is always a good idea to maintain a backup copy of important files to guard against disasters.

Retrieving a File A previously stored document can be loaded into the word processor using the File Retrieve command. This way, you can gradually write a term paper or an article, adding or editing material in stages. As you are working with a document in the word processor, you can import, i.e., bring in, information from other files. This gives you the flexibility to combine portions of text from various documents into a single one.

11.2.5 Printing the Document

The last step in creating a document is to print the text on paper. Word processing programs typically provide the option of viewing the document on the screen prior to printing it. This option is especially useful if the program is not a WYSIWYG word processor (see Section 11.1) and the document is lengthy. By looking at the output on the screen, you can modify the format to suit the requirements prior to generating a hard copy.

When you invoke the Print command, a menu is displayed as shown in Figure 11.17. Among other things, you can select the printer, the number of pages to be printed, the number of copies, and the print quality and the font type and size. Figure 11.18 shows a printed output. The commands for printing a document for two major word processing programs are shown in Quick Reference Table 11.8.

Creating a Text File Instead of sending the output to a printer, you can also send it to a text file. The text file can be printed subsequently on a different printer. For

```
Print

    1 - Full Document
    2 - Page
    3 - Document on Disk
    4 - Control Printer
    5 - Type Through
    6 - View Document
    7 - Initialize Printer

Options

    S - Select Printer          HP LaserJet Series II
    B - Binding                 0"
    N - Number of Copies        1
    G - Graphics Quality        Medium
    T - Text Quality            High




Selection: 0
```

FIGURE 11.17 The Print Command menu (WordPerfect).

example, if you don't have a laser printer attached to your computer, you can create a text file from your document; you can then take the text file to a computer with a laser printer and print out the document.

Thus, using the five-step process, you can create and print term papers, technical reports, research papers, and résumés.

11.3 ADVANCED CONCEPTS

When we write, we often use a dictionary to check the spelling, a thesaurus or synonym finder to find suitable substitutes for words to make the text less repetitious, and a grammar text to avoid grammatical errors. These tools greatly enhance the quality of the written material. Such tools or accessories are increasingly becoming an integral part of word processing programs.

Checking Your Spelling The spell checker in the word processing program can be used to check and correct the spellings in the document. An electronic dictionary with several thousand words is called by the program to check for errors. You can check the spelling of a single word, a page or the entire document. The program will

Software Knowledge You'll Need on the Job
Nicholas Basta

Introduction

Engineers should have a proprietary air when it comes to talking about computers. After all, engineers have been among the <u>foremost</u> users and developers of both minicomputers and mainframes. And they are turning out to be one of the leading users of personal computers.

Today's graduating engineers understand what personal computers can do, and are looked on as the *de facto* PC experts in an office or engineering environment simply because their education has exposed them so much to personal computers. Once on the job, today's graduates can **rapidly solidify** their position by meeting this responsibility head-on. But to do so the new engineer needs to <u>understand the standard types of PC programs</u>, which nearly every PC user must have, and the special programs developed for use in the different engineering fields.

Standard Programs

We come to the point of defining the standard types of PC programs. The initial essential software is the operating system. After the operating system question is settled, you come to the four major software packages that operators use to accomplish different tasks:

i.   a programming language;
ii. a word-processing program;
iii. a database program;
iv. a spreadsheet.

Conclusions

When the PC first appeared, many observers looked on it as a dressed-up calculator. But this view has not prevailed. Instead, engineers are finding that the PC is changing the very nature of their work, removing some of the drudgery (drafting, recordkeeping, repetitive calculations, *etc.*), while extending the range of problems addressable. By the end of this decade, market forecasters are predicting that the ready availability of extremely powerful yet inexpensive PCs will change the face of engineering even more than up to now.

FIGURE 11.18 A printed output.

TABLE 11.8 QUICK REFERENCE TABLE FOR PRINT DOCUMENT COMMANDS

Print	WordPerfect	MS-Word
Document to printer	SHIFT-F7	CTRL-F8
Document to text file	SHIFT-F7-S-E-Port-Other <name>	Print-File <name>

highlight the error and display possible words that are close to the misspelled word and give you the option of editing the misspelled word. In Figure 11.19*a*, the misspelled word "temprature" is highlighted and possible spellings are displayed. Once this is corrected, the checker finds no other errors (Figure 11.19*b*). Since the dictionary is a general-purpose dictionary, it may not contain words specific to a

```
    The temprature is 15° off the set point.

    Their are three types of programs.




    ================================================================================

       A. temperature








    Not Found: 1 Skip Once; 2 Skip; 3 Add; 4 Edit; 5 Look Up; 6 Ignore Numbers: 0
```

(a)

```
    The temperature is 15° off the set point.

    Their are three types of programs.





    Word count: 14      Press any key to continue
```

(b)

FIGURE 11.19 The spell checker—spots only syntax errors. (*a*) Misspelled "temprature" is highlighted. Suggestions displayed at the bottom. (*b*) Misspelling corrected; however, the improper usage of "their" in second sentence goes unnoticed.

particular field. The program typically provides you the flexibility of adding new words to the dictionary.

Syntax versus Semantic Errors Spell checkers are essentially syntax checkers for words. They cannot spot semantic errors or improper usage of words. For example, in Figure 11.19*b*, we know that the word "their" on the second line should have been "there." The checker fails to spot this. Likewise, if the words "off" and "of" in the two sentences were interchanged, the spell checker would not spot the errors. So you should not be lulled into a false sense of security by the spell checker. The spell checker should be used only as an aid, and you should not become dependent on it.

The commands for invoking the spell checker for the two major word processing programs are shown in Quick Reference Table 11.9.

Accessing the Electronic Thesaurus The thesaurus accessory in the word processing program can be used for selecting a synonym to replace a particular word in the document. With the cursor on the desired word, you invoke the thesaurus. The corresponding synonyms are displayed on the screen and you can browse through the words and select the appropriate one or reject all and retain the original word. If no synonyms are found in the electronic thesaurus, the program displays a message.

The commands for invoking the thesaurus for the two major word processing programs are shown in Quick Reference Table 11.9.

Rating Your Writing Software programs are now available for analyzing and rating documents. These programs typically check for grammatical errors (missing verbs, subjects, and objects), readability level of the document (middle school, high school, college, complex), wordiness, long sentences, and excessive repetition of words and phrases. They provide a detailed analysis of the document and suggest improvements. These programs are based on previously defined rules of grammar and writing styles. In short, these programs act like human editors, except that the author is spared the editor's pronouncements. It is important to remember that these programs can only help you become a better writer, they cannot do the writing for you.

Other Useful Features We will now examine some other important features of word processors: the mail-merge facility; creating multiple windows and documents; and headers, footers, and footnotes.

Mail Merge Typically, in an organization, the same type of letter is sent to a large number of individuals. For example, letters announcing an update to a software program are all the same in content, though addressed to different individuals. Word

TABLE 11.9 QUICK REFERENCE TABLE FOR INVOKING ACCESSORIES

Accessory	WordPerfect	MS-Word
Spell checker	CTRL-F2	ALT-F6
Thesaurus	ALT-F1	CTRL-F6

processing programs provide the necessary capability to merge the body of the letter with addresses stored in a different file. This is known as the **mail-merge facility.**

Multiple Windows and Documents Often, it becomes necessary to work with more than one document at the same time. Word processing programs provide windows on the screen in which different documents can be viewed simultaneously.

Headers, Footers, and Footnotes **Headers** are text material often printed at the top of each page before the body of the regular text, while **footers** denote the text printed at the bottom of the page after the text. For example, in a dictionary, the beginning and ending words on a page are displayed at the top of the page. This is known as the "header." The section title displayed at the top of this page is also a header.

A **footnote**, in contrast, is used to qualify a word or sentence in the document. For example, in Figure 11.16*a*, the source of the article is given as a footnote.

Macros As you become proficient in using a word processor, you will find that the number of keystrokes for invoking certain commands is hindering your productivity, and you will want to create shortcuts to these commands. The **Macro facility** in the word processing program lets you chain a sequence of commands and associate this sequence with a particular key.

For example, the Block Move command in WordPerfect requires the following sequence of keystrokes CTRL-F4 1 1, after a block is marked for moving (see Table 11.5). By creating a macro for this sequence of keystrokes and associating it with a key (e.g., M), you can invoke the Block Move command by just pressing ALT-M. It helps to create a set of macros for some commonly used commands, viz., copy, move, retrieve, superscript, subscript, italics, center, etc.

Page Description Languages The recent proliferation of laser printers has brought a new dimension to word processing: **desktop publishing.** Word processing programs can be used with page description languages—languages that specify the page format/layout to the printer—to produce "print-quality" documents, formerly possible only with extensive typesetting equipment, yet considerably faster and cheaper. Once the text is created in the word processor (without any of its formatting features), it is processed through a software program, such as LaTEX, in which the formatting features and/or commands are added. The finished document is then printed on the laser printer.

SUMMARY

We introduced the concepts of word processing and presented a five-step sequence for creating and printing a document using word processing programs. We then discussed commands for manipulating blocks of texts (copying, moving, deleting, and saving); searching for and replacing character strings; improving the appearance of the document; and printing it. We looked at tools in word processors that assist in creating better documents: spell checker, thesaurus, and writing analyzers. Finally, we covered some other useful features of word processors, such as the mail-merge facility, using multiple documents and windows, and page description languages.

INDEX

@prefix, 86–87
@@, 199
@CHOOSE, 170–171
@FV, 198
@IF, 200–201
@IRR, 198
@PI, 87
@PMT, 198
@PV, 198

ABS key in spreadsheets, 125
Absolute addressing of cell, 122–124
Ada, Augusta, 24
Add:
   fields in database, 378–380
   records in database, 376–378
Aiken, Howard, 24
Algorithm, 28, 31
ALTAIR 8800, 25
Analytical engine, 24
AND operator, 19–20, 310, 386
Apollo, 5
Append Variable Names feature in TK, 265
Apple II, 25

Application software, 4, 22, 39–40
   (*See also specific entries, for example:* Spreadsheet; Word processor)
Arithmetic-logic unit (ALU), 3
Arithmetic operations, precedence of:
   in programming language, 46–47
   in spreadsheet, 86–88
   in TK, 224–226
Arrays, 59–62
Arrow keys:
   in spreadsheets, 77
   in TK, 214
   in word processors, 398
ASCII (American Standard Code for Information Interchange), 13–14
Assembler language, 36, 41
Assignment statement, 46

Babbage, Charles, 24
Backsolving, 235–236, 313
Bar graph, 140

BASIC, 22, 38, 211
Beam:
   cantilever, 262
   simply supported, 262
Beam design, 261–262
Bending moment diagram, 275, 287
Binary addition, 16–17
Binary decimal conversion, 14–15
Binary division, 19
Binary multiplication, 18–19
Binary number system, 13–15
Binary numbers, 3
Binary subtraction, 17
Bit (binary digit), 5
Blank field in TK, 250
Block of cells (*see* Range of cells)
Block command in spreadsheets, 128
Block operations on text (*see specific entries, for example:* Copy command)
Boldface text, 417–418
Boolean algebra, 19–20
Boolean operators, 19, 298, 310, 386

425

Bootstrap loader, 20
Bricklin, Dan, 25, 109, 260
Bubble sort, 57
Bugs, 42, 48
Built-in functions:
   and engineering economics, 191
   in programming language, 47
   in spreadsheet, 86–87
   in TK, 224, 297–298
Byte, 5

C language, 22, 38
CAD (computer-aided design), 2
CAE (computer-aided engineering), 2
CALC key, 118, 186
Calculation unit, 228, 268
   (See also Display unit)
Calculator function in TK, 231
Call statement, 65
   in TK, 329, 336
Cancelling a command:
   in spreadsheet, 102–103
   in TK, 250
   in word processor, 401
Case sensitivity, 87
Cash-flow diagram, 192–194
Cell, 74
   aligning label in, 84
   blank, in formula, 88
   formula evaluation, 87–88
   protecting, 135, 178–179
   reference errors, 115
   walking labels in, 84
Cell address, 75
Cell entries, 74
Cell pointer, 75
Cell width, 84, 93–95
Center text, 416
Central processing unit (CPU), 5, 20
Character variable, 45
Chemical reaction, 348–350
Chemical reactor, mass balance in, 180–181
CIRC indicator, 116, 185
Circular reference, 115–123
   avoiding, 188–189
   direct, 115

   indirect, 11
   and iteration, 183–184
Circularity in spreadsheets (see Circular reference)
Clock speed, 5
Coefficient of variation, 63
Combinations, number of, 335
Command menu:
   in database program, 372–374
   in spreadsheet, 91–92
   in TK, 248–249
   in word processor, 397
Command processor, 21
Command structure (see Command menu)
Communication software, 22
Compaq, 25
Compiler, 37–38
Computers:
   anatomy of, 3
   evolution of, 23–24
   in product development, 1–2
   stored program, 24
   (see also Personal computer)
Conditional rules in TK, 305, 308–309
Constants, 45
Constraint resolution system, 231
Control statement (see Decision-making statement)
Control unit, 3
Copy command:
   in spreadsheets, 119–126
   in TK, 250
   in word processors, 409–411
CP/M (control program for microprocessors), 21
CPU (central processing unit), 5, 20
Cray Research, 5
Cubic interpolation in TK, 300
Cue (see cursor)
Cursor, 75, 212, 397
Cursor movement commands:
   in spreadsheets, 79
   in TK, 212–21
   in word processors, 399
Curtis, Diane, 260
Customized menus using-macros, 175–176
Cut-and-paste text, 407

Data General Eclipse, 5
Data security, 369–371
Data structure, 58–59
Database, 363–364
   adding records, 376–378
   creating, 366–368
   deleting records, 379–380
   editing, 368, 376
   employee, 364
   field, 365–366
   file, 365
   major programs, 371
   pack command, 379
   Periodic Table of Elements, 364, 371
   print, 387
   record, 365
   saving, 368, 376, 379
   searching, 369–370, 381–387
   size limits of programs, 371
   sorting records 368–370
   telephone directory, 364
   using in TK, 304, 359
   utilizing, 368–369
   worksheet for design, 373
Date functions, 86
dBASE IV, 373n.
Debugging:
   program, 48
   spreadsheet template, 114–119
   TK model, 272, 348
   (See also Program tracing; VEC matrix)
Decimal binary conversion, 14–15
Decimal number system, 13–15
Decision-making statement, 51–53, 307–308
Declaration statement, 45–46
Declarative language (see Very high level language)
Defensive programming, 69
Definite iteration, 54, 60
Delete:
   block of text, 411
   columns in spreadsheet 152–153
   fields in database, 379–380
   records in database, 379
   rows in spreadsheet, 152–153
   rows in TK, 250
Δ network, 321

De Morgan's theorem, 19
Dictionary, 25, 419–420
Digital Equipment Corporation, 5
Digitizing tablet, 11–12
DIMension statement, 59
Dimensional analysis, 34, 270
Direct Solver, 212, 221
   workings of, 233–236
Discount rate, 195
Disk:
   capacity, 9
   drive, 9
   floppy, 3, 8
   hard, 3, 10
   optical, 10
   sectors, 8
   tracks, 8–9
Display formula command, 113–116
Display unit, 228, 268
   (See also Calculation unit)
Dive command in TK, 258, 277
DO loop, 56
Document program code, 68
Dot matrix printer, 12
DOS (disk operating system), 21

E format, 129
Edit:
   spreadsheet entry, 90, 125
   in TK, 219, 237
   word processing document, 397, 403
Electric circuit analysis, 313–325
Electrolysis example in TK, 356–359
Engineering design, 261
Engineering economics, 191
ENIAC, 24
EPROM (erasable programmable read-only memory), 7
Equation solver, 208, 211
   (See also TK Solver Plus)
Equation solving, 208, 211, 223
Erase cell names, 137
Erase command, 149–150
ERR message in spreadsheets, 85, 88

Errmsg function in TK, 309–310, 346
Errors:
   cell reference, 115
   computational flow, 115
   logic, 68–69, 115
   user-defined, in TK, 303–304, 341–342
   when using spreadsheets, 115
   when using TK, 220, 226–227
ESCape key, 102–103
   (See also Cancelling a command)
Eureka, 211
Excel, 92n.; 114n.
Exit command (see Quit command)

Factorials:
   iterative computation of, 335, 338–339
   recursive computation of, 339–341
Field length, 366
Field name in database, 365–366
Field types in database:
   character, 365
   date, 365
   memo, 365
   numeric, 365
Field width (see Field length)
File command (see Retrieve command; Save)
Files in database, 365
Fill List command, 279
   options in, 281
Financial functions, 86
   (See also specific entries, for example: @IRR)
Fixed disk (see Hard disk)
Flat panel display, 12
Floppy disk, 3, 8
Footers, 423
Footnote, 423
FOR-NEXT loop, 55
   in TK, 337, 352
Format statements, 65

Formatting:
   cell entries, 92
   disk, 8
   numbers (see Numeric formatting)
Formula:
   display command, 113–116
   evaluation order in spreadsheet, 87–88
FORTRAN, 22, 38, 46, 211
Frankston, Bob, 25, 109, 260
Function key overlay:
   in database programs, 372, 375
   in spreadsheets, 78
   in TK, 215, 219
   in word processors, 401, 405
Function Sheet, 213, 301–302

Given function, 305–307
Global command, 94
Global Sheet, 213, 244
GOSUB statement, 66
GoTo function key:
   in spreadsheets, 78
   in TK, 212, 215, 337
Go To statement, 53, 337
Graph types:
   bar, 140
   line, 142–143
   pie, 143
   stacked-bar, 140–141
   X-Y, 142
Graphics software, 22
Graphs in spreadsheets:
   customizing, 145
   naming, 147
   plotting, 147
   printing, 148
   steps in creating, 143–147
   in TK, 281, 287–288
   viewing, 147
Greatest common divisor (GCD), Euclidean algorithm for, 345–347
GUI (graphical user interface), 10

Hard disk, 3, 10

Hardware, 3
   (*See also specific entries, for example:* CPU; Printer)
Headers, 423
Heat transfer and TK, 361
Helenius, Fred, 260
Help facility:
   in database program, 372
   in spreadsheet, 103–104
   in TK, 250–251, 342
   in word processor, 401–404
Hiding:
   columns in spreadsheets, 136
   rows in spreadsheets, 136
High-level language, 37, 41
Hollerith, Herman, 24
How-can analysis, 211, 238
Hyperbolic functions, 86
Hyphenation in word processor, 416

IBM, 5, 21
Ideal gas, behavior of, 111
IF-THEN-ELSE statement:
   in programming language, 51–53
   in spreadsheet, 200–201
   in TK, 309
Impact printer (*see* Dot matrix printer)
IMSAI, 25
Indefinite iteration, 54, 60
Index in database, 388–392
Industrial waste, processing of, 348
Information processing, 4, 13
Ink-jet printer, 13
Input-output devices, 10–12
Input-output manager, 21
Input statement, 46
Inserting columns, 97, 100–102
Inserting rows, 94–97, 236
Integrated circuit, 5
Intel, 5
   i860, 7
   I-80486, 7
Interactive tables in TK, 283, 294–295, 353–356
Internal memory, 7

Interpreter, 37–38
Investment alternatives:
   annual cost method, 195–197
   future worth method, 195–197
   present worth method, 195–197
Italicize text, 417–418
Iteration:
   convergence, 184
   divergence, 184
   indefinite, 54, 60
   in spreadsheets, 180
      caveat using, 187
      limitations of, 203
      specifying number of steps, 186–188
Iterative Solver, 212
   Newton-Raphson method in, 244
   selecting guess values and, 245
   selecting guess variables for, 245
   using, 243–245
   working of, 244–245
Iterative Solving, 241–242
   need for, 242
   and Rule functions, 323–324, 332
   trial-and-error method in, 242
   when to use, 248
   (*See also* Iterative Solver)

Jacquard machine, 23–24
Jayaraman, Sundaresan, 235*n*., 260
Jobs, Steve, 25
JOIN operation in relational database, 392–394

Key fields in database:
   primary, 368
   secondary, 368
Keyboard, 3, 10–11
Kildall, Gary, 21
Kilobyte (K), 8
Konopasek, Milos, 235*n*., 259, 260

Label-align command, 128–129
Label entry, 83, 86

Label-prefix command, 128–129
Laser printer, 12
Laws of motion, 32
Letter-quality printer, 12
Light pen, 3, 10–11
Line graph, 142–143
Linear interpolation in TK, 300
Lisa, Apple, 25
List in TK:
   associating, 277
   declaring, 277
   disassociating, 277
List function in TK, 297
   creating, 300–301
   domain of, 297
   mapping in, 299–300
   range of, 297
   using in rules, 299, 303
List Sheet, 277
List Solving, 275–277
   steps in, 277
   triggering, 279–281
List Subsheet, 279
Load file (*see* Retrieve command)
Logic variable, 45
Loop:
   in BASIC, 55
   in FORTRAN, 56
   nesting of, 61
   in TK, 337, 352
Lotus 1-2-3, 92*n*;, 114*n*.

Machine language, 36–37, 41
Macintosh, Apple, 25
Macros:
   in spreadsheets:
      creating, 164–165
      customized menus using, 175–176
      debugging, 166–167
      definition of, 164–165
      documenting, 166
      entering, 166
      naming, 166
      programming keywords, 178
      testing, 166
   in word processor, 423

Mail merge, 422–423
Mainframe, 4–5
Margins, 413–414
Mars, playing on, 90
Mass balance calculations, 180–181
MathCAD, 211
Mathematica, 211
Matrix manipulation in TK, 361
Mayer, Melinda, 260
Mean, 63
Mean time between failures (MTBF), 162
Memory:
  primary, 7
  random-access (RAM), 7–8
  read-only (ROM), 7
  secondary, 7
Menu (see command menu)
Message area in TK, 212
Methodology:
  for database program, 366–369
  for equation solver, 216–223
  for spreadsheet, 78–83
  for word processor, 396–397
Microcomputer (see Personal computer)
Microprocessor, 5
  growth of, 25
  manufacturers of, 5
Microsoft Word, 399n.
Minicomputer, 4–5, 24
MITS, 25
Mode indicator in spreadsheets, 75
Model:
  library of in TK, 360–361
  mathematical, 30–32
  (See also TK Solver Plus)
Monitor (see Video display terminal)
Motorola, 5
Mc-68040, 7
Mouse, 3, 10–11
Move command:
  in spreadsheet, 133–135, 139
  in TK, 250
  in word processor, 409–412

MS-DOS (Microsoft Disk Operating System), 21
MTBF (mean time between failures), 162
Multiplan, 109
Multitasking, 21

Naming:
  cells, 130–131, 137, 168
  graphs, 147
Nelder-Mead simplex algorithm, 338
Neumann, John von, 24
Newton-Raphson method, 244
NeXT, 10, 25
Nonimpact printer (see Laser printer)
NOT operator, 19–20, 310, 386
Number representation in TK, 342–343
Numeric Format Sheet, 288, 290, 292
Numeric formatting:
  in spreadsheet, 130–133
  in TK, 288, 292
Numeric keypad keys:
  in spreadsheets, 77–78
  in TK, 212–214
  in word processors, 398–399
Numeric variable, 45

Object code, 48
Octal numbering system, 15–16
O'Dell, Jim, 260
Ohm's law, 54
One-dimensional array, 59
One's complement, 18
Operating system, 20
  (See also specific systems, for example: UNIX)
Optical disk, 10
OR operator, 19–20, 310, 386
Output statement, 48

Pack command in databases, 379
Page description language, 423

Page layout (see Formatting)
Parallel circuit, 314
Parsing rules, 225
Pascal, Blaise, 23
Pascal, 22, 38
PDP-8, 24
Periodic Table of Elements:
  in database, 364, 371
  in TK model, 349–351
Peripheral devices (see specific entries, for example: Laser printer)
Personal computer, 4–5
  manufacturers of, 5
Physics and TK, 361
Pie chart, 143
Pixel, 12
Playing the computer (see Program tracing)
Plot Sheet, 287–288
Plotter, 13
Precedence of operations (see Arithmetic operations)
Primary key, 368
Primary memory, 7
  (See also RAM; ROM)
Prime number, 28
Printer, 12
  (see specific entries, for example: Dot matrix printer)
Printing:
  in database program, 387
  in programming, 49
  in spreadsheet, 103–107
  in TK, 250–253
  in word processor, 418–420
Problem solving, 27
  framework for, 30–31
  general methodology for, 28–30
  mechanisms in TK, 212
  methodology for equation solver, 216–221
  methodology for spreadsheet, 78–82
  requirements of tool, 262
  role of computers in, 35–36, 39
  role of humans in, 39

Procedure functions in TK, 335, 344–345
  control statement in, 337
  creating, 335, 338–339
  defining, 336
  input variable, 336
  invoking, 336
  loop construct in, 337
  output variable, 336
  parameter variable, 336
  passing values to, 337–338
  recursion in, 336–337
  syntax of, 337
ProDOS, 21
Program, 3
  bugs in, 42, 48
  steps in development of, 41–42
Program tracing, 48–49
Programming:
  good practices in, 67–68
  modular approach to, 62, 68–69
Programming languages:
  BASIC, 22, 38, 211
  C, 22, 38, 211
  FORTRAN, 22, 38, 46, 211
  hierarchy of, 37
  Pascal, 22, 38, 211
  in TK, 336–338
PROJECT operation in relational database, 392–394
Projectile trajectory, 71–72
PROM (programmable read-only memory), 7
Prompt/errror line in TK, 212
Propagation of solution in TK, 232
Proportional spacing, 415
Protecting cells, 135, 178–179
Pseudocode, 41
Punched card, 3

Quadratic equation, roots of, 42
Quality control, 161
Quattro, 92n., 114n.
Quattro Pro, 110
Query, 381–387

Quick reference command tables:
  for database programs:
    create, 376
    edit, 380
    print, 390
    records, 376
    save, 379
    search, 386
    sort, 383
    use or retrieve, 376
  for equation solver:
    slash, 250
    storage, 258
  for spreadsheets:
    cell labels, 185
    changing cell width, 95
    copy, 123
    create text file, 107
    cursor movement, 79
    delete, 153
    erase, 150
    file combine, 159
    formula display, 116
    graph, 147
    insert columns, 97
    insert rows, 97
    move, 139
    name, 137
    numeric format, 133
    print, 107
    quit, 109
    recalculation, 118
    retrieve, 107
    save, 107
    set number of iterations, 188
    titles, 157
    window, 155
  for word processors:
    block delete, 403
    copy block, 409
    cursor movement, 399
    move block, 409
    page layout, 414
    print text, 420
    retrieve file, 406
    save file, 406

search string, 407
spell checker, 422
text display control, 418
thesaurus, 422
Quit command:
  in spreadsheet, 107–109
  in TK, 259

R (Relationship) graph, 263–265
RAM (random-access memory), 7, 8
Range of cells, 103, 126
Range command (see Block command)
Read-only memory (ROM), 7
Recalculation:
  method of:
    automatic, 118, 172, 186
    manual, 118, 172, 186
  order of:
    columnwise, 118
    natural, 118
    rowwise, 118
Records in database, 365
  (See also Database)
Recursion, 339
  in TK, 336–337, 340
  limits of, 337, 342, 345
Reed, David, 260
Reflex, 373n.
Relational databases, 392–394
Relational operators, 52, 309, 386
Relative addressing, 119
Reliability analysis, 161
REMark statement, 44
Repeating a character, 99–100
Repetitive computation (see Loop)
Reserved words in programming language, 45
Retrieve block of text, 411
Retrieve command:
  database file, 376
  spreadsheet template, 107
  TK model, 254, 257
  word processing document, 418

Return command in TK, 258
RETURN statement, 67
Reverse search string, 404
RISC (reduced-instruction-set computer), 7
ROM (Read-only memory), 7
Rule function in TK, 317
   advantages of, 325
   argument variable, 317
   defining, 317–318
   invoking, 317
   and iterative solving, 323–324, 332
   parameter variable, 317
   passing values to, 317
   result variable, 317
   saving, 321
   when to use, 318, 344–345
   working of, 319–321
Rule Sheet, 212–213
Runge-Kutta procedure, 338

Save:
   block of text, 411
   database, 379
   spreadsheet template, 106–107
   TK Function Sheet, 254–257
   TK model, 253–257
   TK Unit Sheet, 254–257, 271
   word processing document, 403, 406, 418
Scientific format, 129
Scroll bars, 77, 212
Search:
   in database, 369–370, 381–387
   and replace string, 404
Search string function, 404
Secondary memory, 7
Sectors in disk, 8
Select command in TK, 216, 258
SELECT operation in relational database, 392–394
Series circuit, 313
Shearing force diagram, 275, 287
Sheets in TK, 212–214, 259
Simpson's rule, definite integrals, 338

Slash key (/) 91, 248–249
Software Arts, 260
Solution strategy (*see* Algorithm)
SolverQ, 211
Sorting:
   database, 368–370, 380–383
   numbers, 56–57
Source code, 48
Source range (*see* Copy command)
SP (solution path) matrix, 233–235, 239–244
Spell checker, 419–422
Spreadsheet:
   backspace key in, 83
   basics of, 72–75
   cells in, 74
      (*See also* Cell)
   command line in, 75
   control panel in, 75
   formula evaluation, 86–88
   graphs in (*see* Graphs in spreadsheets)
   history of, 109
   iteration in (*see* Iteration, in spreadsheets)
   limitations of, 202
   macros in (*see* Macros, in spreadsheets)
   mode indicator, 75
   moving around in, 77–78
   prompt line in, 77
   screen display, 76
   template, 71
   what-if analysis, 72, 82–83, 90, 127
Spring design, 207
Stacked-bar graph, 140–141
Standard deviation, 62–63
Statement number, 44
Statistical functions, 86
Statistics, 62–63
Status line in TK, 212
Steinberg, Seth, 260
Step mapping in TK, 300
Stoichiometry, 348
String variable, 45

Subprogram, 65
Subroutine, 65
Subscript, 417–418
Subsheets in TK, 258–259
Sun Microsystems, 5, 25
Supercomputer, 4–5
Superscript, 417–418
Switch key, 214
Symbolic value, 295
System software, 3–4, 19–20

Table mapping in TK, 299
Table Sheet, 282–285
Tables:
   in spreadsheets, 132
   in TK, 281
      defining, 282–283
      interactive, 283
      multiple, 286
Tape, 3, 10–11
Target range (*see* Copy command)
Template:
   combining, 158–159
   creating, steps in, 80–82
   document, 82
   errors in, 82
   extracting, 159–160
   layout, 112
   what-if analysis, 82–83, 90
Termination statement, 48
Text file creation:
   in spreadsheet, 107
   in TK, 253
   in word processor, 420
Text justification, 415
Thermodynamics, 111
Thesaurus, 25, 422
Time value of money, 192
Titles, freezing, in spreadsheet, 154–157
TK (*see* TK Solver Plus)
TK Solver Plus:
   backsolving in, 235–236
   calculation unit, 228, 268
   command structure, 248
   Direct Solver, 212, 221
   display unit, 228, 268

errors when using, 220, 226–227
history of, 259–260
Iterative Solver, 212, 243–245
model:
 development of, 217–222
 documentation of, 220–221
moving around sheets, 212–216
sheets, 212–214, 259
SP matrix, 233–235, 239–244
subsheets, 258–259
VEC matrix, 233–235, 263, 266
workings of, 231–235
(*See also* Equation solving)
Tracks in disk, 8–9
Transistor, 7
Transistor control circuit, 162
Triangle, elements of, 326
 rule function in, 326–328
Trigonometric functions, 86
 degree version of, 326
Tuple, 392
Two's complement, 18

Underlining text, 416–417
Unit conversions in TK, 270–271
 errors, 228–229, 274
 indirect, 269
 inverse, 269
Unit names, case sensitivity of, 270
Unit Sheet, 213, 268
Units of measurement, 34
Units library in TK, 271

UNIX, 21
User-defined function in TK, 293–294
Utility software, 22

Value entry, 85
Variable Sheet, 212–213
Variable Subsheet, 279
Variables:
 character, text or string, 45
 logic, 45
 numeric, 45
Variance, 63
VEC matrix, 233–235, 263, 266
Very high level language, 37–38, 41
Video disk, 10
Video display terminal (VDT), 12
VisiCalc, 25, 109
VP Planner, 110
Walking label, 84
What-if analysis, 72, 82–83, 90, 127
Wheatstone Bridge circuit, 321
WHILE loop, 60
Window command:
 in spreadsheets, 152–155
 in TK, 249–250
Windows, Microsoft, 21
Wood, properties of, 294
Word, 5
Word processing, 395
 steps in document creation, 396–397
 (*See also* Word processor)
Word processor:
 advantages of, 396
 carriage return in, 401

deleting text in, 401
editing text in, 403
entering text in, 397, 401
formatting document in, 413–418
indenting text in, 401
insert mode in, 397
macros in, 423
margins in, 413–414
moving around screen of, 397–399
printing document using, 418–420
ruler line in, 397
screen display of, 397
spell checker, 419–420
stand-alone, 395
status line in, 397
typeover mode in, 397
Word wrap-around, 401
WordPerfect, 399*n*.
Worksheet (*see* Spreadsheet)
Workstation, 5
Wozniak, Steve, 25
Writing analyzer, 422
WYSIWIG, 396

X-Y graph, 142
XENIX, 21

Y network, 321–322
Y-Δ transformation, 321–325

Z-80, 21
Zilog, 21

Also Available from McGraw-Hill
Schaum's Outline Series in Engineering

Most outlines include basic theory, definitions, and hundreds of solved problems and supplementary problems with answers.

Titles on the Current List Include:

Acoustics
Advanced Structural Analysis
Basic Equations of Engineering Science
Computer Graphics
Continuum Mechanics
Descriptive Geometry
Dynamic Structural Analysis
Engineering Economics
Engineering Mechanics, 4th edition
Fluid Dynamics, 2d edition
Fluid Mechanics & Hydraulics, 2d edition
Heat Transfer
Introduction to Engineering Calculations
Introductory Surveying
Lagrangian Dynamics
Machine Design
Mathematical Handbook of Formulas & Tables
Mechanical Vibrations
Operations Research
Programming with C
Programming with Fortran
Programming with Pascal
Reinforced Concrete Design, 2d edition
Space Structural Analysis
State Space & Linear Systems
Statistics & Mechanics of Materials
Statics & Strength of Materials
Strength of Materials, 2d edition
Structural Analysis
Structural Steel Design, LFRD Method
Theoretical Mechanics
Thermodynamics, 2d edition
Vector Analysis

Schaum's Solved Problems Books

Each title in this series is a complete and expert source of solved problems containing thousands of problems with worked out solutions.

Related Titles on the Current List Include:

3000 Solved Problems in Calculus
2500 Solved Problems in Differential Equations
3000 Solved Problems in Electric Circuits
2500 Solved Problems in Fluid Mechanics & Hydraulics
1000 Solved Problems in Heat Transfer
3000 Solved Problems in Linear Algebra
2000 Solved Problems in Mechanical Engineering Thermodynamics
2000 Solved Problems in Numerical Analysis
700 Solved Problems in Vector Mechanics for Engineers: Dynamics
800 Solved Problems in Vector Mechanics for Engineers: Statics

Available at your College Bookstore. A complete list of Schaum titles may be obtained by writing to: Schaum Division
        McGraw-Hill, Inc.
        Princeton Road, S-1
        Hightstown, NJ 08520